Crosscutting Concepts
Strengthening Science and Engineering Learning

Crosscutting Concepts

Strengthening Science and Engineering Learning

Jeffrey Nordine and Okhee Lee, Editors

Arlington, Virginia

Dedication

Aaron Rogat initiated this book. His creative thinking and forethought provided a road map for the project and laid the foundation for the series of collaborations that ultimately led to the chapters that follow. A talented scientist and science educator, Aaron dedicated his career to improving science teaching and learning. Aaron left us too soon, but he had meaningful impact on those who had the privilege to work with him.

Rachel Ledbetter, Managing Editor
Andrea Silen, Associate Editor

ART AND DESIGN
Will Thomas Jr., Director

PRINTING AND PRODUCTION
Colton Gigot, Senior Production Manager

NATIONAL SCIENCE TEACHING ASSOCIATION
1840 Wilson Blvd., Arlington, VA 22201
www.nsta.org/store
For customer service inquiries, please call 800-277-5300.

24 23 7 6

Library of Congress Cataloging-in-Publication Data

Names: Nordine, Jeffrey, editor. | Lee, Okhee, 1959- editor.
Title: Crosscutting concepts : strengthening science and engineering learning / Jeffrey Nordine and Okhee Lee, editors.
Description: Arlington, VA : National Science Teaching Association, [2021] | Includes bibliographical references and index. |
Identifiers: LCCN 2020055448 (print) | LCCN 2020055449 (ebook) | ISBN 9781681407289 (paperback) | ISBN 9781681407296 (pdf)
Subjects: LCSH: Science--Study and teaching--Standards--United States. | Engineering--Study and teaching--Standards--United States.
Classification: LCC Q183.3.A1 C757 2021 (print) | LCC Q183.3.A1 (ebook) | DDC 507.1073--dc23
LC record available at *https://lccn.loc.gov/2020055448*
LC ebook record available at *https://lccn.loc.gov/2020055449*

Contents

Preface

The Role of Crosscutting Concepts in Three-Dimensional Science Learning

Helen Quinn

This book is designed to enrich and expand your understanding and use of each of the crosscutting concepts defined by *A Framework for K–12 Science Education* (the *Framework*). That *Framework* introduces these concepts and defines them as one of the three dimensions of science learning. The premise of this book is that you, as a teacher, parent, or mentor, can better guide students' three-dimensional science learning, including learning to understand and use these concepts as you enrich your own understanding of them and see examples of their use in both real-world problem solving and classroom contexts.

I chaired the committee that developed *A Framework for K–12 Science Education*. When, in the course of that work, we formulated the idea of three-dimensional science learning, we were seeking a description of teaching and learning that would resonate in the field. We wanted a term that would remind everyone to attend to three important aspects of science learning in their instructional planning. The goal was to avoid some of the failures of previous attempts to encourage and support effective science teaching through the formulation of standards for science learning.

As we introduced the crosscutting concepts (CCCs), we knew that certain "big ideas" or "themes" (AAAS 1993) or "unifying concepts and processes" (NRC 1996) had been chosen and described as important in prior documents intended to guide science education. However, they were rarely taken up by curriculum developers, and their use in classrooms was limited at best. Our challenge was to describe them better and to come up with a formulation in which they could not be forgotten. We did this by calling the science and engineering practices (SEPs), crosscutting concepts, and disciplinary core ideas (DCIs) each a separate dimension of science learning and by stressing that learning should be three-dimensional. We hoped to achieve the goal of integrating the learning about and use of these practices and crosscutting concepts into science learning. The aim was to make them useful to students, not to simply add to the list of things students must learn and remember.

Looking at the situation now eight years later, I see that our success so far has been limited. Many states have adopted standards based on, or derived from, our *Framework*. A shift to two-dimensional learning is clearly underway. One state has even adopted a two-dimensional approach for their standards. Teachers across the country are beginning to see that students' science learning is deepened and enriched when students engage in the full cycle of science and engineering practices while simultaneously learning and applying the disciplinary core ideas to explain phenomena or develop designs. However, it seems to me that most teachers and curriculum designers still struggle to incorporate the dimension that we call crosscutting concepts. Disciplinary core ideas are seen as the usual "content," and the practices are seen as the doing of science or engineering, so why is this other set of concepts needed?

What are the crosscutting concepts (CCCs)? And why are they needed? I find that the description that resonates most for teachers is one that we did not use in the *Framework*. The CCCs are conceptual tools for examining unfamiliar situations and finding an approach that helps develop understanding. Each CCC is a lens for looking at a problem. Each lens highlights a particular perspective and thus leads its user to ask productive questions that arise from that perspective. Productive questions here mean questions that are useful and effective in guiding and expanding thinking and thus aid in sense-making and problem-solving efforts.

Scientists use these lenses all the time because we have somehow learned that they are effective. When I talk to scientists about the CCCs, they generally agree that they use all of them and that they are most useful when confronting an unfamiliar problem or situation. All too often, we expect students to discover these lenses for themselves, without ever explicitly discussing their use. Three-dimensional learning should provide students with experiences in using these concepts as they seek to build models and explain phenomena or design solutions to problems; time to reflect on the use of these concepts is also important. This experience of use and reflection makes the usefulness of the crosscutting concepts an explicit element of students' science learning.

Each lens is a tool to be taken up as needed and used to enrich the SEPs and DCIs as they are applied to develop designs to solve engineering problems or to explain phenomena. A tool is not useful if it is unfamiliar, so students must develop familiarity with these tools and need to be guided to use them in multiple contexts. Eventually with use,

CCCs should, like the SEPs, become part of the toolkit that a student freely calls on and uses when confronted with any unfamiliar science phenomenon or engineering problem.

If these things are tools, why did we call them *concepts*? The answer is that they are *conceptual* tools. We could call them conceptual frames; the idea of a lens is that it helps us frame a problem in a particular way, directing our attention to aspects relevant to that framing. However, we don't start out by teaching students that a given concept is a lens; instead, we start by giving them experiences where the lens is useful, suggesting they use it, and then asking them to reflect on how it helped them approach the phenomenon or problem. Only as they become familiar with the lens do we begin to talk about it abstractly as a tool they might use again in other circumstances.

Being told to look for patterns is not useful unless the student has begun to develop the idea that the patterns they observe in any given system are something that their model for the system must reproduce and explain. This is learned not by talking about noticing patterns but by first doing activities that ask students to use patterns and find relationships between them. For example, a third-grade class might be looking at the native plants in their region and given photos of the plants and a map that shows where each type of plant is likely to be found. Then the students are asked how they could organize the plants into groups. Some teams of students decide to group the plants by the size of their leaves. Others may choose to group the plants by the locations where they are found. Each has found a pattern and used it to group the plants. (If the teacher sees a possible grouping that none of the students has chosen, he or she can suggest it as an alternative way to carry out the task.) Now the teacher can ask two groups of students who have used different strategies to discuss whether there is any relationship between their grouping methods. Imagine their excitement when they discover that the plants by the stream generally have larger leaves than those on the open hillside. Now they have found a pattern that begs for an explanation and also generates many interesting questions to explore: Why is it that these two completely different ways of grouping the plants actually were quite similar in their outcomes? What else is different about the hillside environment and the streamside one? What advantages and disadvantages are there to having big leaves? What could be the advantage for smaller leaves on the hillside? What do leaves do for the plant anyway?

There is a deep interplay between developing the crosscutting concepts as tools and engaging in the SEPs, always, of course, with the goal of explaining some phenomenon or designing a solution to a problem using science ideas. In some cases, a pairing between a particular concept and a particular practice is obvious and explicit: We cannot develop or use models effectively without having some concept of systems and system models, nor can we construct explanations or design solutions without calling on our understanding of the concept of cause and effect.

Four of the crosscutting concepts—patterns, relationships between structure and function, conditions for stability or change in a system, and conserved qualities (flows and cycles of matter and energy into, out of, and within a system)—provide the conceptual basis for asking productive questions about one's model and how well it represents these essential aspects of the system. These questions can help us refine our model and deepen our understanding of the system and the phenomena occurring within it.

Notice that in the discussion above, I have slightly renamed several of the crosscutting concepts in order to stress what is essential about them. For example, I renamed the CCC of cause and effect: mechanisms and explanation, labeling it instead as "mechanisms of cause and effect." I find that too many people lose track of what comes after the colon, which is in fact the essence of the CCC. The same is true for energy and matter: flows, cycles, and conservation. It is not the idea of energy or matter that is crosscutting here. Instead, what is crosscutting is the additional concept of conserved quantities, namely that something that cannot be produced or destroyed must be supplied and disposed of, and its availability—or lack of it—limits what can occur. Hence, it is very useful to understand how a system functions in order to track how the conserved quantity or substance flows into, out of, or within the system. For this reason, I renamed the CCC "tracking conserved quantities (flows and cycles of matter and energy into, out of, and within a system)." Likewise, with the CCC of stability and change, the question is not whether a system is stable or changing. Rather, the question is under what conditions or over what timescale is it stable, and what changes in conditions lead to what changes in the system. Therefore, I renamed this concept "conditions for stability or change."

I suggest the aforementioned name shifts based on my experiences working with teachers to help them use these concepts. The names we gave them in the *Framework* work once you know how to use the concepts, but I find renaming them helps point the way to begin using each concept.

The CCC of patterns invokes a particular type of observation: namely, looking for patterns in the form or behavior of the system. This concept, which we could likewise rename "recognizing and explaining patterns," can be useful in that it is connected to every practice. In particular, there are questions one can ask about the patterns in a phenomenon or system that link to every practice: What questions do I have about the patterns I have noticed? Does my model reproduce these patterns? What do I need to investigate about this system to understand this pattern better?

The CCC of scale, proportion, and quantity comes into play as one seeks to define any quantitative relationships in developing a model and to test and refine the model through the practices of (a) planning and carrying out investigations (which, of course, involves observations and measurements to be recorded) and (b) analyzing and interpreting data (which is often but not always quantitative). Note that this crosscutting concept cheats; it introduces not one concept but three, linked together by the fact that all involve quantitative thinking, as well as units of measurement.

Thinking about relationships of scale, proportion, and quantity in a system typically involves us in the practice of using mathematics and computational thinking. Measurement requires that students decide (a) on what scale to model the system, (b) what to measure about it, and (c) what units of measurement to use for those quantities. These are critical for developing models and planning investigations. Measurement also makes the mathematics of proportion in science something more than that of fractions in mathematics. This is because in science one can take ratios of quantities with different and incommensurate units and define entirely new quantities, as is the case with speed—a ratio of distance traveled to time elapsed. Students need support to see the critical role that the different types of units play; unlike feet and meters, which are both units of length and can be freely converted, distance units and time units bear no relationship to one another.

Many readers have interpreted the *Framework* and the *Next Generation Science Standards* as a call for an approach to the teaching of science that is integrated across all disciplines. There is, however, little agreement in the field about what it means for a curriculum to be "integrated." Does "three-dimensional" imply that the curriculum is integrated? Are units designed around explaining phenomena necessarily integrated science? Let me define a unit as "integrated" if its learning goals include core ideas from more than one disciplinary area and as "discipline-focused" if all the core ideas addressed are from the same discipline. Either way, it could be taught in the three-dimensional approach, where students are applying the DCIs they are learning and using the SEPs and the CCCs in order to explain an overarching phenomenon that provides the central core of the unit.

Certainly, both the SEPs and the CCCs highlight what is common across all areas of science. Beyond that, the sciences today are much more interconnected than they were when high school science was divided into three courses—biology, chemistry, and physics—and even the experts saw little connection between them. Today, even high school biology contains a large segment of biochemistry, and no serious biologist thinks there is a "life force" (*vis vitalis*) that is outside of physics and chemistry. Chemistry functions by the same quantum physics as materials science. And Earth systems science requires expertise from geology, meteorology, oceanography, as well as physics, chemistry, and biology to understand the complex interconnectedness of the geosphere, biosphere, atmosphere, and hydrosphere.

The *Framework* does stress that whatever the course structure, whether discipline-focused or cross-disciplinary, science overall needs to be taught in such a way that students are supported in building connected knowledge across the disciplines and use their knowledge from one discipline in the context of another when and where it is relevant.

Topics that play a role across all disciplines need to be discussed in such a way that students can connect what they learn in one disciplinary context with that in another. Consider the teaching of energy, for example. No matter the order in which their courses are offered, students should be supported in connecting their physics understanding about energy to what they are learning in chemistry, and they should be able to connect both chemical and physical ideas about energy to language about energy used in biology or Earth science courses or units. To apply the crosscutting concept of conservation of energy and matter, one needs a single approach to energy (and to matter) that begins to be developed in middle school and is applied similarly across all high school courses. This does not argue for any particular organization of courses or units, but it does argue for curriculum design and course planning that looks at more than a single course and its particular disciplinary goals.

I do not think you can integrate or apply ideas that you have not met. The core ideas of each discipline need attention and must be developed as such, even when this work occurs in an interdisciplinary context. Detailed physical science ideas may be relevant to a larger real-world problem, for example in Earth systems science. But if the ideas are to be first introduced in that context, the unit will also require some experiments and activities that look more like traditional discipline-based school science to develop the relevant core ideas effectively. Curriculum design around real-world problems requires careful planning to include the relevant smaller scale activities that support learning the disciplinary ideas well, and these activities must be introduced in a way that students see they are indeed connected to the larger question. A unit can be designed to introduce core ideas from more than one discipline, or from only one of them. What makes it three-dimensional is that the students are seeking to explain a phenomenon that is relevant and interesting to them by (a) using the science they are learning to develop models of the relevant system, (b) engaging in multiple science practices, and (c) using crosscutting concepts as they develop and refine those models to produce a model- and evidence-based explanation of the phenomenon. In fact, most often, the evidence is used to refine and then support the model, and then the model provides the reasoning that connects this evidence to the explanation.

So far, this discussion has viewed the crosscutting concepts from the perspective of how scientists use them and how they can be used in the classroom to enrich and inform student work to develop explanations of phenomena. They also play a role in developing engineering designs. As with the SEPs, the role of and language around the use of each CCC is somewhat different in the engineering context than in the science one. There is a chapter in this book that discusses engineering uses, so I will only give a couple of examples here. Engineers design systems, which may be objects, collections of objects, or processes. Clearly, the CCC of systems and system models is critical for system design. Engineered systems are governed by the same rules of physics, chemistry, and even biology; therefore, CCCs such as the tracking of conserved quantities or the conditions for stability and instability are equally applicable to designed systems as they are to natural phenomena. Their use in the design process most often is in asking questions like: the following: How can I improve my design? How can I make it function well under a broader range of conditions (conditions for stability)? How can I make it use less fuel to the do the same job (tracking flows of matter and energy)?

In this latter context, it is worth noting that engineers talk about inputs and outputs with a somewhat different meaning than the science concept of inflows and outflows to the system. In engineering, an input is something that must be provided (and hence paid for) in order for the system to operate, and an output is either a product or a task that the machine is intended to make or do. An inflow of oxygen from the air is rarely counted as an input, and outflows of waste products such as exhaust gases are not generally included in the term *output*, even though they are clearly both things that must be considered in machine design. It is helpful for students to experience and use the terms *input* and *output* and to become aware of both the overlaps and the differences with the terms *inflow* and *outflow* as they apply this concept.

To support three-dimensional learning, we need to consider how to elicit use of all three dimensions not only in instruction but also in assessment tasks, especially as we design formative assessment. Decisions about which CCC and SEP to highlight in a lesson or unit must be made during the design of the instructional sequence. Leaving this decision to be made only during classroom instruction runs the risk of allowing these important concepts to slip into the background. Therefore, the design of an instructional unit, or an extended curriculum plan that includes multiple units, should include the intentional and explicit use of particular crosscutting concepts, and related SEPs, within each unit. However, as students become familiar with these tools, they may call on others not specifically stressed in the curriculum plan.

With regard to assessment, this too should be a part of curriculum design—which outcomes are intended and how they will be measured should inform the content and approach of the unit. The crosscutting concepts are a particular challenge for external- or test-based assessment in that the use of the tool is not necessarily visible in the finished product or response on the test. I can pose a problem for which a particular crosscutting concept would be a powerful tool, yet not be able to tell with certainty from looking at the solution achieved whether or not the student called on that tool. For summative assessment, this may not matter much; a good solution to a problem is good however it was reached. If one does wish to know how students have used a particular crosscutting concept, then the test tasks must explicitly elicit that information. However, in formative assessment, one does need to know to what extent students are able to take up and use the appropriate tools. In particular, we need to know whether students are using the crosscutting concept that is being developed and used in a unit in order to make decisions about further instruction. As stated above, this is difficult to see from short assessment tasks. Planned and documented classroom observations of students at work on projects can provide some of the missing information, not necessarily for grading purposes but in a way that can guide further instructional choices. Formative assessments need not be designed to assign grades; they should be designed to inform subsequent instruction. This allows a wider range of methodologies than those for summative assessment where a grade is to be assigned.

In the chapters of this book, each crosscutting concept is explored in depth and its usefulness across the sciences is highlighted. Each chapter also seeks to illustrate how that concept might be used in the classroom. However, it is easy to lose sight of what is common about these concepts in exploring the richness of each one of them. Here is one commonality to keep in mind: The metaphor of the lens for viewing phenomena or problems and asking productive questions about them is a useful one that spans all of the crosscutting concepts and links them together as conceptual tools.

Another commonality is that the use of any of these concepts is not specific to any one discipline of science; they are all useful across all disciplines, and many of them apply well beyond the natural sciences and are useful in many problem-solving contexts. Hence, students who develop the ability and the disposition to use these tools are better positioned to apply their science knowledge to everyday situations. This ability develops, as do most, through practice and effort, with coaching and guided reflection on that effort. Students need multiple opportunities to use each of the crosscutting concepts in many different contexts, and they also need time and support to reflect on what they did and how it helped them understand the problem they were tackling. In other words, the crosscutting concepts are concepts students must learn to use appropriately, rather than concepts to be taught as abstractions.

The same can be said of the other two dimensions; students learn the practices by engaging in them and reflecting on that work, and they learn science concepts by using them to help explain phenomena. The three dimensions work together to build a "knowledge for use" of science and how it functions. The expectation is that students who experience such learning will be better able to apply their learning in new contexts than those who have just learned disciplinary ideas as things to remember.

This book will enrich your thinking about each of the crosscutting concepts and prompt you to think of many questions related to the perspective brought by each. My hope is that it also helps you see them each as a powerful tool for student learning and recognize them as a class of concepts with some similar uses, even though each of them is distinct. Use of these concepts as lenses interweaves with and supports student engagement in the practices and the application of their growing knowledge of disciplinary concepts. Learning and assessment tasks that ask students to use them in powerful ways can help students recognize their importance. The goal of three-dimensional science learning is that students will take this learning out of the classroom and into their lives, using and expanding it as they meet issues and opportunities or challenges where it can be helpful to them. Once a tool enters their conceptual toolkit, it becomes theirs, and they can use it whenever and wherever they choose.

Among your goals as a teacher, I expect that you want not just to "cover" the required material but also to provide your students with tools for life and lifelong learning. Both the science and engineering practices and the crosscutting concepts can be such tools, useful well beyond the science classroom. Providing students with the multiple experiences they need to master the use of these tools and add them to their personal toolkit requires well-designed three-dimensional curricula and teaching approaches over multiple years. This book can help you construct the experiences that deliver such learning.

Acknowledgments

This material is based in part on work supported by the National Science Foundation (Grant No. DUE-1834269).

References

American Association for the Advancement of Science (AAAS). 1993. *Benchmarks for science literacy*. New York: Oxford University Press.

National Research Council (NRC). 1996. *National science education standards*. Washington, DC: National Academies Press.

Acknowledgments

The editors wish to thank Christian Nordine for her extensive contributions in preparing and copy editing the initial manuscript. We further wish to thank the manuscript reviewers, whose valuable feedback much improved the final book. Finally, we wish to thank the NSTA editors who have expertly guided the production of the book.

About the Editors

Jeffrey Nordine is the associate professor and deputy director of physics education at the Leibniz Institute for Science and Mathematics Education (IPN) in Kiel, Germany. His research focuses on the design, implementation, and effects of coherent science instruction. In particular, he studies how the energy concept might be taught in order to strengthen connections across science disciplines and to support future learning about energy-related contexts, both in and out of school. He was an award-winning physics teacher and dean of instruction for mathematics and science in San Antonio, Texas, and he was the chief scientist for the San Antonio Children's Museum (The DoSeum). Beginning in August 2021, he will be an associate professor of science education at the University of Iowa.

Okhee Lee is a professor in the Steinhardt School of Culture, Education, and Human Development at New York University. Her research involves integrating science, language, and computational thinking with a focus on English language learners. Her latest research addresses the COVID-19 pandemic and social justice. She was a member of the *Next Generation Science Standards* (*NGSS*) writing team and served as leader for the *NGSS* diversity and equity team. She was also a member of the steering committee for the Understanding Language Initiative at Stanford University. She became a fellow of the American Educational Research Association (AERA) in 2009, received the Distinguished Career Contribution Award from the AERA Scholars of Color in Education in 2003 and the Innovations in Research on Equity and Social Justice in Teacher Education Award from the AERA Division K Teaching and Teacher Education in 2019, and was recognized by the National Science Teaching Association (NSTA) Distinguished Service to Science Education Award in 2020.

Contributors

The following list contains the contact information for the book's contributors.

Preface

Helen Quinn, Stanford University, *quinn@slac.stanford.edu*

Chapter 1

Jeffrey Nordine, Leibniz Institute for Science and Mathematics Education (IPN); Kiel, Germany, *nordine@leibniz-ipn.de*

Okhee Lee, New York University, *olee@nyu.edu*

Chapter 2

Joseph Krajcik, CREATE for STEM, Michigan State University, *krajcik@msu.edu*

Brian J. Reiser, Northwestern University, *reiser@northwestern.edu*

Chapter 3

Marcelle Goggins, Research Improving People's Lives, *mgoggins@ripl.org*

Alison Haas, New York University, *ams728@nyu.edu*

Scott Grapin, University of Miami, *sgrapin@miami.edu*

Rita Januszyk, Gower District 62; Willowbrook, Illinois, *ritajanuszyk@gmail.com*

Lorena Llosa, New York University, *lorena.llosa@nyu.edu*

Okhee Lee, New York University, *olee@nyu.edu*

Chapter 4

Kristin L. Gunckel, University of Arizona, *kgunckel@arizona.edu*

Yael Wyner, The City College of New York, City University of New York, *ywyner@ccny.cuny.edu*

Garrett Love, North Carolina School of Science and Mathematics, *garrett.love@ncssm.edu*

Chapter 5

Tina Grotzer, Harvard Graduate School of Education, *tina_grotzer@harvard.edu*

Emily Gonzalez, Harvard Graduate School of Education, *emily_gonzalez@harvard.edu*

Elizabeth Schibuk, Conservatory Lab Charter School; Dorchester, Massachusetts, *eschibuk@conservatorylab.org*

Chapter 6

Cesar Delgado, North Carolina State University, *cesar_delgado@ncsu.edu*

Gail Jones, North Carolina State University, *mgjones3@ncsu.edu*

David Parker, The Outdoor Campus, South Dakota Game, Fish and Parks, *David.Parker@state.sd.us*

Chapter 7

Sarah J. Fick, Washington State University, *s.fick@wsu.edu*

Cindy E. Hmelo-Silver, Indiana University, *chmelosi@indiana.edu*

Lauren Barth-Cohen, University of Utah, *lauren.barthcohen@utah.edu*

Susan A. Yoon, University of Pennsylvania, *yoonsa@upenn.edu*

Jonathan Baek, Honey Creek Community School; Ann Arbor, Michigan, *jbaek@hc.wash.k12.mi.us*

Contributors

Chapter 8

Charles W. (Andy) Anderson, Michigan State University, *andya@msu.edu*

Jeffrey Nordine, Leibniz Institute for Science and Mathematics Education (IPN); Kiel, Germany, *nordine@leibniz-ipn.de*

MaryMargaret Welch, Seattle Public Schools; Seattle, Washington, *mmwelch@seattleschools.org*

Chapter 9

Bernadine Okoro, Ephesus Media; Washington, D.C., *bernadine.okoro75@gmail.com*

Jomae Sica, Beaverton School District; Beaverton, Oregon, *jomae_sica@beaverton.k12.or.us*

Cary Sneider, Portland State University, *carysneider@gmail.com*

Chapter 10

Brett Moulding, Partnership for Effective Science Teaching and Learning, *mouldingb@ogdensd.org*

Kenneth Huff, Williamsville Central School District; Williamsville, New York, *khuff@williamsvillek12.org*

Kevin McElhaney, Digital Promise, *kmcelhaney@digitalpromise.org*

Chapter 11

Joi Merritt, James Madison University, *merritjd@jmu.edu*

Kristin Mayer, Kentwood Public Schools; Kentwood, Michigan, *kristin.mayer@kentwoodps.org*

Chapter 12

Jo Ellen Roseman, American Association for the Advancement of Science (retired), *roseman.joellen@gmail.com*

Mary Koppal, American Association for the Advancement of Science (retired), *mrk1346@yahoo.com*

Cari Herrmann Abell, BSCS Science Learning, *cabell@bscs.org*

Sarah Pappalardo, Howard County Public Schools, *sarah_pappalardo@hcpss.org*

Erin Schiff, Howard County Public Schools, *erin_schiff@hcpss.org*

Chapter 13

Ann E. Rivet, Teachers College, Columbia University, *rivet@tc.columbia.edu*

Audrey Rabi Whitaker, Academy for Young Writers; Brooklyn, New York, *rabi@afyw.org*

Chapter 14

Christine M. Cunningham, Pennsylvania State University, *ccunningham@psu.edu*

Kristen B. Wendell, Tufts University, *kristen.wendell@tufts.edu*

Deirdre Bauer, State College Area School District; State College, Pennsylvania, *dmb13@scasd.org*

Chapter 15

Erin Marie Furtak, University of Colorado Boulder, *erin.furtak@colorado.edu*

Aneesha Badrinarayan, Learning Policy Institute, *abadrinarayan@learningpolicyinstitute.org*

William R. Penuel, University of Colorado Boulder, *william.penuel@colorado.edu*

Samantha Duwe, Aurora Public Schools; Aurora, Colorado, *srduwe@aurorak12.org*

Ryann Patrick-Stuart, Aurora Public Schools; Aurora, Colorado, *repatrick-stuart@aurorak12.org*

Chapter 16

Emily C. Miller, PBL Science Connections; University of Wisconsin-Madison, *emilycatherine329@gmail.com*

Tricia Shelton, National Science Teaching Association, *tshelton@nsta.org*

Chapter 17

Jeffrey Nordine, Leibniz Institute for Science and Mathematics Education (IPN); Kiel, Germany, *nordine@leibniz-ipn.de*

Okhee Lee, New York University, *olee@nyu.edu*

Ted Willard, Discovery Education, *twillard@discoveryed.com*

PART I
Introduction to Crosscutting Concepts

Chapter 1

Strengthening Science and Engineering Learning With Crosscutting Concepts

Jeffrey Nordine and Okhee Lee

On a nature walk during science class, Maria notices something funny about the wildflowers in the park by the school. The park is filled with beautiful trees and wildflowers, and Maria notices that the flowers seem to grow in some areas but not in others. This normally wouldn't catch her attention—after all, she knows that flowers don't grow everywhere. But the flowers seem to make circles around the trees (see Figure 1.1).

Figure 1.1. Wildflowers growing among trees

Maria knows that nobody planted the flowers and that plants don't actively choose where to grow, but it looks almost as if the flowers decided as a group not to grow too close to the trees. Even more puzzling is that other things are growing next to the trees; in this field of brightly colored flowers, every tree is surrounded by a circle of green grass. Maria wonders why. She decides to ask her teacher, and as usual, he doesn't provide an answer but instead asks her to think about how she could design a scientific investigation to find out what she wants to know.

Perhaps without noticing, Maria is using the crosscutting concept of patterns on her nature walk. She notices an unexpected pattern in nature, and this prompts her to ask questions about why it is happening. The pattern is unexpected because she knows something about flowers—they grow from seeds, and seeds certainly don't decide on their own where they are planted. This pattern, combined with her knowledge of plants, prompts her to ask questions that can guide an investigation. If Maria chooses to go on with investigating this phenomenon, she will use her existing knowledge of plant growth to design a test that allows her to establish a cause-and-effect relationship (another crosscutting concept) that could explain why the wildflowers grow in some places but not in others. In both asking scientific questions and designing investigations to answer them, Maria is doing more than using crosscutting concepts—she is engaging in three-dimensional learning to make sense of the phenomenon, as scientists do in their work.

Why Are Crosscutting Concepts Useful?

Before we begin discussing the crosscutting concepts in earnest, we would like to direct your attention to the preface of this book, which was written by Helen Quinn, chairperson of the committee that wrote *A Framework for K–12 Science Education* (the *Framework*; NRC 2012) as the foundation for the *Next Generation Science Standards* (*NGSS*; NGSS Lead States 2013a). The preface provides an account of how the crosscutting concepts came to be included in the *NGSS* and a discussion of what has happened in the time since the *Framework* was released. The preface provides an invaluable and updated perspective on the vision of three-dimensional science learning upon which the *NGSS* and this book are based.

The idea behind three-dimensional learning is that making sense of phenomena involves the blending of three interconnected dimensions of science learning: science and engineering practices (SEPs), disciplinary core ideas (DCIs), and crosscutting concepts (CCCs). SEPs describe how students should be directly involved with how science is done in order to better understand how scientists and engineers do their work and how science knowledge develops (Schwarz, Passmore, and Reiser 2017). DCIs identify the core science principles and theoretical constructs within the disciplines of science and engineering (Duncan, Krajcik, and Rivet 2017). CCCs (the focus of this book) are ways of thinking and reasoning about phenomena that are common across all disciplines of science and engineering.

As we have seen with the example of Maria, the three dimensions of science learning do not work in isolation—they are fundamentally connected to one another. Sure, it is possible to learn only science ideas—for example, that plants grow from seeds—but this fact alone is not useful for piquing Maria's curiosity and guiding investigations. Phenomena and problems are central to science and engineering. The *Framework* clarifies that "the goal of science is to develop a set of coherent and mutually consistent theoretical descriptions of the world that can provide explanations over a wide range of phenomena. Engineering puts the principles of science to work in order to design solutions to problems" (NRC 2012, p. 48). In the science classroom, to make sense of phenomena or to design solutions to problems, students engage in all three dimensions of science learning.

Although CCCs have always been used by scientists, they have received relatively little explicit attention with students compared to SEPs and DCIs. (By *explicit*, we mean presented as a distinct element of student understanding.) Yet, CCCs play a critical role in helping students engage meaningfully in science. By explicitly including CCCs as a dimension of science learning that should be integrated with SEPs and DCIs, the *Framework* and *NGSS* more fully describe what it means for students to be supported in actively doing and learning science.

"By explicitly including CCCs as a dimension of science learning that should be integrated with SEPs and DCIs, the *Framework* and *NGSS* more fully describe what it means for students to be supported in actively doing and learning science."

The central purpose of this book is to unpack CCCs and clarify their role in strengthening science and engineering learning. In this introductory chapter, we (1) address the role that CCCs play in the *Framework* and *NGSS*, (2) unpack how CCCs could support student learning and broaden access to science, and (3) outline the structure of the book.

What Are the Crosscutting Concepts and Where Do They Come From?

According to the *Framework*, CCCs "unify the study of science and engineering through their common application across fields" (NRC 2012, p. 2). CCCs are a set of conceptual tools that, along with SEPs and DCIs, help students explain and predict phenomena. They are "crosscutting" because they apply across science disciplines and are widely used in every science and engineering discipline. The committee that developed the *Framework* identified seven CCCs and described them as follows (NRC 2012, p. 84):

1. *Patterns.* Observed patterns of forms and events guide organization and classification, and they prompt questions about relationships and the factors that influence them.
2. *Cause and Effect: Mechanism and Explanation.* Events have causes, sometimes simple, sometimes multifaceted. A major activity of science is investigating and explaining causal relationships and the mechanisms by which they are mediated. Such mechanisms can then be tested across given contexts and used to predict and explain events in new contexts.
3. *Scale, Proportion, and Quantity.* In considering phenomena, it is critical to recognize what is relevant at different measures of size, time, and energy and to recognize how changes in scale, proportion, or quantity affect a system's structure or performance.
4. *Systems and System Models.* Defining the system under study—specifying its boundaries and making explicit a model of that system—provides tools for understanding and testing ideas that are applicable throughout science and engineering.
5. *Energy and Matter: Flows, Cycles, and Conservation.* Tracking fluxes of energy and matter into, out of, and within systems helps one understand the systems' possibilities and limitations.
6. *Structure and Function.* The way in which an object or living thing is shaped and its substructure determine many of its properties and functions.
7. *Stability and Change.* For natural and built systems alike, conditions of stability and determinants of rates of change or evolution of a system are critical elements of study.

CCCs appear prominently in the *Framework* and *NGSS*, but they are not inventions of the committee. They have consistently been an explicit part of the practice of science, and they were included in previous standards documents. More than 25 years ago, the *Benchmarks for Science Literacy* (AAAS 1993) elaborated on four "Common Themes" across science disciplines (systems, models, constancy and change, and scale). Even before that, these four common themes were identified in the groundbreaking *Science for All Americans* (AAAS 1989).

What is new in the *Framework* and *NGSS* is the extent to which CCCs have been explicitly integrated with SEPs and DCIs. CCCs are not an afterthought; on the contrary, they are critical to the vision of three-dimensional learning. The explicit inclusion of CCCs as a dimension of science learning and their integration into every performance expectation (PE) in the *NGSS* is a recognition that these critical conceptual tools have long been a necessary part of science but have largely remained implicit rather than explicit in science instruction. As a result, many students have been under-supported in their science learning and ability to engage meaningfully in the type of science learning that has long been advocated by science standards and visioning documents (NRC 1996; Olson and Loucks-Horsley 2000). Science instruction that includes an explicit emphasis on CCCs holds the promise of offering better supports for student engagement in scientific thinking.

How Can the Crosscutting Concepts Strengthen Students' Science Learning?

Including CCCs explicitly as a dimension of science learning reflects a recognition that they are a critical part of science and that students need better support in using these conceptual tools effectively. The precise role that CCCs play in students' science learning is an active area of research and debate within the research community, with some going so far as to question the value of including the CCCs as a dimension of science learning in the *NGSS* (Osborne, Rafanelli, and Kind 2018). While the discussion continues, there is substantial agreement that making CCCs more explicit in science instruction holds the promise of better supporting students in three key ways: (a) integrating CCCs with SEPs and DCIs to make sense of phenomena, (b) using CCCs to make connections across science disciplines, and (c) using CCCs as lenses on phenomena (Fick, Nordine, and McElhaney 2019; Rivet et al. 2016; Science SCASS States 2018).

Integrating Crosscutting Concepts With Science and Engineering Practices and Disciplinary Core Ideas

Too often, science instruction focuses on conveying scientific ideas to students without deeply involving them in the sensemaking process. When students' knowledge of science principles is the focus of instruction and assessment, it is easy to lose sight of the idea that science ideas by themselves are not particularly useful. For example, when students learn about heat transfer, it is common to emphasize that heat can be transferred via convection, conduction, and radiation. However, the ability to recount and define these transfer mechanisms does little to help students use what they know about heat transfer to decide how best to keep their drink cold on a hot day. To use DCIs productively, students must engage in SEPs, and CCCs play an indispensable role in connecting these two dimensions.

To illustrate the value of CCCs in the process of doing science and using science ideas, consider the role of the CCC of cause and effect (see Chapter 5) in a scientific experimentation. Controlled experiments, in which only one variable is changed between an experimental and a control condition, are inherently designed to establish cause-and-effect relationships between the variable that is changed and the outcome that is measured. In a controlled experiment, cause-and-effect relationships are relatively straightforward because any change in the outcome must be caused by the variable that is changed. For example, one common school science experiment is investigating the effect of light on plant growth by placing one plant of a particular type in the sunlight and another plant of the same type in a closed cabinet. Students readily accept that the difference in growth is caused by light, and an understanding of photosynthesis further strengthens this causal connection by providing a mechanism to explain why sunlight leads to plant growth.

Cause-and-effect relationships that appear to be straightforward are often difficult to establish, especially when controlled experiments are not possible. For example, it is not possible to run experiments in which the effect of single factors that influence global temperatures can be investigated; however, it is still important to try to establish whether certain factors cause global temperatures to change. In this case, it is important to gather a large amount of data and look for patterns (see Chapter 4). Recognizing and describing patterns is critical in science, but the existence of patterns alone does not provide sufficient grounding to make the sorts of causal connections that are important for explaining and predicting phenomena. For example, as the COVID-19 pandemic began to unfold, some people began to notice a pattern of overlap between the location of confirmed COVID-19 cases and the location of newly installed 5G cellular towers. Images like the one in Figure 1.2 began to circulate online.

Figure 1.2. Graph showing 5G cellular coverage (top) and the approximate location of COVID-19 cases (bottom) in spring 2020

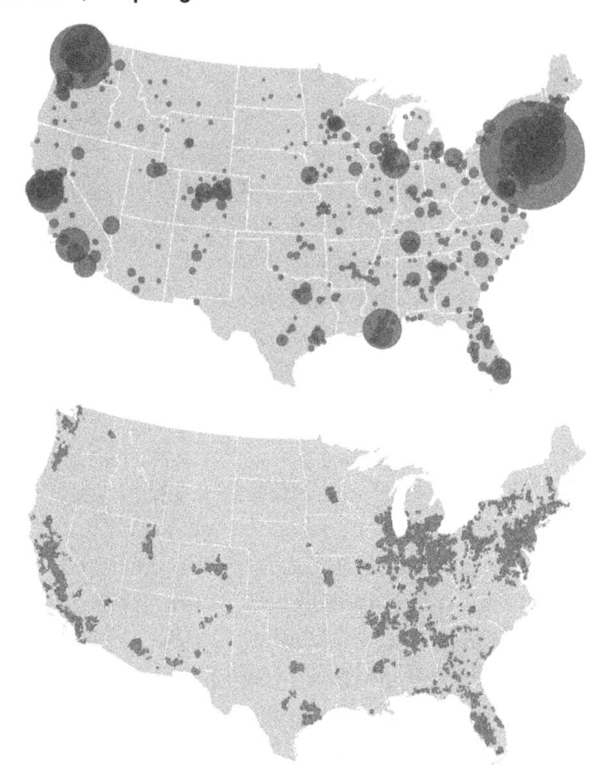

The similarity in the pattern of 5G networks and COVID-19 cases was used by some to support a claim that 5G towers were causing illnesses reported as COVID-19—ignoring that both 5G towers and COVID-19 cases are both more likely to be found in areas with large, dense populations. To establish causality, one must do more than find a pattern in data. Scientific principles must be used to establish a causal mechanism for why one factor should influence another—this is why mechanism and explanation are included in the CCC of cause and effect, appearing after the colon. In the example of COVID-19, scientific ideas about the transmission of viruses and the symptoms they can cause provide a necessary causal mechanism to explain why patterns of COVID-19 cases are connected to population density rather than 5G tower distribution. In summation, to make causal claims about how and why phenomena occur, one must make connections between SEPs, CCCs, and DCIs.

Determining causality is certainly not the only area in which three-dimensional learning is required. Three-dimensional learning emphasizes that the process of doing science relies more generally on different dimensions of conceptual tools—SEPs, CCCs, and DCIs. For example:

- To develop and use scientific models of phenomena (SEP), students must carefully define the system that is represented by the model (CCC) and use one or more core ideas or elements of core ideas (DCIs) to specify how the various components of the system interact (CCC).
- To make measurements during an investigation, students must use standard units to quantify and relate changes in observed quantities (CCC) and to analyze those data (SEP) in light of at least one core idea (DCI).

Including CCCs as an explicit part of three-dimensional learning recognizes the critical role they play in scientific thinking. When science instruction explicitly focuses on CCCs, students are better supported in using their knowledge to explain and predict phenomena.

Using Crosscutting Concepts to Make Connections Across Science Disciplines

CCCs were identified because they are used across science and engineering disciplines. This broad usage is perhaps easier to see for some CCCs than others. For example, it is not difficult to see that all scientists and engineers look for and describe patterns. Pattern recognition is a core element of how people make sense of the world, from learning the rules of language to recognizing that a full Moon occurs about once every month. It is no surprise that patterns play a central role in every science discipline. On the other hand, it can be less clear to see connections across science disciplines in terms of how the CCC of energy and matter: flows, cycles, and conservation is used. For instance, in physics class, energy is commonly discussed as a conserved quantity that can be either "potential" or "kinetic"; in biology, energy is often discussed as a resource that is stored within ATP molecules and used as life functions are carried out. These different ways of discussing energy make sense within the context of the phenomena under consideration, but such differences obscure similarities in how key energy ideas are used more consistently across disciplines. The CCC of energy and matter: flows, cycles, and conservation stresses that energy (and matter) is a conserved quantity that can never be created or destroyed—only manifested in different ways or transferred between systems. Thus, as students learn about potential and kinetic energy in physics or ATP in biology, they would benefit from more explicit discussion of how these ideas connect to the conservation of energy and its transfer between systems. (For elaboration of these ideas, see Chapters 8 and 12.). Every CCC is used across science and engineering disciplines, but students need support in the form of explicit instruction in order to see the connections across disciplines and to learn to use CCCs across a wide range of phenomena and problems.

Explicit instruction about CCCs may help students see similarities in how problems can be approached across a range of disciplinary contexts and develop a more consistent and coherent understanding of science and engineering. A central assertion of the *Framework* is that explicit instruction about CCCs will "help provide students with an organizational framework for connecting knowledge from the various disciplines into a coherent and scientifically based view of the world" (NRC 2012, p. 83). As previously noted, evaluating this claim is an active area of research, but there is ample evidence that students commonly develop disjointed and disconnected ideas about science (NRC 2007). CCCs may help students see new connections. Furthermore, just because today's scientists may not use and discuss ideas, such as energy, in overtly similar ways doesn't mean that making these connections more explicit won't help the scientists and engineers of tomorrow. Some of the most promising and impactful work in science and engineering is occurring across the boundaries of the traditional disciplines, and making sense of most socioscientific issues (e.g., sustainable resource usage, genetically modified foods) requires ideas from several disciplines. Including CCCs as an explicit part of science instruction recognizes the interdisciplinary nature of the problems and issues that will likely be faced by future generations.

Explicit instruction about CCCs may help students see similarities in how problems can be approached across a range of disciplinary contexts and develop a more consistent and coherent understanding of science and engineering.

> Among the most powerful ways that CCCs help students learn science is by providing multiple lenses through which they can make sense of phenomena.

Using Crosscutting Concepts as Lenses on Phenomena

Among the most powerful ways that CCCs help students learn science is by providing multiple lenses through which they can make sense of phenomena. As lenses on phenomena, CCCs provide different perspectives for thinking about how and why phenomena occur. Thus, one way to put different CCC lenses into practice is to think about different questions or prompts that align with each CCC (Penuel and Van Horne 2018). Table 1.1 shows some examples of questions that students might consider when using the various CCC lenses to make sense of phenomena and engage in engineering design. These questions are taken or adapted from a list provided by the aforementioned Helen Quinn, chairperson of the *Framework* committee and author of the preface for this book.

Table 1.1. Questions that may guide students in using the CCCs as lenses on phenomena

Crosscutting Concept	Example Questions
Patterns	• What do I notice after careful observation? Do any features or patterns emerge that are interesting? • What patterns or relationships do I see in the data? • Can I explain the causes of these patterns?
Cause and Effect: Mechanism and Explanation	• Does it seem likely that some events or changes are causing others? • What is the mechanism by which one change causes another? • How can I design a system to cause the desired effect?
Scale, Proportion, and Quantity	• What should be measured or quantified in order to better understand this phenomenon? • How can we describe the relationships between different measured quantities? • Can I use a scale model or prototype to test my design?
Systems and System Models	• What is involved in this phenomenon? What is not? • What are the components of the system under consideration? What are the relationships between them? • How could I represent this system with a diagram or model?

Continued

Table 1.1. *(continued)*

Crosscutting Concept	Example Questions
Energy and Matter: Flows, Cycles, and Conservation	• What matter flows into or out of the system? • What energy transfers occur into or out of the system? • What physical or chemical changes happen within the system?
Structure and Function	• What shapes or structures can I observe within this system? • What role do these shapes or structures play in how the system functions? • What features of the shape or structure of this design are desired by the user?
Stability and Change	• Under what range of conditions does this system operate effectively? • What changes would cause the system to become unstable or fail? • How can I improve the stability of my design?

Different CCCs provide complementary lenses on phenomena and problems. By looking at the same phenomenon in different ways, students can expand the richness of their sensemaking and generate new questions about how and why things happen, which generate scientific investigations to answer these questions.

To illustrate the power of using different lenses to think about the same phenomenon, consider how students typically learn about watersheds. Many students learn that natural land features create dividing lines between land areas, which separate whether precipitation collects in one body of water or another. However, in learning only about watersheds from a landforms perspective, they may miss out on developing a much deeper and more useful understanding of the Earth.

Consider the critical role that systems thinking plays in watersheds (Fick 2018). A systems lens involves identifying inputs and outputs to the watershed system, as well as the identification of subsystems within the watershed. Using the systems lens, students are prompted to think explicitly about how water flows into and out of the system, and whether there are subfeatures within a watershed that may lead to different water features (e.g., two tributaries that join each other in a larger river). Fick (2018) has shown that such explicit systems instruction can enrich student discussion and learning about watersheds.

Now consider watersheds from the lens of energy and matter flows, and apply this lens to a watershed that is fed by the melting of snow and ice. The rate at which water flows through the watershed is directly influenced by how energy from the Sun is transferred at different rates to ice and water due to differences in how they absorb sunlight. Furthermore, more direct light from the Sun (and thus higher energy transfer to the watershed) increases the rate of evaporation from a body of water, which influences both the volume of water in water reservoirs and the flow rate through rivers and streams. Using an energy and matter lens can help students understand why the same body of water can look so different at different times of year.

Finally, consider watersheds from the lens of structure and function. Rather than focusing on defining a watershed as a land area that feeds certain water reservoirs, a structure and function lens emphasizes how the specific structural features of an area lead to different dynamics. For example, a deep lake with a small surface area has a much slower evaporation rate and a more stable volume than a shallow lake with a large surface area. A structure and function lens helps students generate insights that may be missed by simply focusing on defining watersheds in terms of land areas.

Watersheds are just one example of how using different CCCs to interpret the same phenomenon can lead to new insights that enrich students' understanding. Student learning about the phases of the Moon, ecosystems, electric circuits, the periodic table, and more (the list goes on and on!) can all be enhanced by applying a variety of CCCs as lenses on phenomena.

Broadening Access to Science and Inclusion of All Students in the Science Classroom

Broadening access to science is a central theme of the *Framework* and *NGSS*, hence "All Standards, All Students" as the vision of the *NGSS* (NGSS Lead States 2013b). As we have seen through various examples in this chapter, CCCs provide important conceptual tools for students to productively connect SEPs with DCIs as they make sense of phenomena and problems. Likewise, scientists and engineers commonly use CCCs to make such connections in their work. Through the *Framework* and *NGSS*, explicit teaching of CCCs could promote access to science and inclusion in the science classroom for students who otherwise might not have exposure to such opportunities.

CCCs provide a conceptual lens (or multiple lenses) to wonder and ask questions about phenomena. Think back to the opening of this chapter, which describes a nature walk taken by Maria's science class. During the walk, Maria notices something funny about the wildflowers in the park by the school. She wonders why this phenomenon is occurring. She uses the CCC of patterns and the CCC of cause and effect, which serve as lenses on the phenomenon and provide an entry point into learning science. Traditionally, the science classroom expects students to come to science, as students are expected to learn about canonical knowledge of science. In contrast, phenomena—especially in local contexts that are real and relevant to students, such as the phenomenon that Maria notices—allow students to realize that science has a bearing on their lives and future careers and that they are scientific thinkers (Lee 2020). Moreover, the science classroom traditionally assumes students, especially those who have been underserved, bring little or limited prior knowledge as defined in terms of canonical knowledge of science. On the contrary, all students bring to the science classroom their funds of knowledge from home, community, and school, which can serve as intellectual resources to learn science. For Maria, her prior knowledge and her use of CCCs as a lens on the phenomenon make her well equipped to develop an understanding of science.

CCCs are critical for using SEPs with DCIs together to make sense of phenomena. Out of wonder and curiosity, Maria wants to figure out why the flowers seem to grow in some areas but not in others—she notices a pattern that prompts her to ask a question. She engages in SEPs, as scientists do in their work. After asking a question, she plans and carries out an investigation to explore cause-and-effect relationships, and she develops and revises her model until she constructs an explanation of the phenomenon and communicates her explanation with others. Through engagement in SEPs and use of CCCs, she develops an understanding of DCIs and begins to learn more about the phenomenon that piqued her curiosity.

“Teachers can make CCCs explicit to promote science learning for all students, especially those who do not see science as real or relevant to their lives or future careers.”

Teachers can make CCCs explicit to promote science learning for all students, especially those who do not see science as real or relevant to their lives or future careers. After first noticing the flower pattern, Maria asks her teacher about it. He doesn't provide her with a definitive answer but instead asks her to think about how she can design a scientific investigation to find out. He could capitalize on his students' prior experiences in home, community, and school. Building on their prior experiences, he could guide them to engage in SEPs and develop an understanding of DCIs to explain the phenomenon. Over time, he could scaffold CCCs so students learn to use CCCs independently across science disciplines and across contexts.

CCCs have traditionally been implicit in the science classroom. According to the vision of the *Framework* and *NGSS*, CCCs could serve as conceptual tools for all students to engage in three-dimensional learning. By making CCCs explicit in the science classroom, teachers could provide all students with opportunities to make sense of phenomena and problems, as scientists and engineers do in their professional work.

Summary

CCCs have the potential for strengthening students' science and engineering learning in three ways: integrating with SEPs and DCIs to support sensemaking and problem solving, supporting connections between contexts and disciplines, and serving as a set of lenses through which to view phenomena and problems.

Three-dimensional learning emphasizes that making sense of phenomena (in science) and solving problems (in engineering) requires the blending of three dimensions of conceptual tools: SEPs, CCCs, and DCIs. Although the role of CCCs has long been implicit, the *Framework* and *NGSS* make this role explicit. Explicitly supporting students' use of CCCs as they make sense of phenomena and solve engineering problems helps support all students in more fully participating in science and engineering and in learning with deeper understanding.

The wonderful thing about CCCs, and why they have been identified as "crosscutting," is that they are very broadly applicable across a diverse range of disciplinary contexts. Try it for yourself—pick a CCC, look around you, and think how it can be applied to interpret different things you see. You may notice that you focus on features you may normally have missed. Now pick a different CCC. You'll likely find that applying different CCC lenses helps you notice things you might not have before. One of the most powerful features of CCCs is that they provide students with a set of complementary lenses through which they can interpret and learn about the world around them. This book is about leveraging the power of CCCs to strengthen students' science learning.

Structure of This Book

This book is divided into four sections. Part I addresses foundational issues that undergird the CCCs and how they might change classroom instruction, strengthen students' engagement in science, and broaden access and inclusion of all students in the science classroom. Part II includes an in-depth look at each CCC. Each chapter is written by a team of science educators who bring experience in both research and K–12 classroom teaching, and each includes a discussion of why the CCC is a powerful conceptual tool, how the CCC is used across disciplines, challenges that students face in learning about the CCC, and exemplary teaching strategies. Whereas Part II focuses on each CCC individually, Part IV focuses on how CCCs can strengthen the teaching of canonical topics in science (i.e., nature of matter, plant growth, and weather and climate) and the engagement of students in engineering design. Finally, Part 4 addresses issues that relate to using CCCs to enhance and enrich science instruction, including implications for student assessment and professional collaboration.

Throughout the book, chapter authors illustrate examples of CCC-informed instruction drawn from their experience in the classroom. These illustrations of practice fall into two categories: Instructional Applications and Classroom Snapshots. (Note that all student names in Classroom Snapshots have been changed for privacy.) We draw a distinction between Instructional Applications and Classroom Snapshots in order to highlight the role the CCCs play in both instructional planning and real-time facilitation of classroom activities.

The central goal of this book is to clarify the nature and potential of CCCs to strengthen science and engineering teaching and learning. We hope this book provides you with insights into the design and implementation of effective three-dimensional science instruction that leverages the power of CCCs. As you read, we hope three core messages become clear. First, CCCs are conceptual tools that, when used along with science and engineering practices and disciplinary core ideas, provide lenses for making sense of phenomena and problems. Second, CCCs should be made accessible to all students, as they hold great promise for both broadening access to science and strengthening outcomes for a diverse group of science and engineering learners. Third, there is a powerful opportunity—in fact, a critical need—for teacher leadership in illustrating what high-quality CCC-informed instruction looks like. The research literature on the teaching and learning of CCCs is limited, and although the ideas presented in this book are based on years of experience in teaching and research, there is still much to be done to better understand the opportunities and challenges of CCC-informed instruction.

We hope this book serves as a guide for designing and implementing truly three-dimensional instruction for all students, and we hope it further serves as a call to action both to re-envision what science and engineering instruction can look like and to actively contribute to the growing community of educators who are seeking to better understand the potential of CCCs for strengthening science and engineering teaching and learning.

References

American Association for the Advancement of Science (AAAS). 1989. *Science for all Americans*. New York: Oxford University Press.

American Association for the Advancement of Science (AAAS). 1993. *Benchmarks for science literacy*. New York: Oxford University Press.

Duncan, R. G., J. S. Krajcik, and A. E. Rivet, eds. 2017. *Disciplinary core ideas: Reshaping teaching and learning*. Arlington, VA: NSTA Press.

Fick, S. J. 2018. What does three-dimensional teaching and learning look like?: Examining the potential for crosscutting concepts to support the development of science knowledge. *Science Education* 102 (1): 5–35. *https://doi.org/10.1002/sce.21313*.

Fick, S. J., J. Nordine, and K. W. McElhaney, eds. 2019. Proceedings of the summit for examining the potential for crosscutting concepts to support three-dimensional learning. University of Virginia. *http://curry.virginia.edu/CCC-Summit*.

Lee, O. 2020. Making everyday phenomena phenomenal: Using phenomena to promote equity in science instruction. *Science and Children* 58 (1): 56–61.

National Research Council (NRC). 1996. *National science education standards.* Washington, DC: National Academies Press.

National Research Council (NRC). 2007. *Taking science to school: Learning and teaching science in grades K–8.* Washington, DC: National Academies Press.

National Research Council (NRC). 2012. *A framework for K–12 science education: Practices, crosscutting concepts, and core ideas.* Washington, DC: National Academies Press.

NGSS Lead States. 2013a. *Next Generation Science Standards: For states, by states.* Washington, DC: National Academies Press. *www.nextgenscience.org.*

NGSS Lead States. 2013b. *NGSS Appendix D: All standards, all students: Making the Next Generation Science Standards accessible to all students.* Washington, DC: National Academies Press. *www.nextgenscience.org.*

Olson, S., and S. Loucks-Horsley, eds. 2000. *Inquiry and the national science education standards: A guide for teaching and learning.* Washington, DC: National Academies Press.

Osborne, J., S. Rafanelli, and P. Kind. 2018. Toward a more coherent model for science education than the crosscutting concepts of the *Next Generation Science Standards*: The affordances of styles of reasoning. *Journal of Research in Science Teaching* 55 (7): 962–981. *https://doi.org/10.1002/tea.21460.*

Penuel, W. R., and K. Van Horne. 2018. STEM teaching tool #41: Prompts for integrating crosscutting concepts into assessment and instruction. STEM Teaching Tools. *http://stemteachingtools.org/brief/41.*

Rivet, A. E., G. Weiser, X. Lyu, Y. Li, and D. Rojas-Perilla. 2016. What are crosscutting concepts in science? Four metaphorical perspectives. In *Transforming learning, empowering learners: The international conference of the learning sciences (ICLS), Volume 2,* ed. C. K. Looi, J. L. Polman, U. Cress, and P. Reimann. 970–973. Singapore: International Society of the Learning Sciences. *https://repository.isls.org/handle/1/356.*

Schwarz, C., C. Passmore, and B. J. Reiser, eds. 2017. *Helping students make sense of the world using next generation science and engineering practices.* Arlington, VA: NSTA Press.

Science SCASS States. 2018. Using crosscutting concepts to prompt student responses. CCSSO Science SCASS Committee on Classroom Assessment.

Chapter 2

How Crosscutting Concepts, Disciplinary Core Ideas, and Science and Engineering Practices Work Together in the Classroom

Joseph Krajcik and Brian J. Reiser

A Framework for K–12 Science Education (the *Framework*; NRC 2012) presents a new and exciting vision of science teaching and learning—one in which students do science by making use of the three dimensions of scientific literacy to make sense of phenomena and develop solutions to design problems. The three dimensions of scientific knowledge—disciplinary core ideas (DCIs), science and engineering practices (SEPs), and crosscutting concepts (CCCs)—work together to support students in figuring out phenomena and building deep and useable knowledge (Pellegrino and Hilton 2012).

DCIs represent the most central ideas of the disciplines of physical science, life science, Earth and space science, and engineering (Duncan, Krajcik, and Rivet 2016). They play a vital role in helping explain a range of phenomena and solving challenging engineering problems. For example, the disciplinary core idea ESS2.C: The Roles of Water in Earth's Surface Processes is a component idea of a larger DCI, ESS2: Earth's Systems. ESS2.C is critical in explaining various phenomena in the Earth sciences, such as the movement of water in the atmosphere and water movement both on and under the ground that helps explain land formations.

Crosscutting concepts, such as cause and effect, structure and function, and stability and change, occur within and across disciplinary boundaries that can help students make sense of phenomena or solve complex problems. We like to think of crosscutting concepts as lenses to examine and explore numerous phenomena. For instance, looking for patterns in streams and riverbanks might help one connect the mechanism of running water with how rocks and soil are deposited more in some spots than in others. Looking at the shape of a riverbank through the lens of structure and function focuses investigators' attention on how the material making up the riverbank may have affected its shape. For instance, even a fast-moving river would have difficulty cutting through granite, whereas a fast-moving river can cut through limestone deposits in a relatively short period of time.

Science and engineering practices (Schwarz, Passmore, and Reiser 2017) are the ways of knowing and doing that scientists and engineers use to investigate and make sense of the natural and designed worlds and to develop solutions to challenging and complex problems. Furthermore, SEPs are the means by which we can build, test, refine, and apply scientific knowledge to explain and make changes in our world. In the effort to explain patterns (a crosscutting concept) that characterize rivers and streams, one might ask what evidence could be collected to support the claim that fast-moving water carries more Earth materials, drawing on the practices of designing and conducting investigations and building an explanation. One might try to build and apply a model to explain why some riverbanks erode more than others.

Although each dimension is important on its own, these three dimensions work together to support researchers and students in figuring out phenomena. This integration of the three dimensions is referred to as three-dimensional learning and supports students in developing deep and useable knowledge they can employ to solve future challenging problems and make sense of new phenomena they experience in their lives.

In this chapter, we examine the essential role that crosscutting concepts play in three-dimensional learning. We focus on the role that CCCs can play in classroom teaching, curriculum materials design, and the work of students in figuring out phenomena or in solving design problems. We explore the following questions: How do the three dimensions of scientific knowledge work together to support learners in making sense of phenomena? What added value do the crosscutting concepts bring to the other two dimensions in making sense of phenomena and solving engineering design problems? How can we include CCCs as tools for *figuring out* phenomena and solving problems, rather than *learning about* the CCCs as objects in and of themselves because the book and teacher say they are important (Schwarz, Passmore, and Reiser 2017)? How explicit should we be in using the crosscutting concepts in instruction?

A Brief Description of Three-Dimensional Learning

A major innovation in *A Framework for K–12 Science Education* (NRC 2012) is that science and engineering practices, disciplinary core ideas, and crosscutting concepts work together to enable learners to figure out phenomena or to find solutions to engineering design challenges. Research from the learning sciences (Brown, Collins, and Duguid

1989; Krajcik and Shin 2014; NRC 2007; Sawyer 2008) shows that content ideas taught in isolation from how learners are to use them results in ideas that learners find difficult to use for future problem solving and sensemaking. Even though an individual might know about an idea, they can't apply it to new situations. Similarly, using science processes or inquiry skills in isolation of science ideas leads to learning how to carry out procedures but not knowing why or when to use those ideas to make sense of the world. Disciplinary core ideas and crosscutting concepts work in conjunction with the science and engineering practices to form deep knowledge that learners can apply to make sense of their world. Moreover, as learners use the three dimensions to make sense of the world, they further develop all three dimensions across time. People cannot learn science ideas isolated from the doing of science, and they cannot learn the practices of science (i.e., the SEPs) in isolation from the ideas of science (i.e., the DCIs and CCCs). Learning how to make use of knowledge will only occur by students using the three dimensions together to make sense of the world.

What Role Should Crosscutting Concepts Play in Three-Dimensional Teaching and Learning?

In the sections below, we describe concrete strategies for using crosscutting concepts, along with science and engineering practices and disciplinary core ideas, in classroom teaching and learning to make sense of phenomena or solve complex problems. Although the *Framework* (NRC 2012) and *Next Generation Science Standards* (*NGSS;* NGSS Lead States 2013a) specify that students should use all three dimensions together in order to use science to make sense of their world and to solve problems, many questions arise about how to orchestrate three-dimensional learning in classroom teaching and in the design of instructional materials. How should the three dimensions work together? How can students use a crosscutting concept in figuring out phenomena? What advantage does the use of crosscutting concepts provide beyond the use of disciplinary core ideas and scientific practices? In this section, we consider how crosscutting concepts can be used in synergy with the other two dimensions, disciplinary core ideas and science and engineering practices, to support learners in the figuring-out process. Science is all about explaining and making sense of the world, and to do so, scientists (including children and learners of all ages who are exploring the world) must use the three dimensions in an integral fashion. We will use two examples in this chapter to illustrate using the three dimensions to explain phenomena. We will highlight the role that crosscutting concepts play, along with the disciplinary core ideas and science and engineering practices, in the sensemaking process. The first example, a composite hypothetical example that stems from the experiences of the authors, highlights the use of crosscutting concepts; the second example, which stems from actual classroom use, illustrates how crosscutting concepts work in a classroom setting.

Example One: Why Does a Puddle Evaporate on a Cold Day?

This first example reflects a composite of experiences the authors had in observing middle school students making sense of an everyday phenomenon—the evaporation of a puddle on a cold day. In this example, we show the important role that several crosscutting concepts (in particular, patterns, cause and effect, and structure and function), along with various SEPs and DCIs, play in helping students make sense of the phenomenon.

The teacher's goal in this scenario is to help her students build understanding toward the following *NGSS* performance expectation[1]:

> MS-PS1-4. Develop a model that predicts and describes changes in particle motion, temperature, and state of a pure substance when thermal energy is added or removed.

To begin, the science teacher has her students make observations about a puddle of water on a cool but sunny day. The students observe a pattern (a crosscutting concept) that puddles of water evaporate outside not only on very hot days but also on cool days. The class asks the question (a scientific practice) "How can this happen?" The students in the class are also curious if the puddle would evaporate faster if the temperature were warmer. One student asks, "Would the puddle still evaporate if it were the same temperature outside but overcast?" The teacher has students make observations about a puddle evaporating as she suspects it will allow them to raise questions, engage in other scientific practices, and use crosscutting concepts and disciplinary core ideas to make sense of the phenomenon. She explicitly asks, "What patterns do you observe?" and "What might be causing the patterns?" Notice how she explicitly uses two crosscutting concepts, patterns and cause and effect, to suggest ways to interrogate the phenomenon.

How could a middle school classroom teacher go about supporting students as they work to explain this phenomenon? To figure out the phenomenon, the students in the class would need to apply (1) disciplinary core ideas from several disciplines, including energy and forces from physical science, (2) several crosscutting concepts, including systems and system models, structure and function, and patterns, and (3) various science and engineering practices, including developing and using models, using mathematics and computational thinking, and analyzing and interpreting data. Notice that in making the initial observation, the students in the class apply a crosscutting concept—patterns. The evaporation of a puddle is a very repeatable phenomenon—even on cool or cold days. If the temperature outside were warmer, the students would notice that the puddle evaporates faster than when it is cooler. Here, students apply a disciplinary core idea (the transfer of energy in an open system) and a crosscutting concept (cause and effect) to build an explanation. We find it critical for teachers to understand how to

1. The *NGSS* use an abbreviation system for referencing the performance expectation. For example, MS-PS1-4 is a middle school (MS) physical science (PS) performance expectation or standard. The numbering system 1–4 refers to the disciplinary core idea. The 1 refers to the disciplinary core idea of matter and interactions, and the 4 indicates it is the fourth middle school standard in this disciplinary core idea.

use crosscutting concepts strategically to focus students' questioning and investigation. Teachers need this explicit knowledge about the role crosscutting concepts can play and facility with the meaning and utility of each crosscutting concept. For students, explicit declarative knowledge about crosscutting concepts is less important than being able to use these concepts as a tool. For example, learners need to know we need to represent cause-and-effect relationships when we build and evaluate models and that looking for causes is a general part of figuring out phenomena in science. They do not need to know that cause and effect is labeled as a crosscutting concept or be able to recount and define all the crosscutting concepts when prompted.

To help the students develop an explanation of what causes the puddle to evaporate, the teacher has them draw and share models of their beginning ideas. She stresses that models are to provide causal mechanisms for why the phenomenon occurs and stresses that relationships need to show cause and effect. In developing and sharing their initial models, students apply a crosscutting concept (i.e., cause and effect), a science and engineering practice (i.e., developing and using models), and initial disciplinary core ideas about the particle nature of matter and the transfer of energy (i.e., when enough energy is transferred to a molecule near the surface of liquid water, the molecule gains enough energy to overcome the attraction of other water molecules; it will then escape and enter the surrounding air as a molecule of water in the gas phase).

The students in the class now wonder how they could gather evidence to support their idea that fast-moving particles will leave the surface of the puddle. They reason that as more and more fast molecules leave the surface of the liquid water, the temperature of the water will decrease. The teacher guides the students in designing an experiment to gather evidence to support their claim. The students, with support from the teacher, decide to do an experiment that involves wrapping a temperature probe in some cloth, dipping the temperature probe into room-temperature water, and observing and collecting data of the temperature change as the water evaporates from the cloth. The class observes that the water temperature continually decreases below room temperature until all the water evaporates. Two crosscutting concepts are involved in making sense of these observations: patterns and energy and matter: flows, cycles, and conservation.

With the teacher's guidance, the students in the class ask, "Why does this happen?" Then they engage in constructing their next model. One group of students draws a model showing that water particles with more energy could break away from the attraction of the other water particles and escape from the surface of the water to the surrounding environment because they are moving faster. As students provide feedback to one another in groups, students in another group ask, "Why is it that some of the molecules are moving faster, and how does that account for a lower temperature?" A student from the first group suggests that the escaping particles cause the temperature of the liquid water to decrease because energy is transferred out of the water system and into the surroundings by the faster-moving water molecules that left the water. These students' questions and comments are touching on two very important crosscutting concepts: structure and function and cause and effect. The teacher guides the students in examining the idea that the structure at the particle level has an important consequence at the macroscopic level (function)—that is, as more and more water particles leave the surface of the cloth, the temperature decreases because the fast-moving particles leave, and slower-moving particles are left on the cloth.

The students suggest a candidate claim for their findings that the water temperature decreases during evaporation. But how could they gather evidence for it? The teacher decides to push the students to see if they can design a way to gather further evidence. One student suggests they conduct the experiment in a closed system (a crosscutting concept) and wonders what would happen if they did an investigation in which they place a temperature probe wrapped in cloth and dipped in water into a plastic bag and then sealed the plastic bag. The students wonder what they would observe and predict that the temperature would stay the same since the water could not evaporate—as it would be contained in the plastic bag. Students collect data from this experimental setup and observe that the temperature does not decrease. Some of the water evaporates, but the rate of evaporation and condensation become equal when the water inside the bag becomes saturated so that the molecules leaving the surface of the water equal those turning into a liquid. The students decide to draw a model of this situation. In drawing their model, they apply another crosscutting concept—stability and change. The model shows that some of the water molecules will escape the surface (change at the micro level), but because the bag will become saturated with water, the probe will stay wet as molecules leave the surface but also condense so that the temperature does not change. (At the micro level, change is constantly occurring as molecules leave and go back into the liquid phase, but at the macro level no change is observed.)

What have the students figured out in this example? They have figured out that water in the puddle will continually evaporate because energy is transferred from the Sun to the water, causing the surface molecules to move faster so they can break away from the other water molecules at the surface of the liquid water. In so doing, students make use of not only various elements of disciplinary core ideas but also the following crosscutting concepts: cause and effect; scale, proportion, and quantity; systems and system models; and structure and function. Students also use various science and engineering practices: asking questions and defining problems, analyzing and interpreting data, planning and carrying out investigations, and developing and using models.

What Role Did the Crosscutting Concepts Play in Explaining the Phenomenon if We Already Have Disciplinary Core Ideas and Science and Engineering Practices?

Why do we need CCCs if we already have DCIs and SEPs? How did the students and teacher make use of crosscutting concepts in helping to develop an explanation of how a puddle can evaporate even on a cool day? To understand a complex phenomenon, learners need to use lenses from various crosscutting concepts in combination with the DCIs and SEPs. First, they used the CCC of patterns to notice that water in a puddle will evaporate even on a cool day and that this pattern happens consistently. Although this pattern of observations describes what the phenomenon does, it does not explain the phenomenon. The teacher suggested drawing a model to characterize at the microscopic level what was causing the temperature to decrease at the macroscopic level. Notice how the teacher explicitly asked the students to use a crosscutting concept, cause and effect (although without necessarily identifying it as a crosscutting concept). The students then drew models (an SEP) to provide a mechanism for why the temperature would drop. Drawing the model facilitated looking at the phenomenon through the lens of cause and effect: mechanism and explanation (a CCC). Students also discussed how what was happening at the microscopic level affects what is observed at the macroscopic level. Here, students were examining the phenomenon by using two additional crosscutting concepts—scale, proportion, and quantity and structure and function. The idea of scale becomes important because the observations the students made at the macroscale are only explainable by what happens at the microscopic level. Students also looked at the phenomenon with respect to open and closed systems and how they affected evaporation. In their experiment to gather evidence that evaporation would not occur in a closed system and that the temperature would remain the same, students looked at the phenomenon through the lens of stability and change. This short scenario provides an example of three-dimensional learning in which learners made use of six of the seven crosscutting concepts as tools to help figure out how a phenomenon works.

In figuring out why this phenomenon occurred, students used specific disciplinary core ideas, science and engineering practices, and crosscutting concepts to develop and use the elements of MS-PS1-4. Although the teacher was supporting her students in building understanding of MS-PS1-4, students also used other SEPs, elements of different DCIs and various crosscutting concepts not explicit in the performance expectation. When figuring out phenomena, students will use multiple elements of DCIs, SEPs, and CCCs. This supports them in building deeper knowledge of all the dimensions so learners can use them in different situations. The three dimensions work together to explain phenomena and help students build knowledge.

Example Two: Students Use Crosscutting Concepts as Lenses to Develop Questions in a Unit on Thermal Energy Transfer

The second example comes from a middle school classroom enacting the OpenSciEd unit *How Can Containers Keep Stuff From Warming Up or Cooling Down?* (Mohan 2019). The teacher, Ms. Bennett, teaches in an urban school in the Midwest. We describe the anchoring phenomenon lesson, in which crosscutting concepts are used in several different ways as critical elements of beginning students' three-dimensional learning about the phenomena of thermal energy transfer. The example illustrates how crosscutting concepts can help guide students' initial explanations of phenomena, uncover key parts of phenomena that need further figuring out, and guide students' development of questions that drive the sensemaking work of the unit. The unit anchor uses the phenomenon of two cups that differ in how well they keep a cold drink from warming up—a "regular cup" and a "fancy cup"—to raise questions about why objects cool off or warm up in general.

The curriculum materials reflect strategies for using crosscutting concepts in the design of the tasks, the tools students use, and the suggestions for teacher guidance. In this unit, the teacher and curriculum materials use three crosscutting concepts in guiding student work. The central crosscutting concept is systems and system models. In using the idea of systems to explain phenomena, students also need to bring in ideas of structure and function and ideas of mechanism, which is an important component of cause and effect. Additionally, we describe the trajectory of students' work in this unit during the anchoring phenomenon and analyze how crosscutting concepts are built into the task design and are evident in the teachers' guidance of students.

At the start of the unit, Ms. Bennett begins with a story about how she has noticed her iced drink warms up more quickly than she would like, making it taste watery as the ice melts. She mentions that she has seen these "fancy cups" in stores that claim this type of cup can keep a cold drink colder for longer. She shows the class two cups, a regular cup and a fancy cup, filled with iced coffee. Students weigh in about whether they think the fancy cup would work better in keeping the drink cold and offer some initial ideas about why they think that. Some students notice the regular cup is made of thin plastic, whereas the fancy cup is made of thicker plastic, and they wonder if the thicker walls or more layers might keep the drink colder. Other students aren't sure there would be much of a difference. The class brainstorms ways to test this claim about the fancy cup and decides to put the same amount of the same liquid at the same cold temperature into the two cups, insert a thermometer in each, and measure the temperature at regular intervals (see Figure 2.1).

Through their measurements, students notice the liquid in the fancy cup warms up less than the regular cup. They also see that the regular cup has water droplets (some of the students refer to this as "sweat" but others call it "condensation") on the outside, whereas the fancy cup does not. In addition, they notice that the outside of the regular cup feels colder than the fancy cup. Ms. Bennett then asks the students to try to *explain* what is going on in this phenomenon. Here is where the crosscutting concepts begin to be explicit. The teacher brings in the crosscutting concept of systems and system models to guide their conversation and the work students will do when they create their models. She says, "Let's think of each cup as a system. When we figure out a system, we want to know the parts and how they work together. So turn to your partner, and talk about these three questions: What are the parts of each cup system? What does each part do in the system? How do the parts work together to keep a drink cold?"

Figure 2.1. The contrasting cups from the anchoring phenomenon

This image shows a fancy cup (right) and a regular cup (left) from the anchoring phenomenon of *How Can Containers Keep Stuff From Warming Up or Cooling Down?* (Mohan 2019).

Next, the students discuss their ideas about the parts of each cup system that are important. To guide the class, Ms. Bennett focuses on the behavior of each cup part, suggesting, "Let's ask ourselves what would happen if this part were not there. Would it change the outcome? If so, the part is probably important." The class decides the important parts of each cup are the walls, the lid, the opening, and the straw. They also decide that whatever the cup contains is also a relevant part of the system, so they include the water inside and air between the water and the lid. Several students suggest that they need to include heat, which they say is entering the cup, while other students suggest that the "cold" is leaving the cup. Some students point out that light is getting into the cup, and that may cause the liquid to warm up. A few students think the air outside the cup might matter, so the outside air needs to be in the system.

Ms. Bennett suggests the students consider these ideas and make an individual model to explain how the cups differ. Here again, she uses crosscutting concepts to guide their work with the modeling practice. She uses the crosscutting concept of systems and system models to keep the class focused on what they need to explain. In this discussion, she also brings in a related CCC, structure and function, which is often used in conjunction with systems. She tells the students, "For each part you want to argue is important in your system, make sure you say what the part does—its *function*. For example, if you think the lid is important, think about how the lid works and how it functions similarly or differently in the two cup systems." Students individually develop their initial models.

Students share their models with others at their table and look for similarities and differences. Ms. Bennett lets the students know that they will work tomorrow on putting their ideas together. The next day, students meet in a "Scientist Circle" (Michaels and O'Connor 2012) to compare their initial models and attempt to reach consensus on an initial class model. Ms. Bennett emphasizes that their class model needs to explain their findings about the fancy cup and regular cup. To do this, their model should explain how the temperature change happens in each cup system and why it happens differently in the two systems. Students realize that this is just their initial model that represents their starting ideas before they start figuring out new science as part of this unit. As such, they expect this initial model will need to change as they figure out more of the science.

Ms. Bennett suggests they start with the components of the system that their model should include. They begin with the regular cup, and Ms. Bennett acts as a recorder at the front of the classroom, drawing and labeling what the class decides should be in the model. Students agree that the major structures in the cup system that their model must include are the cup walls, the cup lid, the hole in the top, the straw, the liquid, air inside the cup, and air outside the cup. Although there are some disagreements among students about which of these is most important, students agree that factors of light, heat, and cold are worth investigating, so they represent three possible factors in their initial model to explain temperature change in the liquid: light entering the cup, heat entering the cup, and cold leaving the cup. Students agree that there is something about the thicker cup walls in the fancy cup that affects these factors, leading to less light and less heat entering the cup and less cold leaving the cup. The consensus model from Ms. Bennett's class is shown in Figure 2.2.

Figure 2.2. Consensus cup model

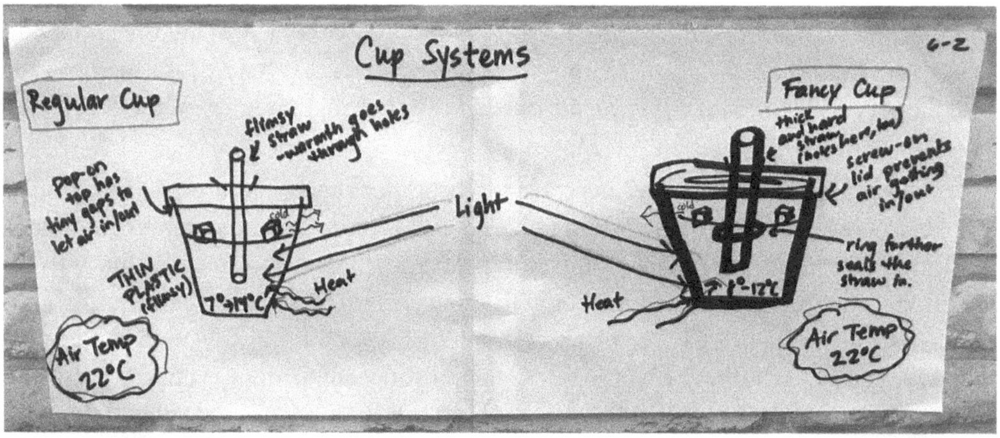

The initial class consensus model from Ms. Bennett's classroom to explain how temperature change happens in each cup

Ms. Bennett is pleased with this consensus model. She feels that students did a nice job of using the crosscutting concept of systems and system models to organize it. Of course, this initial model leaves open the key explanation of *how* these elements of the system work together to produce different results. Although students were confident that the thicker walls would interfere in some way with the cold, heat, or light, they were not exactly sure how. Furthermore, they were not sure how to explain whether it was the heat or the cold that was "moving" between different materials. Ms. Bennett knows these issues are what students will be figuring out and explaining at the particle level through their investigations in the next few weeks.

The previous day, Ms. Bennett had suggested it would be helpful to consider other phenomena where things cool down or warm up. She directed students to look around their homes to find other systems designed to maintain the temperature of something inside. She reminded them to focus on systems that work due to the materials they're made of rather than systems that require electricity (such as air conditioners or heaters). After developing the class consensus model, students discuss their examples. Here, we see a third use of the crosscutting concept of systems and system models to guide students' work. Ms. Bennett invites students to share their examples using two guiding questions:

- What are the parts of the system?
- How do they work together to keep something from changing temperature?

As students share their examples, Ms. Bennett also poses a question related to the CCC of structure and function, asking students to consider whether these containers are similar to or different from one another, depending on how their parts keep the object inside cold or hot. Examples and relevant structures students cite include a blanket (where thick layers of fabric trap heat), a sweater (where lots of thick fabric traps heat), a cooler (where thick layers of plastic prevent heat from coming in), skin (where "many layers" keep in body warmth), a fireplace (surrounded by thick bricks that keep the heat from getting to the walls), and a house (where insulation inside walls keeps heat in and cold out).

Ms. Bennett then says, "Do you think the ideas in our class model for the fancy cup might apply to some of these other objects, too?" She goes on to ask students to take the consensus model and revise it in order to explain one of the related phenomena they cited. Ms. Bennett reminds them that their goal in developing scientific models is to explain a whole class of phenomena, and so this will be a good test of what parts of their consensus model might be promising. She stresses again that students' models should focus on how the parts of the system work together to slow the process of the object inside changing temperature. Students then develop and share their models, comparing their proposed model-based explanations of how the blanket, cooler, fireplace, or other containers affect the warming up or cooling off of the objects inside. A few promising ideas emerge from these discussions that the students feel are worth pursuing further. They want to investigate the following components of the fancy cup system: the type of material used, the thickness of the walls or lid, the use of insulated or vacuum layers, and the idea of trapping or blocking heat. Equally important, students have accumulated many questions throughout the discussions about how the elements of the cup system and the other examples of insulation systems (coolers, fireplaces) influence the movement of heat.

Ms. Bennett feels the class is ready for the next important step. She brings in another crosscutting idea, cause and effect: mechanism and explanation, to push on what has been missing from the students' ideas so far. She says, "We have a lot of ideas about how this *might* work, but we do not yet agree on how the parts of the system actually *cause* these different results. In science, we need to be able to tell the cause-and-effect story, step by step, about what is happening to the objects. In this case, that means explaining what is happening to the heat or cold and why the different kinds of layers and materials seem to change this process. That's our goal for the investigations we are going to do."

After helping focus the students on this goal, Ms. Bennett asks them to write out questions they now have about cause and effect in these systems. Again, crosscutting concepts are important in her framing of how the students should think about focusing on this next science and engineering practice, asking questions and defining problems: "I want you each to come up with questions to figure out how and why a system can work to maintain the temperature of the stuff inside it." She encourages them to develop questions that "(1) we can answer through investigation and (2) will help us explain how these things work the way they do." She has the class develop a driving question board that reflects all the questions they come up with, organized in clusters of related questions (e.g., Nordine and Torres 2013; Singer et al. 2000; Weizman, Shwartz, and Fortus 2010). Each student offers a question, linking it to another student's question that they feel is related. Similar categories of questions emerge across Ms. Bennett's two sixth-grade classes. A sample of the clusters of questions are shown in Figure 2.3 and Table 2.1.

Figure 2.3. Driving question board

The board shows student-generated questions resulting from their experience with the anchoring phenomenon, organized by the class in a driving question board.

Table 2.1. Sample questions generated by students and grouped by clusters

Cluster	Student Questions
Air	• If there are little holes in the fabric of the sweater, wouldn't the hot air escape? • Does the [temperature] of the air affect how fast or slow something stays cool? • How does air get hot? • Wouldn't hot air escape from holes in the blanket?
Heat	• Does the hotness or the coldness move around? • How long can heat stay in a blanket? • How does heat get through thick plastic? • How do some things get hot or cold?

Continued

Table 2.1. *(continued)*

Cluster	Student Questions
Holes/Gaps	• How do sweaters keep you warm when they have three big holes? • Why does the heat go through the plastic when there are no holes? • If there are air holes in both, why does the fancy cup stay colder longer? • Do gaps make things cooler or hotter?
Layers	• Does the thickness of the object matter in keeping the object cold/warm? • How many layers of metal are needed to keep cold or warm air in a hydroflask or thermos?
Type of Material	• Does metal heat up faster than plastic? • Does the material always matter to keep stuff hot or cold?
Light	• Do some materials bounce off light while others don't? • How does light change the air temperature? • How does light change the temperature of water?

Ms. Bennett is pleased that the class's work with the anchoring phenomenon has set up a very productive set of questions she can now use to connect to the investigations she has planned for the unit. Over the next six weeks of this unit, the crosscutting concepts of systems and system models and cause and effect continue to play a guiding role as students build more and more elements of their evolving models of thermal energy transfer. As their empirical investigations proceed, they figure out that the same properties of objects that keep an object warm also keep it cold. They figure out that when an object warms up or cools off, it happens without the mass changing. Thus, they focus on how energy transfers (rather than on how mass moves) from inside the system to outside. They use their ideas of particles too small to see (from fifth grade) to try to explain what could be happening to those particles. They have concluded that the number of particles is not changing as the temperature changes and develop a model of temperature as the average kinetic energy of the particles. Then, they extend this model to figure out how energy can transfer within a material, or between two materials, eventually developing a model of energy transfer through particle collisions and realizing that "cold does not flow to warmer areas."

In this example, we have seen several different ways that instructional materials and teaching strategies can use crosscutting concepts as critical components to focus students' engagement in science and engineering practices and to support their work with the disciplinary core ideas. In the next section, we review the roles that the CCCs played across the two examples of sensemaking and draw out some general recommendations about how CCCs should be used in three-dimensional teaching and learning.

The Role of Crosscutting Concepts

In this section, we look across the two examples and return to the questions that motivated this chapter. What is the role of crosscutting concepts in three-dimensional learning? How should the three dimensions work together? What advantage does use of a crosscutting concept provide beyond the use of disciplinary core ideas and science and engineering practices? Crosscutting concepts are essential in supporting students in making sense of phenomena and an integral aspect of three-dimensional learning (Fick 2018; Krajcik 2015).

Crosscutting Concepts Help Guide Students' Engagement in Practices

The students in both examples learned a lot about how to develop models and use them to explain phenomena. Naturally, developing a model to explain a phenomenon, like a sweater keeping someone warm on a cold day, requires building and using disciplinary core ideas, such as how energy transfers through adjacent materials and how layers of material can slow that process down. However, engagement in this practice equally depends on crosscutting concepts, particularly in guiding learning when disciplinary core ideas are not yet sophisticated enough to fully explain the phenomenon. In both examples, we saw how the teacher plays a critical role in guiding students as they construct models and use crosscutting concepts in sensemaking. In example one, which focused on why puddles evaporate on cold days (p. 22), we saw how students made use of several crosscutting concepts to support making sense of the phenomenon. The teacher made some of the CCCs more explicit than others. In example two, which focused on thermal energy transfer (p. 26), we saw how crosscutting concepts played a focusing role, helping students identify what was important to include in their model. As students observed the cup warming up, they saw this as a change within a system. This helped them focus on what the components of the system were and how the parts could work together to behave differently in different variations of the system.

The crosscutting concepts also played an evaluative role that helped students reflect on the adequacy of their explanations. For instance, in later lessons of the thermal energy unit, the idea of cause and effect helped students see that they needed to go beyond statements of a general relationship (e.g., "heat moves from hotter to colder objects") to a mechanistic, step-by-step causal account in terms of the movement of particles. This allowed them to explain *why* heat moves from hotter to colder objects in terms of random movements of particles that on average lead to slower particles gaining speed and faster particles losing speed.

Crosscutting Concepts Develop Over Time

The crosscutting concepts are dynamic tools that become more sophisticated as students engage in three-dimensional learning across time (NRC 2012). In Ms. Bennett's classroom, students employed CCCs to develop more sophisticated disciplinary core ideas about particle motion and energy transfer (PS1-4), moving from "cold objects warm up in a warm room" to the particle-level explanations that explain energy transfer in warming up and cooling down through the transfer of kinetic energy by way of particle collisions. These advances in SEPs and DCIs also help students build more sophisticated approaches to using the crosscutting concepts themselves. For example, Ms. Bennett's students learned that developing cause-and-effect explanations of a physical system meant they needed to go down to the particle level (i.e., explaining how the particles that make up liquid in a cup and the particles that make up the solid cup walls interact) and could not stay at the macro level (i.e., focusing on the cold liquid and the double walls of the cup). Similarly, using the particle model to apply energy transfer through particle collisions also has implications for how students think about systems and system models in new contexts. In explaining how everyday objects warm up or cool down, students develop a strategy for analyzing where the matter is going in a system, whether it can escape the system or not, and how energy is transferred through matter interactions. These strategies are part of a more sophisticated set of tools for using the crosscutting ideas of systems and system models in new contexts. It also connects students' ideas of systems to their developing ideas about another crosscutting concept—energy and matter: flows, cycles, and conservation. Students begin to see that they can make progress in figuring out how systems behave by tracking the flow of the matter and energy into, out of, and within the systems.

Crosscutting concepts, like the other two scientific knowledge dimensions, develop over time as students build on their initial ideas. As students continue to use, challenge, and refine these concepts, they develop more sophisticated versions of the concepts (Duschl 2012). For instance, third-grade students can use ideas related to cause and effect. They know that when they push a toy car and it collides with another toy car, the second toy car will move (unless it is much heavier). In part, they develop this idea of a toy car causing a second car to move because they recognize a pattern (another CCC). They can also see the connection between heating water and evaporation due to the recognition of a pattern. Their ideas of causality will become more sophisticated over time as they experience and try to figure why other phenomena occur. The sophisticated use of the CCC of cause and effect at the middle school level is evident in the two examples presented in this chapter. For instance, in the first example, we show students building a model to explain that collisions at the microscale can account for observables at the macroscopic level.

In summary, as learners progress through grade levels, their use of crosscutting concepts becomes more sophisticated (Duschl 2012; NGSS Lead States 2013b), as do the science and engineering practices (Schwarz, Passmore, and Reiser 2017) and the disciplinary core ideas (Duncan, Krajcik, and Rivet 2016). Furthermore, knowledge of each dimension becomes more sophisticated as students use the various ideas to make sense of a new phenomenon or to solve a new problem. In this way, the knowledge students develop becomes deeper and more useable over time (NRC 2012).

Crosscutting Concepts Play a Key Role in Applying Ideas From One Science Context to Another

Crosscutting concepts are particularly important in helping students bring prior knowledge to bear when encountering new phenomena. Think back to the example of Ms. Bennett's students, who developed crosscutting ideas through their unit on thermal energy. During this unit, they used a system model approach to explain phenomena by identifying elements, figuring out how they interact, and tracing the flow of matter and energy through that system. They saw that for physical systems, they need to go down to the particle level of interaction to explain the observable level (e.g., hot liquids cool off, warm liquids warm up, both to reach equilibrium with the room temperature). Now imagine that the next scientific challenge students take on is in a very different context. The class leaves the subject of chemistry and starts an Earth science unit. The anchoring phenomenon involves a surprise hailstorm where large balls of hail fall on a town during what was shortly before a bright, sunny day. It is obviously too warm outside for ice to form, so how could hail (which is clearly frozen water) fall from the sky?

In the anchoring phenomenon for this weather unit (Novak 2019), students start by considering how the approach they used in the prior unit might help them here; at the same time, they bring beginning disciplinary core ideas with them along with various science and engineering practices. They realize that a storm is a complicated event, with various parts that interact over time. Furthermore, they quickly see that their understanding of general systems can be a strategy to help them start to make sense of the storm. The students consider the elements at play. They identify air, water in the air, light from the Sun, the transfer of energy from the Sun, and the temperature of the matter as relevant components. They start to ask how these elements might be interacting and how this could lead to a change in the system over time. The class decides to use the idea of tracking matter and energy flow through a system (a crosscutting concept) and the transfer of energy (a disciplinary core idea) as a way to try to explain, step by step, how a pleasant, sunny day changed into a stormy day, with lots of wind, darker clouds, and frozen water falling from the sky.

As with the prior thermal energy unit, the idea of systems and system models helps focus students' initial models, explanations, and development of questions (SEPs). However, with their more sophisticated sense of systems and system models, which is connected to the idea of flow of matter and energy, students have a head start on pursuing lines of inquiry that will help them explain the hailstorm phenomenon. Indeed, placing the thermal energy unit prior to the weather unit in the OpenSciEd materials (OpenSciEd 2020) was done on purpose. In this middle grades sequence, students first worked on developing the DCI of thermal energy transfer, the CCC of systems and system models, and the SEPs of developing and using models and defining problems in a simpler phenomenon context (i.e., the cups in a room). The new more complex context of weather phenomena helped them further extend both the DCIs and CCCs to more complex systems. Learners still make use of their ideas regarding particles colliding, but now the particles are colliding in a much larger system (i.e., air masses moving across the land). The crosscutting concepts of characterizing the system and tracking the flow of matter and energy help students apply these particle ideas, providing the organizing frame of identifying what is in the system, figuring out how the elements interact, and tracking the matter movement and energy transfer through those interactions.

Thus, the crosscutting concepts not only play a guiding role but also can help students frame their work as they take on new phenomena. Although students will need to develop further understanding of the target DCIs in order to fully explain the new phenomena, the crosscutting concepts enable them to develop an initial analysis that helps guide their subsequent investigations. For instance, in the previous example, the students embarking on the weather unit have not yet figured out how air masses interact, how water moves through the atmosphere, or how this leads to temperature changes, phase changes in water, and precipitation. Nevertheless, they quickly recognize that they have a system of interacting parts, and they know they need to follow the matter as it moves at the particle level and figure out how that transfers thermal energy in order to explain how the system behaves. The examples in this chapter illustrate the key role that each dimension—be it the DCIs, SEPs, or CCCs—can play in supporting learners' progression on the other two dimensions. Indeed, studies of middle school students engaging in science and engineering practices—particularly those related to modeling, argumentation, and explanation—reveal that students can increase their sophistication of developing mechanistic accounts, apart from specific disciplinary core ideas. That is, students learn something about mechanistic explanations that can then help them make progress in making sense of new phenomena, prior to building the specific disciplinary core ideas they need to fully explain the phenomena (Krist, Schwarz, and Reiser 2019).

Crosscutting Concepts Can Help Problematize Students' Explanations, Pushing Them to Go Deeper

Another aspect of using CCCs to help guide students' work is that they challenge students to realize when they need to go deeper in their models and explanations. An important element of engaging students in science is helping them find limitations in explanations that seem sensible but do not get at the underlying cause of the phenomena they are analyzing (Reiser et al. 2021). In the first example presented in this chapter, learners knew that puddles of water "dry up." But does that mean the water disappears or that something else happens to it? Perhaps a student has learned about the water cycle and knows that water somehow goes into the air and can come back out again. That may be a sufficient explanation in grades 3–5, but in middle school, we want learners to figure out that the reason water can go into the air is because energy is transferred to the particles of water, whereupon they gain enough kinetic energy to overcome the attraction of the other molecules and go into the gaseous state. Also, notice how in the first example students were guided by the teacher to design a new experiment to find out what would happen to the temperature of the cloth within a closed system. This experience pushed students to think further about what is happening at the molecular level and how it affects observable properties. Allowing students to experience and investigate new phenomena and apply the developing disciplinary knowledge, crosscutting concepts, and science and engineering practices can push students to go even deeper.

In the second example featuring the thermal energy unit, part of the guiding work that the crosscutting concept did was to problematize students' partial explanations and help identify what part of the phenomenon needed to be explained. Students quickly determined that layers of material—such as double-walled cups, double-paned windows, and layers of clothing—are helpful in "keeping heat in," but the crosscutting concepts helped reveal the limitations in that explanation. Using the CCCs of systems and system models and cause and effect, the teacher pushed on students' initial explanations to help them realize they didn't have the full story yet. As students worked on the whole-class model, the teacher helped capture areas of agreement. By encouraging the continuation of the investigations, she also helped the class see how their explanations didn't go deep enough, using prompts such as these:

- "How does the thicker layer help keep heat in?"
- "Let's think about what's happening to the matter or the energy. Does the layer prevent matter from escaping? Does heat leave the system somehow? If so, how?"
- "Can we tell a cause-and-effect story, step by step, that explains what is happening to the matter and energy at each step? It looks like we are in agreement that the layers are important. But we need to push deeper to figure out exactly how layers work to keep heat or cold in."

Throughout the rest of the unit, the teacher continued to use the crosscutting concepts of systems and system models and cause and effect to push students' intuitive ideas and help them construct a particle-level explanation. The teacher pushed students to realize that "heat" results from the movement of molecules.

Crosscutting Concepts Should Be Explicit So Students Use Them as Tools for Sensemaking

In these examples, the crosscutting concepts were made explicit so students could use them as tools. In the unit on evaporation (example one), the teacher explicitly said that student models needed to provide a mechanism (cause and effect) for why the water evaporated. The teacher also pushed students to think about how what was happening at the molecular level affected the observable properties of the system (structure and function). In the unit on energy transfer (example two), the teacher told students it would be helpful to bring in the idea of systems and system models at the start of the unit. Similarly, at the beginning of the weather unit, she asked students how they could use their approach from the previous unit (i.e., looking at temperature change in cups as a systems phenomenon) to view the new phenomenon of how the weather changes.

This explicit use contrasts with alternative approaches, such as the guiding role of crosscutting concepts being apparent only to teachers and curriculum designers. Implicit use of CCCs, however, will not allow users to develop deeper, more sophisticated knowledge of the CCCs they can use in new situations. The examples we present suggest that it can be natural and helpful for the teacher to make this crosscutting frame an explicit focus of the conversation. The students know they are using systems as a thinking tool, but they do not necessarily know that systems is labeled as a crosscutting concept. Learning crosscutting concepts is a critical knowledge domain of science that students need to develop to make sense of new phenomena or to solve new, challenging problems. Teachers need to explicitly support students in learning and using the crosscutting concepts as tools for sensemaking.

However, it is important to stress that the crosscutting concepts should not be brought in simply because the teacher knows the CCCs are something students have to learn about. Rather, the teacher and the task design should employ them, when needed, as tools. The CCCs provide strategies for students to think about questions and problems and to make progress in sensemaking. The CCCs are *not* ends in and of themselves. So, for example, in the anecdotes provided in this chapter, the teachers were not doing lessons on patterns or on systems in the abstract. Instead, these CCCs provided a framework students could use to guide their questioning of the phenomena, direct their model construction, and evaluate their explanations.

Crosscutting concepts are an integral component of three-dimensional learning. Realizing, for instance, that the phenomenon of a puddle evaporating needs a cause to occur begins to push learners to think about what causes the evaporation. Thinking of how to explain the phenomenon existing in systems suggests next steps for students to take: they need to identify the elements, trace how they interact, and so on. It's not simply the teacher calling out what students are studying as a system and answering some questions about it. The crosscutting concepts, like the other two dimensions, should serve as tools students can use to make sense of phenomena or solve problems.

Conclusion

The two examples we describe in this chapter show how the three dimensions of scientific knowledge—disciplinary core ideas, crosscutting concepts, and science and engineering practices—work together as tools to support students in making sense of the world and in developing deep and useable knowledge of the three dimensions (Fick 2018; Krajcik 2015). Notice how each of the examples starts with a phenomenon to drive student learning and to prompt them to wonder why the event occurs. Students then use the three dimensions to figure out how the phenomenon occurs—rather than to learn content. It is through the figuring-out process that all students build deeper and more useable knowledge of all three dimensions of scientific knowledge. In the process of making sense of a phenomenon, student knowledge of each dimension becomes deeper and more sophisticated. Each of the dimensions plays a critical role in supporting students as they make sense of phenomena. As teachers, we need to make sure we explicitly help learners in developing knowledge of each of the dimensions.

The innovation of *A Framework for K–12 Science Education* is that it brings the science and engineering practices together with explanatory ideas from the science disciplines and ideas that cut across these disciplines (i.e., the CCCs). The crosscutting concepts play a unique and important role. As we pointed out throughout this chapter, we find it critical for teachers to be explicit about using specific crosscutting concepts; however, we see no value in a student knowing that structure and function, for example, is a crosscutting concept. Rather, the value of crosscutting concepts for learners is in using them to examine a phenomenon or to solve a problem. Students need to know how and when to apply the lenses of the various crosscutting concepts to make sense of phenomena and solve problems.

In this chapter, we illustrated the following unique aspects of crosscutting concepts:

1. They serve as lenses to focus students' engagement in practices to make sense of phenomena.
2. They make explicit the key elements of the disciplinary core ideas at work so students can guide their sensemaking.
3. They problematize students' sensemaking.
4. They are used as tools, rather than as abstract ideas students must learn about.
5. They develop over time to become more sophisticated and useable in different contexts.

However, for CCCs to develop across time or be used as tools for sensemaking, teachers and instructional designers need to make CCCs explicit in instruction and not implicit.

The use of CCCs is a critical tool that works in conjunction with disciplinary core ideas and science and engineering practices to help students make sense of the world.

References

Brown, J. S., A. Collins, and P. Duguid. 1989. Situated cognition and the culture of learning. *Educational Researcher* 18 (1): 32–42.

Duncan, R., J. Krajcik, and A. Rivet, eds. 2016. *Disciplinary core ideas: Reshaping teaching and learning*. Arlington, VA: NSTA Press.

Duschl, R. A. 2012. The second dimension—crosscutting concepts: Understanding *A Framework for K–12 Science Education. The Science Teacher*. 79 (2): 34–38

Fick, S. J. 2018. What does three-dimensional teaching and learning look like? Examining the potential for crosscutting concepts to support the development of science knowledge. *Science Education* 102 (1): 535.

Krajcik, J. 2015. Three-dimensional instruction: Using a new type of teaching in the science classroom. *Science Scope* 39 (3): 16–18.

Krajcik, J., and N. Shin. 2014. Project-based learning. In *The Cambridge handbook of the learning sciences*. 2nd ed., ed. R. K. Sawyer, 275–297. New York: Cambridge University Press.

Krist, C., C. V. Schwarz, and B. J. Reiser. 2019. Identifying essential epistemic heuristics for guiding mechanistic reasoning in science learning. *Journal of the Learning Sciences* 28 (2): 160–205.

Michaels, S., and C. O'Connor. 2012. *Talk science primer*. Cambridge, MA: TERC. *https://inquiryproject.terc.edu/shared/pd/TalkScience_Primer.pdf.*

Mohan, L. (Ed.). 2019. How can containers keep stuff from warming up or cooling down? [Curriculum Materials, Middle School]. OpenSciEd. *www.openscied.org/instructional-materials/6-2-thermal-energy.*

National Research Council (NRC). 2007. *Taking science to school: Learning and teaching science in grades K–8*. Washington, DC: National Academies Press.

National Research Council (NRC). 2012. *A framework for K–12 science education: Practices, crosscutting concepts, and core ideas*. Washington, DC: National Academies Press.

NGSS Lead States. 2013a. *Next Generation Science Standards: For states, by states*. Washington, DC: National Academies Press. *www.nextgenscience.org.*

NGSS Lead States. 2013b. *NGSS Appendix G: Crosscutting concepts*. Washington, DC: National Academies Press. *www.nextgenscience.org.*

Nordine, J., and R. Torres. 2013. Enhancing science kits with the driving question board. *Science and Children* 50 (8): 57–61.

Novak, M. (Ed.). 2019. Why does a lot of hail, rain, or snow fall at some times and not others? [Curriculum Materials, Middle School]. OpenSciEd. *www.openscied.org/instructional-materials/6-3-weather-climate-water-cycling.*

OpenSciEd. 2020. Scope and sequence. *www.openscied.org/scope-sequence.*

Pellegrino, J. W., and M. L. Hilton. 2012. *Education for life and work: Developing transferable knowledge and skills in the 21st century*. Washington, DC: National Research Council.

Reiser, B., M. Novak, T. A. W. McGill, and W. Penuel. 2021 (in press). Storyline units: An instructional model to support coherence from the students' perspective. *Journal of Science Teacher Education.*

Sawyer, R. K., ed. 2008. *The Cambridge handbook of the learning sciences.* 1st ed. New York: Cambridge University Press.

Schwarz, C. V., C. Passmore, and B. J. Reiser, eds. 2017. *Helping students make sense of the world using next generation science and engineering practices.* Arlington, VA: NSTA Press.

Singer, J., R. W. Marx, J. Krajcik, and J. C. Chambers. 2000. Constructing extended inquiry projects: Curriculum materials for science education reform. *Educational Psychologist* 35 (3): 165–178.

Weizman, A., Y. Shwartz, and D. Fortus. 2010. Developing students' sense of purpose with a driving question board. In *Exemplary science for resolving societal challenges*, ed. R. E. Yager, 110–130. Arlington, VA: NSTA Press.

Chapter 3

Broadening Access to Science: Crosscutting Concepts as Resources in the *Next Generation Science Standards* Classroom

Marcelle Goggins, Alison Haas, Scott Grapin, Rita Januszyk, Lorena Llosa, and Okhee Lee

Although crosscutting concepts (CCCs) are not new ideas in science education, their inclusion in *A Framework for K–12 Science Education* (the *Framework*; NRC 2012) and the *Next Generation Science Standards* (*NGSS*; NGSS Lead States 2013a) has new implications for science instruction. All students come to school with experiences to make sense of the world around them that relate to CCCs in the *NGSS*. For example, in their everyday lives, students notice patterns, recognize how parts work together as a system, and try to figure out what causes things to happen. Given that students use CCCs in their everyday lives, these concepts can be thought of as resources that students bring to the *NGSS* classroom. Teachers can help students make their thinking explicit as they develop an understanding of CCCs as resources to make sense of phenomena.

A perspective on CCCs as resources is timely in the context of increasing cultural and linguistic diversity of the K–12 student population. In recent years, underrepresented groups in terms of race/ethnicity have become the majority in U.S. public schools, and students classified as English learners represent the fastest-growing subset of the student population. Traditionally, science education has not provided opportunities for students from underserved groups to see science as relevant to their lives or future careers. By viewing CCCs as resources that all students bring to the science classroom, teachers can integrate them into science instruction in ways that build on the students' everyday experiences in their homes and communities. This perspective on CCCs as *resources* makes science real and relevant to all students and allows them to see themselves as scientists from the moment they enter the science classroom.

In this chapter, we describe how a perspective on CCCs as resources is particularly powerful for achieving the *NGSS* vision of "all standards, all students" (NGSS Lead States 2013b). First, we frame CCCs as resources that all students bring to the science classroom. Second, we acknowledge how our perspective builds on and extends the emerging research literature on integrating CCCs into science instruction. Third, we provide classroom examples to illustrate how a perspective on CCCs as resources is enacted by two teachers in a yearlong, fifth-grade curriculum with a focus on English learners. Finally, we conclude with classroom strategies for implementing this perspective on CCCs.

Crosscutting Concepts as Resources With Diverse Student Groups

Broadening access to science is a central theme of the *Framework* and the *NGSS*. Traditionally, it has been expected that students come to the science classroom to learn canonical science knowledge. Moreover, it has traditionally been assumed that students, especially those from underserved groups, bring little or limited prior canonical science knowledge with them. It is imperative that science be made real and relevant to all students. Utilizing CCCs as resources is one way to do this.

CCCs have previously been thought of as "common themes" (AAAS 1989) and "unifying concepts and processes" (NRC 1996) that are present in different science disciplines; however, they were not emphasized in science standards, which was problematic from the perspectives of both science and equity. From the perspective of science, these "themes" and "concepts" became secondary to science content or inquiry in both the research literature and classroom implementation. As a result, CCCs did not figure prominently in science instruction, especially with student groups that were traditionally underserved in science education (NGSS Lead States 2013b).

In contrast to previous standards, the *NGSS* explicitly integrates seven CCCs into the standards, which is an advance from the perspectives of both science and equity. From the perspective of science, by including CCCs alongside science and engineering practices (SEPs) and disciplinary core ideas (DCIs) as part of three-dimensional learning, the *NGSS* elevates the status of CCCs. From the perspective of equity, the *NGSS* posits the importance of CCCs for all students. Specifically, the *NGSS* states that "explicit teaching of crosscutting concepts enables less privileged students, most from non-dominant groups, to make connections among big ideas that cut across science disciplines. This could result in leveling the playing field for students who otherwise might not have exposure to such opportunities" (NGSS Lead States 2013b, p. 6).

To address CCCs in relation to diverse student groups, we propose that teachers view them as resources that students use in their everyday lives to make sense of the world and that they bring to the science classroom to make sense of phenomena. By capitalizing on students' funds of knowledge from their homes and communities (González, Moll, and Amanti 2005), including their everyday experiences with CCCs, teachers demonstrate value for students' cultural and linguistic resources. With English learners in particular, we posit that a view of CCCs as resources invites these students to use all of their meaning-making resources, including everyday language, home language, and multiple modalities, in the science classroom (Lee et al. 2019b).

Emerging Literature on Crosscutting Concepts

The *Framework* defines CCCs as concepts "that unify the study of science and engineering through their common application across fields" (NRC 2012, p. 2). The research literature on CCCs has been limited (Osborne, Rafanelli, and Kind 2018) and is only beginning to emerge (e.g., Fick, Nordine, and McElhaney 2019). Our perspective on CCCs as resources builds on and extends this limited literature by considering CCCs from an equity perspective (Goggins et al. 2019). Specifically, our perspective is informed by the theoretical ideas of Lave and Wenger (1991), who conceived of learning as participation in communities of practice. In such communities of practice, all members are viewed as legitimate and recognized for bringing individual resources that contribute to the collective functioning of the community. Initially, these resources may be invisible[1], thus allowing for "smooth entry into practice" (Adler 2000, p. 214) as students use their everyday experiences for initial meaning-making. Over time, these resources are made visible so they can be more intentionally "used [to] extend practice" (Adler 2000, p. 214). The dual functions of invisibility and visibility allow all students' resources to be used as individual resources for meaning-making from the outset and to become collective resources of the classroom over time.

Based on this theoretical grounding, we propose viewing CCCs as resources that all students bring to the science classroom community of practice and that teachers can build on and make visible across science disciplines and over the course of instruction. This perspective on CCCs has three key strategies for teachers:

1. All students come to the science classroom with intuitive ideas about CCCs that can serve as resources that develop into knowledge they learn to use more intentionally (Fick, Arias, and Baek 2017). Thus, as teachers leverage these intuitive ideas about CCCs, they guide students in using CCCs to make sense of phenomena. Over time, teachers can build on and make visible students' intuitive ideas about CCCs. In the science classroom, all students bring their funds

1. Lave and Wenger's (1991, p. 103) use of the term *invisible* is not intended in a pejorative sense (e.g., to indicate "missing" or "absent") but rather to indicate "unproblematic interpretation and integration [of resources] into activity." In contrast, they use the term *visible* to indicate "extended access to information" about how and why a resource is used in a particular way.

of knowledge about CCCs from their homes and communities (see Chapter 8 in Fick, Nordine, and McElhaney 2019). This perspective on CCCs as resources calls for a shift in instruction from a deficit perspective (i.e., students from underserved groups come to the science classroom with limited sensemaking resources) to an asset perspective (i.e., students from underserved groups come to the science classroom with sensemaking resources).

"This perspective on CCCs as resources calls for a shift in instruction from a deficit perspective (i.e., students from underserved groups come to the science classroom with limited sensemaking resources) to an asset perspective (i.e., students from underserved groups come to the science classroom with sensemaking resources)."

2. Teachers can provide opportunities for students to use CCCs across science contexts and disciplines. Meaningful use of CCCs in different disciplines allows all students to formalize their intuitive ideas about CCCs. Rather than associate a particular CCC with a specific discipline, students should view CCCs as resources they can flexibly draw on to make sense of phenomena in any discipline.
3. Teachers can guide students in using CCCs intentionally when presented with unfamiliar phenomena so their understanding and use of CCCs becomes more sophisticated across grade levels, grade bands, and K–12 education. For students to progress from an intuitive use of CCCs to one that is more intentional, teachers can design coherent instructional sequences that help students recognize how and when CCCs are useful resources for sensemaking of phenomena.

To summarize, building on CCCs as resources during instruction makes students' intuitive ideas about CCCs visible (Strategy 1), provides opportunities for students to apply CCCs across science disciplines (Strategy 2), and guides students in using CCCs intentionally over the course of instruction (Strategy 3). These strategies come together to support equity by viewing students' everyday and home experiences as sensemaking resources (Strategy 1), extending students' resources to other contexts in order to showcase their value for sensemaking (Strategy 2), and making the resources explicit for all students so the classroom community of learners use the resources collectively to make sense of phenomena (Strategy 3).

Classroom Examples

This section provides classroom examples from the implementation of Science and Integrated Language (SAIL), a yearlong, fifth-grade curriculum aligned to the *NGSS* with a focus on English learners. The SAIL curriculum bundles the 16 fifth-grade performance expectations in the *NGSS* into four units that address physical science, life science, Earth science with engineering embedded, and space science. Each unit in the curriculum focuses on a local phenomenon that is real and relevant to students (Lee 2020; Lee et al. 2019a).

- Unit 1: What happens to our garbage? (physical science)
- Unit 2: Why did the tiger salamanders disappear? (life science)
- Unit 3: Why does it matter if I drink tap water or bottled water? (Earth science)
- Unit 4: Why do falling stars fall? (space science)

To develop the curriculum, we worked closely with teachers in an urban school district who field-tested the curriculum and provided feedback on how to improve the curriculum to better meet the needs of all students, particularly English learners. In this section, we provide examples from two classrooms where we field-tested our curriculum over three years. In one school, 25% of the student body were English learners, and 89% of students qualified for free or reduced-price lunch. In the other school, 24% of the student body were English learners, and 77% of students qualified for free or reduced-price lunch.

The following classroom examples each highlight the three strategies previously described. First, they illustrate how teachers capitalize on students' intuitive ideas about CCCs to make sense of phenomenon (Strategy 1). Second, they illustrate how a perspective on CCCs as resources applies across science disciplines (Strategy 2). Third, they illustrate students' learning progressions in using CCCs more intentionally from the first unit to the final unit of the school year (Strategy 3). For each of the classroom examples, we provide a description of the classroom instruction and then offer our commentary with a focus on these three key strategies for CCCs. In the examples, we include excerpts from the SAIL curriculum to illustrate how the curriculum is purposefully designed to promote the perspective on CCCs as resources.

Patterns in Garbage Materials

Description

The first classroom example shows how a teacher capitalized on students' everyday experiences with the CCC of patterns.

CLASSROOM SNAPSHOT 3.1

On the first day of science instruction in the school year, fifth-grade students walked into their classroom and immediately saw something unusual: piles of garbage from their school cafeteria on tarps. The teacher divided the class into groups of four or five students with varying levels of English proficiency in each group and assigned each group to a pile of lunch garbage. In preparing the garbage materials, the teacher ensured there were no hazardous materials included. (See Figure 3.1 for the safety measures the teacher took for this garbage sort.)

Figure 3.1. Safety measures taken for the garbage sort

Garbage Sort Safety Guidelines
• When assembling the piles of garbage for the activity, the teacher made sure not to include broken glass or sharp objects. • The teacher ensured the garbage had as little liquid as possible. • The teacher directed students to wear plastic gloves, plastic aprons, and protective goggles and to use tongs for handling the garbage. • The teacher directed students to wash their hands after handling the garbage. • If students had allergies (e.g., to nuts or mold), the teacher consulted the school nurse before proceeding with the garbage sort.

Wearing gloves, aprons, and goggles and using tongs to move the garbage materials around, students made observations of the materials. Then the teacher described their task, explaining that the groups of students would sort their garbage piles into smaller piles or categories. Since students had to agree with their group members about which sorting categories to use, the groups were communicating patterns they found in the garbage. The teacher guidance and prompting was minimal for groups making decisions about the categories. Teacher prompts such as "Why are you grouping these materials together?" allowed students to express the underlying reason for why they sorted the materials in a particular way. (See Figure 3.2 for possible teacher prompts related to the CCC of patterns.) Groups sorted the garbage materials based on how the materials looked or what they had been used for before being discarded. For example, one group sorted materials by color and texture, whereas another group sorted materials into the three categories of utensils, bowls, and food, recognizing that different materials had different purposes before being thrown away. The teacher listened to groups' rationales for their garbage categories, looking for students' use of the CCC of patterns. While listening, the teacher recognized that his students were

Continued

Classroom Snapshot 3.1 *(continued)*

already using the CCC of patterns by identifying similarities and differences in the garbage materials.

Figure 3.2. Possible teacher prompts to probe for students' intuitive use of patterns before making the CCC explicit

Possible Teacher Prompts Related to Patterns
1. How did your group decide which materials go together?
2. What is similar about the materials in each category?
3. What is different about the materials in each category?
4. If you were given a new material, how would you know which category it belonged to?

After talking with each group about similarities and differences in the garbage materials, the teacher brought the class together to discuss their observations. Each group shared its categories of school lunch garbage. In this discussion, the teacher made students' use of the CCC of patterns explicit by telling the class how scientists look for and find patterns of similarity and difference in their observations, which can lead scientists to ask new questions or find new ways to organize their data. At the end of the lesson, the teacher commended the students for using patterns, as scientists do, to categorize the garbage materials. He also suggested that the class keep in mind this concept of patterns when investigating other phenomena in the future. For homework, students identified patterns of similarity and difference in their home garbage materials. Here, students were able to use their intuitive understanding of patterns more intentionally to make sense of the garbage phenomenon.

Commentary

In this classroom example, the teacher capitalized on students' everyday experiences with the CCC of patterns as a resource to begin making sense of the phenomenon of garbage. He listened to how groups decided on their categories, and students were able to use their intuitive ideas about patterns based on their everyday experiences. After providing students with the opportunity to use patterns based on their everyday experiences, the teacher made the use and purpose of the CCC of patterns visible for students. The teacher connected students' intuitive use of CCCs to the work of scientists and encouraged all students to see themselves as scientists from the very beginning of the school year (Strategy 1).

This classroom example comes from the first day of instruction in a physical science unit. By starting the unit with students' intuitive ideas about the CCC of patterns and by making students' use of the CCC of patterns visible, the teacher laid the foundation early in the year so this CCC could be used to make sense of phenomena in other science disciplines (e.g., space science) in future instructional units (Strategy 2).

Finally, the teacher guided students to use the CCC of patterns more intentionally through specific probing. As shown in Figure 3.2, the first probe ("How did your group decide which materials go together?") is an open-ended question intended to elicit how students intuitively used the CCC of patterns. The second and third probes ("What is similar about the materials in each category?" and "What is different about the materials in each category?") prompted students to identify similarities and differences in their observations, which is an important element of the CCC of patterns in fifth grade (NGSS Lead States 2013c, p. 4). Finally, the fourth probe ("If you were given a new material, how would you know which category it belonged to?") presented a hypothetical scenario to promote students' more informed use of the CCC. By following this sequence of probes, the teacher was able to move students from a more intuitive to a more intentional use of the CCC (Strategy 3).

The teacher's perspective on CCCs as resources was especially beneficial to the English learners who were able to use all of their meaning-making resources to make sense of the phenomenon (Lee et al. 2019b). In this example, the opportunity to use everyday language in combination with gestures (e.g., saying "Put that one here!" while pointing to specific garbage materials) at the beginning of instruction before progressing to the more specialized language of the *NGSS* performance expectation (i.e., "Distinguish materials by patterns in their observable properties") enabled English learners to participate meaningfully from the outset. This perspective on CCCs departs from a more traditional approach of introducing specialized language (e.g., patterns) at the beginning of instruction before students have experienced and developed an understanding of science concepts. In his instruction, the teacher embraced the notion that language is a product of, not a precursor to, "doing" science (Lee et al. 2019b) by recognizing how students' everyday language, related to the CCC of patterns, could serve as an entry point to science learning.

Systems of Garbage Disposal in the School, Home, and Community

Description

The second classroom example illustrates how the teacher capitalized on students' everyday experiences with a different CCC, systems and system models, to figure out how garbage was disposed of in their school, home, and community.

CLASSROOM SNAPSHOT 3.2

After sorting their school lunch garbage in the first lesson of the unit, students began to wonder where all of the garbage would go once it left the classroom. The teacher called on several students to share their initial ideas. Student responses included "the garbage can," "garbage trucks," and "landfills." The teacher wrote student responses on sticky notes and posted them on the board, asking students how the responses were related to one another. Student responses provided connections between the ideas listed on the sticky notes. For example, students said, "The janitor takes the garbage outside of the school" and "The garbage bin gets dumped into the garbage truck." The class connected the sticky notes with arrows to show how the different parts worked together to transport garbage from the classroom to the landfill.

Next, the teacher assigned groups to develop their own models of garbage disposal in either the home or community. Each group wrote the different parts on sticky notes (e.g., garbage can, dumpster, garbage truck) and used arrows between the sticky notes to show how the parts were related to one another. As the groups worked, the teacher asked probing questions and provided feedback on how the parts worked together. (See Figure 3.3 for possible teacher prompts related to the CCC of systems and system models.)

Figure 3.3. Possible teacher prompts to probe for students' use of systems and system models before making the CCC explicit

Possible Teacher Prompts Related to Systems and System Models
1. Where do you put your garbage at home?
2. Where in your neighborhood do you throw out garbage?
3. What would happen if a part, like the garbage truck, were missing?
4. How would the garbage end up in the landfill?

After circulating to each group, the teacher asked groups to place their models on the board in the front of the room. Students identified similarities and differences in the systems, which allowed the teacher to reinforce the CCC of patterns from the previous lesson, and students noticed that all of the garbage ended up in the landfill. In this discussion, the teacher described what the class developed as "models

Continued

Classroom Snapshot 3.2 *(continued)*

of several garbage disposal systems." He also explained to the class that scientists identify parts, or components, of systems and identify how those components work together, or interact. The interactions among the components enable the system to carry out functions that the individual components cannot. The teacher commended the students for using systems, as scientists do, to figure out where garbage goes when it is disposed. The teacher also suggested that the class keep in mind this concept of systems when investigating other phenomena in the future.

Commentary

In this classroom example, the teacher used students' everyday experiences with the CCC of systems and system models as resources to make sense of how garbage in the school, home, and community end up in a landfill. Similar to the first example, the teacher made students' intuitive ideas about the CCC of systems and system models visible by making the CCC explicit after students had experience using the CCC (Strategy 1). For English learners in particular, the opportunity to use all of their meaning-making resources (e.g., saying "This one comes first!" while rearranging the position of the sticky notes in their model) at the beginning of instruction before progressing to more specialized language (e.g., "Identify the components and interactions of the garbage disposal system.") provided access to science learning.

Using the CCC of systems and system models in the physical science unit exposed students to the CCC in a particular discipline. Students' classroom experiences with the CCC of systems and system models could then be extended in future units on different science disciplines. Furthermore, using this second CCC, in addition to the CCC of patterns, provides an example of how different CCCs may be used to make sense of one phenomenon in a particular discipline. This promotes a more flexible approach to using CCCs as resources for sensemaking, as multiple CCCs can be used to make sense of the same phenomenon (Strategy 2).

The teacher extended students' understanding of the CCC of systems and system models by probing their thinking on how the components of the system interact. As shown in Figure 3.3, the probes move from questions that ask students to name individual components of a system relevant to their everyday lives (e.g., "Where do you put your garbage at home?") to questions that probe students' thinking about how the components work together as a system (e.g., "What would happen if a part, like the garbage truck, were missing?"), which promotes more sophisticated use of the CCC over the course of instruction (Strategy 3).

Patterns and Systems in the Night Sky

Description

In the final unit of the year—the space science unit—students confronted the following driving question: Why do falling stars fall? They used the CCC of patterns to distinguish the properties of falling stars from stars such as the Sun. Students handled real falling stars (meteorites) and watched videos of falling stars at different times of the year. They wondered why they could see specific falling stars at certain times of the year. This final classroom example shows how students use their prior experiences with stars to help scaffold their understanding of falling stars before constructing physical models later in the unit.

CLASSROOM SNAPSHOT 3.3

The teacher began the first class of the unit by asking students if the sky looks the same every night of the year. Students shared out their experiences of seeing different objects in the night sky at different times of the year. The questioning valued students' intuitive ideas about finding the pattern of changing objects in the sky over the course of a year. The teacher passed out a data table listing the annual falling star showers (groups of falling stars), the constellations they are named after, and the dates for when the falling star showers occur. She prompted partners to talk about what the data table revealed about falling star showers. One English learner excitedly exclaimed that she and her partner already knew what to do: They were going to look for patterns in the data. The student demonstrated agency in learning science, as she recognized the power of patterns in the data to make sense of the phenomenon.

As partners talked, the teacher listened to how students made sense of the data. She recognized and shared with the class that partners found two different patterns. Some students looked for patterns within a single year, whereas others noticed patterns over multiple years. To make students' use of the CCC of patterns more intentional, the teacher asked students what each pattern could help them figure out. Through a class discussion, students concluded that patterns *within a single year* could help them figure out why they see different falling stars at different times of the year, whereas patterns *over multiple years* could help them figure out whether they would see the same falling stars at the same times next year.

Next, the teacher projected Stellarium, a free, open-source virtual planetarium program (*https://stellarium.org*), to show the night sky from the schoolyard over the course of the year. Working in groups, students made predictions about when they

Continued

Classroom Snapshot 3.3 *(continued)*

would see certain constellations. As they made observations of the night sky, they collected data about constellation positions during different months. As students shared their thinking about why they would see specific falling stars at certain times of the year, the teacher circulated to groups to prompt their use of the CCC of patterns. (See Figure 3.4 for possible teacher prompts related to patterns in space.)

Figure 3.4. Possible teacher prompts to probe for students' different use of patterns in space before making the CCC explicit

Possible Teacher Prompts Related to Patterns
1. What did you observe for each constellation over one year?
2. What did you observe for each constellation over multiple years?
3. What do these observations tell us about why we see different constellations at night during different times of the year?
4. What predictions can you make from these observations?

Finally, the teacher brought the class together for a discussion of students' observations. Groups shared their thinking about the patterns in the data they collected from Stellarium. Students noted that they only saw the constellations in some of the months each year (e.g., Leonids falling stars are visible in November but not May), but it was the same months every year (e.g., Leonids were visible in November over multiple years). To close the lesson, students used the patterns they had identified to write their predictions about which constellation and falling star shower they would see in November two years in the future and to record their initial ideas about *what caused* these patterns.

The teacher began the next class period by asking students to refer back to their initial ideas based on the Stellarium constellation data about why they only saw specific falling stars at certain times of the year. She then prompted students to think about how they could test these ideas using a physical model in the classroom. The teacher called on students to share what components they should include in their models, reminding them of the different systems they studied over the course of the year (e.g., the garbage disposal system). She wrote students' suggestions on the board. The class came to a consensus to include Earth, the Sun, and two different falling star showers: the June Bootids falling star shower and the November Leonids falling star shower. Then, the teacher presented students with supplies for developing a physical model to test their ideas: a polystyrene foam ball and a pencil to

Continued

Classroom Snapshot 3.3 *(continued)*

represent Earth, a toothpick in the Earth to represent the school's location, a light bulb in the center of the classroom to represent the Sun, and images of falling stars hanging from coat hangers to represent the two falling star showers.

The teacher directed students to share their ideas with their group members about why they would only see specific falling stars at certain times of the year. After passing out supplies for the physical model to each group, the teacher listened to students as they shared and tested their initial ideas. Students were eager to test out their ideas using the physical model, noticing that if they changed one component (e.g., how Earth moves around the Sun), the change resulted in different interactions (e.g., Earth interacted with different groups of falling stars at different times of the year). The teacher prompted groups to think about how changes to their system model related to their everyday and classroom observations of the night sky. (See Figure 3.5 for possible teacher prompts related to systems and system models in space.)

Figure 3.5. Possible teacher prompts to probe for students' use of the CCC of systems and system models in space

Possible Teacher Prompts Related to Systems and System Models
1. How are the components in your model interacting?
2. How do the interactions explain why we only see falling stars at certain times of the year?
3. How does knowing that Earth moves through the debris of the falling stars help us understand Earth's movements?
4. What does your model predict?

After observing each group, the teacher gathered the students together so the class could formulate an answer to the lesson question: "Why do we only see falling stars at certain times of the year?" In sharing their models, most groups noted that in order to see falling stars at different times of the year, Earth must move around, or orbit, the Sun. The teacher asked students about how the components of their models interacted. Groups showed different shapes of orbits, resulting in a discussion about how they might modify their system models to better support their ideas about Earth's orbit around the Sun. Some students added a new component, a third falling star shower, to better show how Earth passes different groups of falling stars at different times of the year.

Continued

Classroom Snapshot 3.3 *(continued)*

In the final section of the unit, students watched a video of a meteor falling to Earth, prompting them to ask what caused the meteor to fall. Students then obtained information about gravity's effect on objects and revised their models to explain that gravity pulls the falling stars toward Earth.

Commentary

This classroom example represents the three key strategies of implementing CCCs as resources. First, students used their everyday and classroom experiences with the CCC of patterns as a resource to answer the lesson question, "Why do we only see falling stars at certain times of the year?" (Strategy 1). Building on their experiences seeing different objects in the night sky at different times of the year, students analyzed and interpreted the data table about falling star showers and their observations of the night sky over multiple years in Stellarium.

Second, students used the CCCs of patterns and systems and system models in different science disciplines—from sorting garbage and developing the garbage disposal system in physical science to analyzing constellation data in space science (Strategy 2). This lesson came at the end of the school year after students had used patterns and systems and system models multiple times to make sense of phenomena in different science disciplines, which allowed students to move from more concrete observations (e.g., developing a system model of components they see every day) to more abstract observations (e.g., modifying a system consisting of components and interactions at a scale too big to see.)

Finally, the classroom example highlights students' progression with using the CCCs of patterns and systems and system models over the course of the school year—from using CCCs intuitively to build on their everyday experiences to using CCCs more intentionally as scientists do (Strategy 3). In this example, students intuitively used the CCC of patterns based on their everyday and classroom experiences *and* intentionally said they were going to use the CCC of patterns to make sense of their data. Whereas students' use of patterns in the first unit focused mainly on identifying similarities and differences in observations, their use of patterns in the final unit was much more sophisticated. Here, they considered how different patterns were useful for answering different questions about phenomena. Students also used the CCC of systems and system models in a more sophisticated manner in the final unit. Whereas students focused on naming system components in the first unit, they focused on changing the interactions of the system components in the final unit, which their teacher prompted with questions such as those in Figure 3.5. Additionally, students demonstrated a more sophisticated understanding of a system by including multiple interactions in their models, including Earth's daily rotation around its axis and Earth's orbit around the Sun and through falling star debris.

Students' intentional use of the CCCs suggests their understanding of how CCCs may be useful in science, building on intuitive use from their everyday and classroom experiences. Furthermore, students demonstrated a deep understanding of using CCCs as resources at the end of the school year when they used two CCCs to explain a single phenomenon. In the final classroom example, students used patterns to make sense of the constellation data, which in turn made students wonder what interactions of the system could be causing the observed patterns. This example contrasts with the first two examples in which students used the CCCs separately, and it represents a progression in students' use of CCCs as resources. Here, not only did students draw on such resources from their homes and communities and from previous experiences in the science classroom, but they also understood when and how to use these resources.

For the English learners in the examples, the progression from an intuitive understanding to an intentional use of the CCC of systems and system models also represents a progression in sophistication of language. As their understanding of when and how to use the CCC became more sophisticated, these English learners communicated their understanding in more sophisticated ways. The students moved from naming individual components of the garbage disposal system (e.g., "garbage truck") to explaining the interactions of the components in the space system (e.g., "How does knowing that Earth moves through the debris of the falling stars help us understand Earth's movements?"). Although this learning progression applies to all students, it is particularly important for English learners who are valuable members of the classroom community based on the merit of their ideas, even if using less-than-perfect English.

Summary

To carry out the perspective of CCCs as resources in the classroom, we recommend three key strategies for curriculum design and classroom implementation. First, we propose intentionally designing classroom investigations and activities that provide opportunities for students to use resources from their everyday experiences with CCCs. To do this, CCCs may be introduced in the context of local phenomena (e.g., garbage, falling stars) that students have experience with in their everyday lives (Lee et al. 2019a). In this way, the phenomena act as scaffolds for introducing CCCs at the beginning of the year and for extending students' use of CCCs over the course of instruction. The intentional design around local phenomena allows students to draw on CCCs as resources so local phenomena and CCCs are mutually supportive for making students' ideas visible.

Second, we recommend that teachers and curricula consider "look fors" that help teachers recognize students' intuitive use of CCCs regardless of how that use is communicated. For instance, in the first classroom example, the teacher knew to look for different ways in which students sorted their garbage (e.g., by color, use, or material type). Once teachers ensure students are using a given CCC, they can use targeted probes (Figures 3.2–3.5) to make students' ideas visible and make the CCC in use explicit (Grapin et al. 2019).

Finally, we suggest scaffolding students' use of CCCs over time. Some CCCs are more intuitive to students (e.g., patterns), especially at the elementary level, than others (e.g., energy and matter in terms of flows, cycles, and conservation). To promote all CCCs as resources, designing instruction around students' use of more intuitive CCCs at the beginning of the school year lays the foundation for students to better use all CCCs over time. By explicitly naming the CCCs students use intuitively at the beginning of the year, students are made aware of the resources they bring to the classroom and see themselves doing the work of scientists. In turn, students develop more intentional use of CCCs over time as they make sense of different disciplines and different phenomena. By explicitly building on students' intuitive ideas, CCCs can act to both broaden participation and strengthen science learning for all students.

"All students come to school with experiences from their homes and communities they can use as resources to make sense of phenomena in the real world and the science classroom. This perspective on CCCs as resources promotes student participation and inclusion and allows students to see themselves as scientists."

All students come to school with experiences from their homes and communities they can use as resources to make sense of phenomena in the real world and the science classroom. This perspective on CCCs as resources promotes student participation and inclusion and allows students to see themselves as scientists.

References

Adler, J. 2000. Conceptualising resources as a theme for teacher education. *Journal of Mathematics Teacher Education* 3 (3): 205–224.

American Association for the Advancement of Science (AAAS). 1989. *Benchmarks for science literacy*. New York: Oxford University Press.

Fick, S. J., A. M. Arias, and J. Baek. 2017. Unit planning using the crosscutting concepts. *Science Scope* 40 (9): 40–45.

Fick, S. J., J. Nordine, and K. W. McElhaney, eds. 2019. Proceedings of the summit for examining the potential for crosscutting concepts to support three-dimensional learning. Charlottesville, VA: University of Virginia. *http://curry.virginia.edu/CCC-Summit*.

Goggins, M., A. Haas, S. Grapin, L. Llosa, and O. Lee. 2019. Integrating crosscutting concepts into science instruction. *Science and Children* 57 (2): 56–61.

González, N., L. C. Moll, and C. Amanti. 2005. *Funds of knowledge: Theorizing practices in households, communities, and classrooms*. Mahwah, NJ: Erlbaum.

Grapin, S., A. Haas, L. Llosa, and O. Lee. 2019. Using discipline-specific probes with English learners in the science classroom. *Science and Children* 57 (4): 36–43.

Lave, J., and E. Wenger. 1991. *Situated learning: Legitimate peripheral participation.* Cambridge, UK: Cambridge University Press.

Lee, O. 2020. Making everyday phenomena phenomenal: Using phenomena to promote equity in science instruction. *Science and Children* 58 (1): 56–61.

Lee, O., M. Goggins, A. Haas, R. Januszyk, L. Llosa, and S. E. Grapin. 2019a. Making everyday phenomena phenomenal: *NGSS*-aligned instructional materials using local phenomena with student diversity. In *Culturally and linguistically diverse learners and STEAM: Teachers and researchers working in partnership to build a better path forward*, ed. P. Spycher, and E. Haynes, 211–228. Charlotte, NC: Information Age Publishing.

Lee, O., L. Llosa, S. E. Grapin, A. Haas, and M. Goggins. 2019b. Science and language integration with English learners: A conceptual framework guiding instructional materials development. *Science Education* 103 (2): 317–337.

National Research Council (NRC). 1996. *National science education standards*. Washington, DC: National Academies Press.

National Research Council (NRC). 2012. *A framework for K–12 science education: Practices, crosscutting concepts, and core ideas.* Washington, DC: National Academies Press.

NGSS Lead States. 2013a. *Next Generation Science Standards*: For states, by states. Washington, DC: National Academies Press. *www.nextgenscience.org*.

NGSS Lead States. 2013b. *NGSS Appendix D: All standards, all students: Making the* Next Generation Science Standards *accessible to all students.* Washington, DC: National Academies Press. *www.nextgenscience.org*.

NGSS Lead States. 2013c. *NGSS Appendix G: Crosscutting concepts*. Washington, DC: National Academies Press. *www.nextgenscience.org*.

Osborne, J., S. Rafanelli, and P. Kind. 2018. Toward a more coherent model for science education than the crosscutting concepts of the *Next Generation Science Standards*: The affordances of styles of reasoning. *Journal of Research in Science Teaching* 55 (7): 962–981.

PART II
The Seven Concepts

Chapter 4

Patterns

Kristin L. Gunckel, Yael Wyner, and Garrett Love

Classroom Snapshot 4.1 highlights the role that patterns play as a crosscutting concept (CCC) in doing, learning, and teaching science.

CLASSROOM SNAPSHOT 4.1
Zebras, Congers, and Sandpipers

"Look, Mr. Lopez! Some stripes on the zebra go up and down, and some go sideways," said third-grade student Laymon after reading a book on animals. "Tigers also have stripes," said Georgia, as she was putting her book away on the shelf. Mr. Lopez, who was beginning a science unit on animal diversity, had provided the class with books about animals to hook their interest, and the students had spent the past half hour perusing the books. Mr. Lopez had suggested that as they looked through the books, they make note of interesting things they observe about the animals featured. He thought Laymon and Georgia had made some interesting observations about the phenomenon of animal skin colorations. He gathered his third-grade class together and asked the students what other patterns they had noticed on animal skins. Enrique raised his hand to say that some animals had spots. "Like what animals?" Mr. Lopez asked. "Like cheetahs and leopards," answered Mia. Mr. Lopez went to the computer and began pulling up images of various animals. He found pictures of spotted salamanders, zebras, a spotted sandpiper, and even some

Continued

Classroom Snapshot 4.1 *(continued)*

animals with both stripes and spots, such as the barred sand conger (Figure 4.1), which he projected onto the screen.

Mr. Lopez guided the class in sorting the animals based on their color patterns, and eventually the class drew a Venn diagram of animals with stripes, animals with spots, and animals with both stripes and spots. Mr. Lopez then said, "So what do we want to know about these animals?" The class agreed that they wanted to know why some animals have spots and some have stripes.

Figure 4.1. Patterns on animal skins

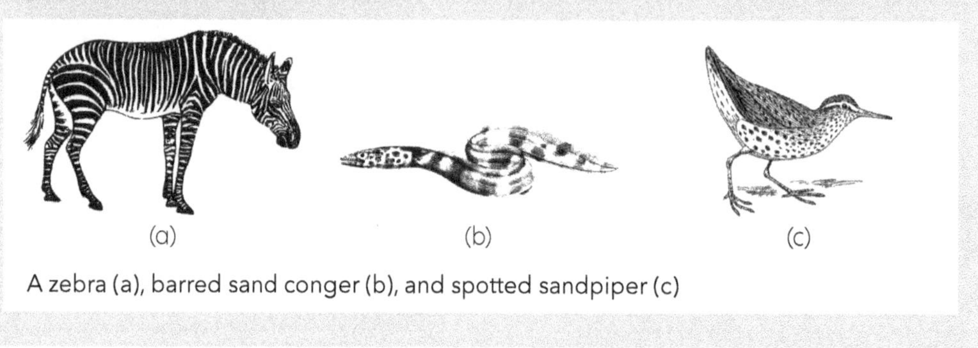

A zebra (a), barred sand conger (b), and spotted sandpiper (c)

One way to think about science is as the art of finding and explaining patterns in the world. At first glance, the world can seem like a chaotic place where events may appear to occur randomly, creating unbalance and confusion. Fortunately, the human mind is adept at noticing patterns; Mr. Lopez's students were quick to notice patterns on animal skins and to begin to wonder about them. Science and engineering rely on tools for identifying, representing, and analyzing patterns to model, explain, and predict events. Teaching science involves supporting students in noticing and using patterns to make sense of phenomena.

In this chapter, we will unpack the role of finding and analyzing patterns in scientific and engineering practices, examine some common types of patterns, and describe some of the tools that scientists use to represent and analyze patterns. We will then look at the teaching of patterns, including some of the challenges students initially experience when looking for patterns, what students should learn about patterns, and how patterns function to support three-dimensional learning. We will end by introducing the Experiences-Patterns-Explanations (EPE) table as a planning tool for supporting students in thinking about patterns.

What Is the Role of Patterns in Science?

The work of science is to build models that can explain and predict observations of the natural world. Hidden and underappreciated in this definition is the role that patterns play as the link between observations of phenomena and explanations (Covitt, Dauer, and Anderson 2017; Gunckel 2010). Understanding this link is an essential part of three-dimensional learning, which brings together crosscutting concepts with disciplinary core ideas (DCIs) and science and engineering practices (SEPs).

Figure 4.2. The Observations-Patterns-Models (OPM) triangle

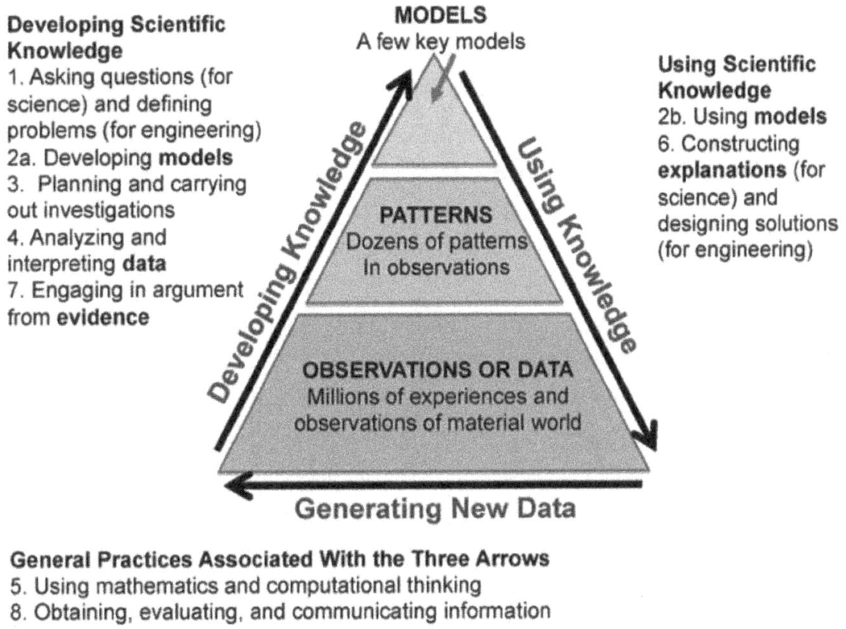

Source: Covitt, Dauer, and Anderson 2017.

Figure 4.2 illustrates the relationships among observations and explanations in science (Covitt, Dauer, and Anderson 2017). At the top of the triangle are *models*, which are abstracted representations of scientific ideas. Models can be children's conceptions or fully developed theories that serve as a basis for explanations and predictions of a given phenomenon. At the base of the triangle are the systematic *observations* (also referred to as *data*) that provide evidence to support models. Models require large quantities of data from multiple sources before they are considered valid. Therefore, the base of the triangle is wider than the top to signify this proportional relationship.

It is important to note, however, that robust models do not arise directly from data. Developing scientific knowledge, represented by the arrow pointing up the left side of the triangle, involves systematically organizing and analyzing observations to identify and verify *patterns*. These patterns are interpreted to propose and test hypotheses and eventually construct models. Models, in turn, are used to explain phenomena and predict patterns, given certain conditions, either natural or engineered. Models serve as a template of "what to look for" and thus greatly improve the capacity for making more careful and informed observations. Using models in this way is represented by the arrow labeled "Using Knowledge" on the right side of the triangle. This process of using knowledge may then generate new data that can lead to the development of new scientific knowledge, as shown by the arrow across the bottom of the triangle. Figure 4.2 also shows the SEPs involved in developing knowledge, using knowledge, and generating new data. Table 4.1 shows how patterns are used as part of the SEPs in science and the classroom.

An Earth Science Example of the Role of Patterns in Science

Plate tectonics is a unifying theory that explains diverse phenomena such as how volcanoes work, where earthquakes happen, and the distribution of rocks of different types and ages on Earth. How geologists developed that theory is one of the classic stories in the history of science. An obvious pattern that many people today recognize is the apparent fit of the continental shelves across oceans. However, when Alfred Wegener proposed in 1912 that South America and Africa were at one time connected, other geologists ridiculed his idea. At the time, there was little evidence beyond the matching outlines of the land massess to support such a preposterous suggestion.

Over time, however, geologists began to notice other interesting patterns in a variety of observations. Groups of scientists separately mapping rocks in North America, South America, Europe, and Africa began to compare their maps and noticed a curious pattern—not only did the outlines of the continents match up, but there were some rock types and ages that matched across continents as well. For example, the Appalachian Mountains in North America matched in rock type and age with the Caledonia Range in England and the Mauritania Atlas Range on the west coast of Africa. Numerous paleontologists working in all of these areas also noticed a similar pattern—there were fossils of the same types of extinct plants and animals that occurred on different continents. This pattern raised the question of how it was possible that, for example, the land reptile called the *Lystrosaurus* could have been in Africa, India, and Antarctica, all with very different modern climates, during the Triassic (Figure 4.3, p. 68). Many other observations by different geologists showed that glacial striations were left on rocks in places where today there are no longer glaciers and that those striations also matched up across continents.

Each of these three types of patterns—matching rocks and mountain ranges, matching fossils, and matching glacial striations—were supported by a vast number of individual observations made by countless geologists over time. They all seemed to support the idea that perhaps the continents were once together after all. But how could vast masses of rock move? What was the mechanism?

Table 4.1. The role of patterns in science and teaching science

Science and Engineering Practice	Role of Patterns in Science	Students Using Patterns in Classrooms
Asking Questions and Defining Problems	Scientists ask questions about patterns and use patterns to identify and define problems.	Students ask questions about patterns they see or use patterns to define problems.
Developing and Using Models	Models represent and serve as the foundation to explain patterns.	Students create representations and models to explain patterns they observe.
Planning and Carrying Out Investigations	Investigations are used to gather data necessary to identify patterns. Investigations can also be designed to test hypotheses about patterns.	Students design and carry out investigations to look for patterns; students identify the measurements that will be necessary to see a pattern.
Analyzing and Interpreting Data	Data are analyzed and interpreted to find patterns and relationships that can be used to explain or predict phenomena.	Students analyze and interpret data to find patterns that need to be explained and/or are predicted by models.
Using Mathematical and Computational Thinking	Mathematical and computational thinking make patterns visible.	Students use mathematical and computational thinking to identify, describe, and critique patterns.
Constructing Explanations and Designing Solutions	Patterns form the basis for explanations; explanations and solutions account for patterns.	Students construct explanations that account for patterns; students use patterns to design solutions to problems identified using patterns.
Engaging in Argument From Evidence	Patterns can be sources of evidence, used to describe evidence, and used to determine what counts as evidence.	Students create arguments about patterns (whether there is a pattern); students use patterns as evidence in arguments.
Obtaining, Evaluating, and Communicating Information	Patterns are effective ways to communicate information.	Students use patterns to communicate ideas, and evaluate and critique the evidence and explanations.

Figure 4.3. Map showing matching continental shelves, rock types, and fossils during the Triassic period

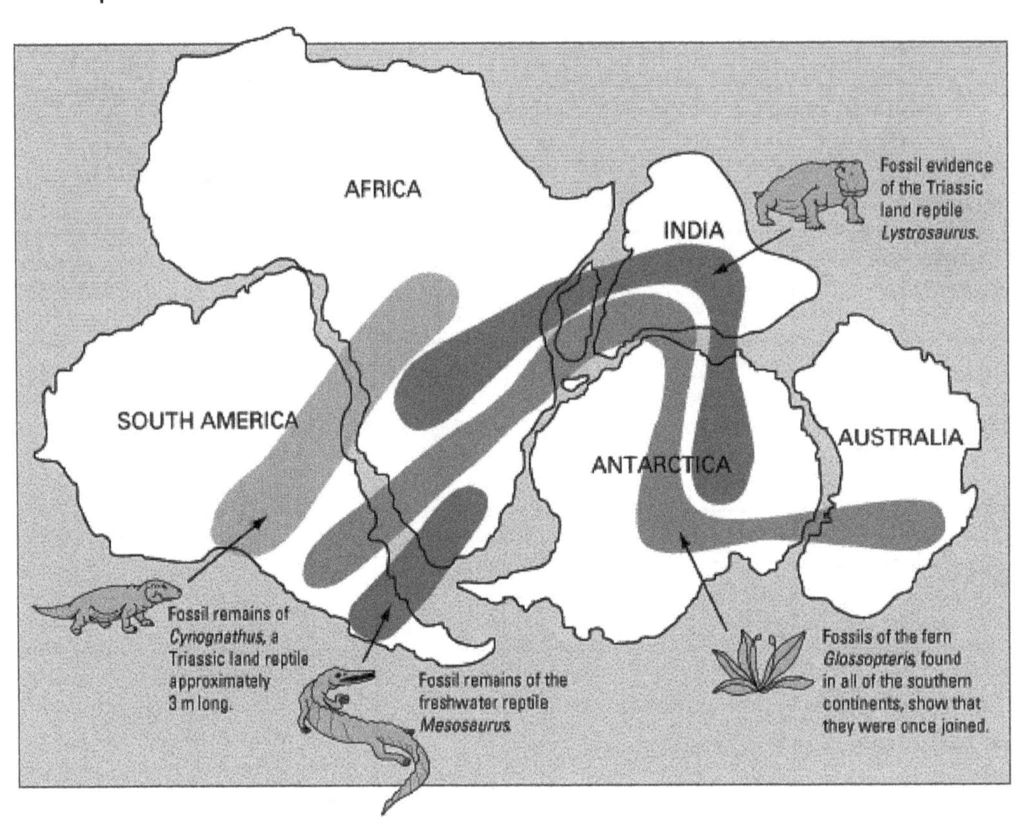

New patterns from other observations emerged. Geologists noticed that there are some places where earthquakes and volcanoes are more numerous. As they located these places on a map, the geologists realized that some large earthquakes and volcanoes were clustered on the edges of pieces of continental crust. At the same time, smaller earthquakes and different types of volcanoes occurred in the oceans (Figure 4.4a). Other geologists began to sample the rocks on the ocean floor, detecting yet another pattern: Rocks of different ages formed symmetrical stripes across the oceans, with younger rocks along mid-ocean ridges and older rocks along the edges of ocean basins. Magnetic minerals in these rocks recorded reversals in the orientation of Earth's magnetic field. When mapped out, these minerals showed a similar pattern, with reversals of the same age mirrored across the ocean ridges (Figure 4.4b).

These new patterns—the location of earthquakes and volcanoes and the patterns in age and magnetism of ocean rocks—were based on millions of observations. Geologists began to realize that some parts of the oceans were expanding along the mid-ocean ridges and other parts were contracting along deep ocean trenches. They began to understand that seafloor spreading along plate boundaries could explain how continents could move around the globe. In what is widely recognized as a paradigm shift, geologists now had a model to explain the patterns they had observed, from the formation of volcanoes along the western edges of North and South America to the occurrence of large earthquakes along these same boundaries; from the matching of glacial striations and mountain ranges to the existence of the same types of fossils on different continents.

Figure 4.4. Patterns observed in Earth's crust

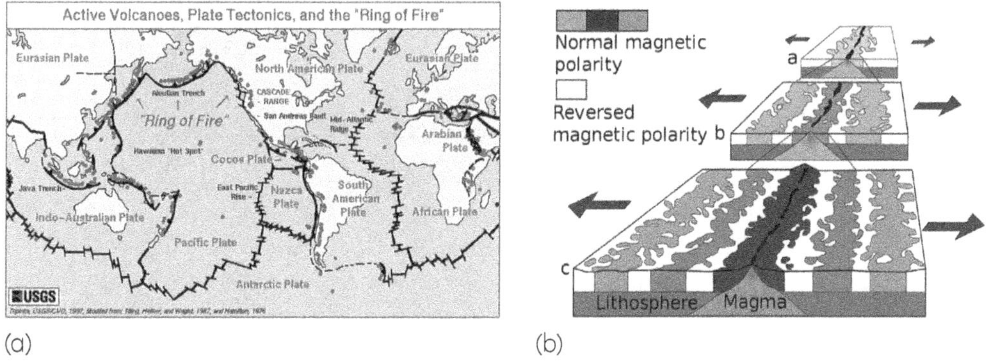

(a) (b)

A map of tectonic plates and volcanoes (a) and an illustration of magnetic seafloor stripes (b)

This new model for how plates move could also be used to explain other phenomena. For example, some scientists wondered how fossils of seashells found on Mt. Everest could have gotten so high above the ocean floor. The model helped geologists recognize many new patterns, including changes in the geochemistry of the rocks of the Himalaya, which indicated a tectonic collision between India and Asia that occurred about 50 million years ago. The scientists were then able to explain how India was once attached to Madagascar but broke away about 80 million years ago and moved northward, eventually colliding with Asia. Oceanic crust was caught up in the process and thrust upward, along with its fossils, to form the Himalaya. Figure 4.5 (p. 70) shows how this story fits into the Observations-Patterns-Models (OPM) triangle for developing and using scientific knowledge. Today, geologists continue to make observations of the rocks in the Himalaya, which raises new questions that they pursue to better understand the history of our Earth.

Figure 4.5. Story of plate tectonics in the OPM Triangle

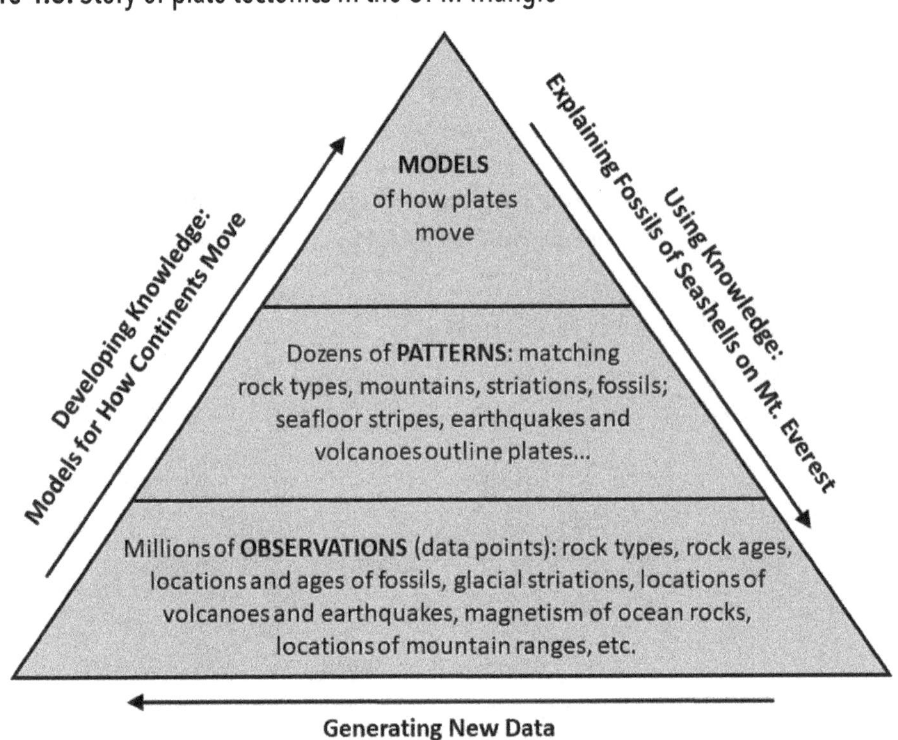

> Many patterns are complex, and a central part of a scientist's job is to convince others that the pattern is real ...

Types of Patterns and Tools for Finding and Communicating Them

Patterns link observations to models by helping scientists find order in phenomena. Randomness is difficult to explain, but once a pattern is noticed, scientists have something to model and use to make predictions. Some patterns may seem simple and obvious and other patterns are more difficult to detect and describe. Many patterns are complex, and a central part of a scientist's job is to convince others that the pattern is real—that under a given set of conditions, the laws of nature would cause the pattern to be repeated.

Scientists use a variety of tools, such as diagrams, graphs, maps, and statistics, to organize data, identify and analyze patterns, and communicate about patterns they find. An effective tool will both help the scientist find new patterns and represent the patterns in a way that others can recognize. Here we describe four common types of patterns and the tools that scientists use to represent them.

Classifications

Living organisms, atomic elements, rock types, body tissues, and cloud types are just some of the natural phenomena that scientists classify. Scientists organize components of the natural world into categories, with each category defined by its own unique characteristics. Within each category, elements that make up the group have similar characteristics and are segregated from other groups that have significant differences. Groups themselves can be divided into subgroups or lumped into more inclusive categories. The hierarchical nesting of groups inside ever larger groups is a structure that gives scientific classification great explanatory potential. For example, evolutionary biologists organize similar interbreeding organisms into entities called species. They then organize species into ever larger groups of most closely related relatives until they've connected all living things in one great group. As such, humans are a species that are part of a larger group called primates, which also includes other animals with grasping hands and forward-facing eyes, like monkeys and apes. All primates are part of an even bigger group called mammals, which includes all animals that have hair and lactate. Each level up in the hierarchy becomes more inclusive until all organisms on Earth come together in one great group, defined as containing cells—a requirement for life that scientists think evolved only one time and then diversified into all the life we see today. This model distills the explanation for the full diversity of life into a single evolutionary "tree."

Chemists often organize the elements by one particular property: atomic weight. The predictive power of patterns was enabled when they created the periodic table, a model that organized the known elements into groups with similar characteristics. "Holes" in the pattern helped identify where to look for unknown elements that chemists later found. This powerful system for organizing observed patterns was not the first or only system scientists developed. The best classification scheme was far from obvious. Scientists settled on the table we use today because it held the greatest predictive and explanatory power.

Classification can be communicated in a number of ways, but any classification tool must have components that communicate two things: differentiation between elements that are different (using, for example, color variation, distance, or dividing boundaries) and connections that highlight similarity (using, for instance, like colors, connecting lines, or proximity). Nonvisual methods like naming conventions (such as the use of *cirro* for "high" and *alto* for "medium height" when describing clouds) can also facilitate associations and help with describing underlying phenomena. Figure 4.6 (p. 72) shows some examples of classification schemes.

Figure 4.6. Examples of scientific tools for visualizing classification patterns

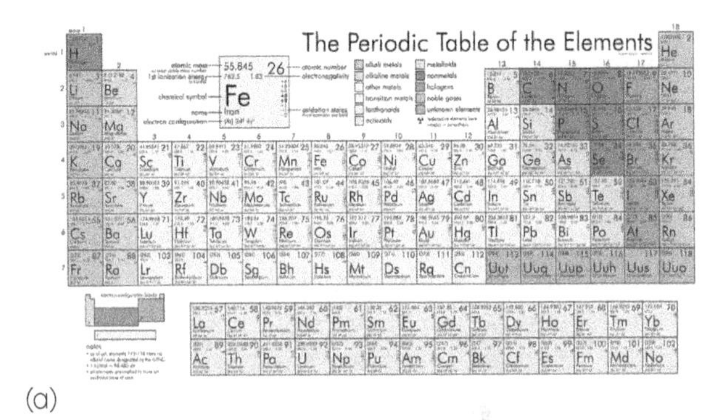

(a)

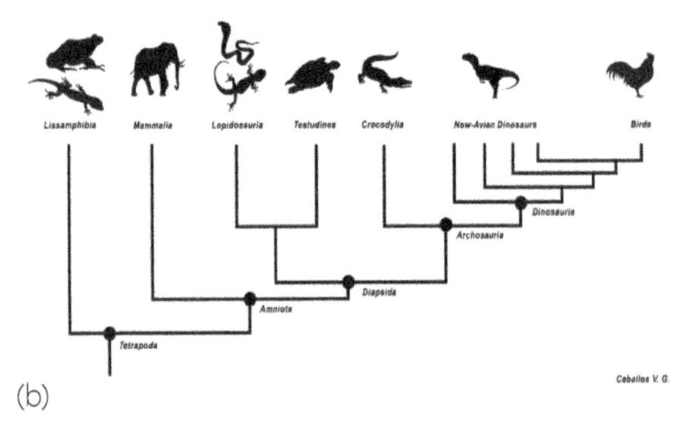

(b)

The periodic table of elements (a) and a cladogram highlighting birds as members of the dinosaur group (b)

Distributions

Scientists are often interested in observing *where* and *when* certain events occur (or where and when they occurred in the past), with the goal of eventually predicting where and when similar events will occur in the future. One way to find these patterns is by mapping distributions. Distributions describe the arrangement of a collection of objects in space or a series of events in time. Discerning patterns within a distribution is useful for identifying relationships between new observations and known phenomena or for creating new explanatory models. For example, identifying the locations of earthquakes has helped reinforce the theory of plate tectonics and establish the location of plate boundaries. Historical hurricane tracks are also used as data for predicting future hurricane paths.

Maps and charts are common and useful tools for representing and analyzing distributions. Integral to any such representation is an element that communicates the context in either space (e.g., using map boundaries or gridded coordinates such as latitude and longitude) or time (e.g., using a scaled timeline). Observations can be individual events, as in the earthquake example, or even classifications. Consider, for example, a map of the distribution of biomes on Earth, which serves to show a relationship between biomes and latitude. This map not only reinforces the relationship between temperature and biome, but closer inspection reveals wet and dry biome regions that correspond to circulation patterns in the atmosphere. Now consider a distribution of fossils mapped on a geologic timeline. This can explain the ages and evolution of different lineages of life. Figure 4.7 shows examples of distributions.

Figure 4.7. Examples of scientific tools for visualizing distribution patterns

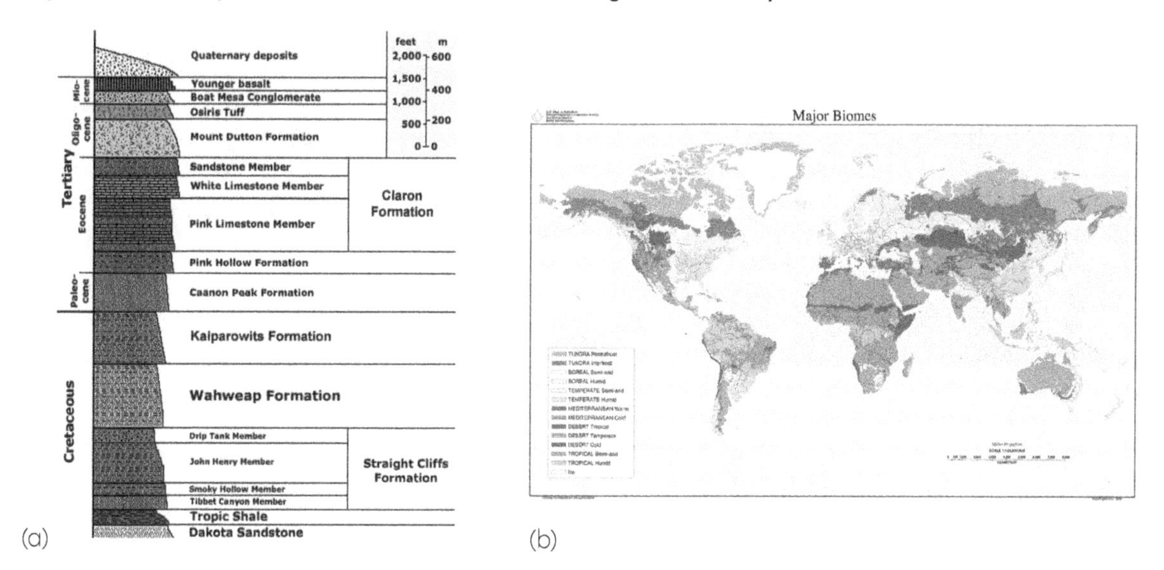

A stratigraphic column of Bryce Canyon National Park showing the distribution of rock types through time (a) and a map of major biomes of the world (b)

Cycles

Patterns that repeat are called cycles. Examples include the phases of the Moon; the movements of water, carbon, and nitrogen through the environment; the birth, growth, and reproduction of living organisms; and high and low tides. Cycles are characterized by the repetition of a particular condition or state, typically at different points in time, although alternatively at different places in space (e.g., consider ripples in windblown sand). An assumption of any cycle is that because the repetition has been observed, one can expect to observe it again at a predictable time or place.

Tools for representing cyclical patterns include sequences of photographs or drawings, diagrams, and graphs. The critical element for communicating a cycle is some indication of the repeated state—a repeated image, a similar location in a graph, or, typically, a sequence of arrows that connect an image back to itself. Figure 4.8 (p. 74) shows common representations of cyclical patterns.

Figure 4.8. Examples of cycles and tools for representing them

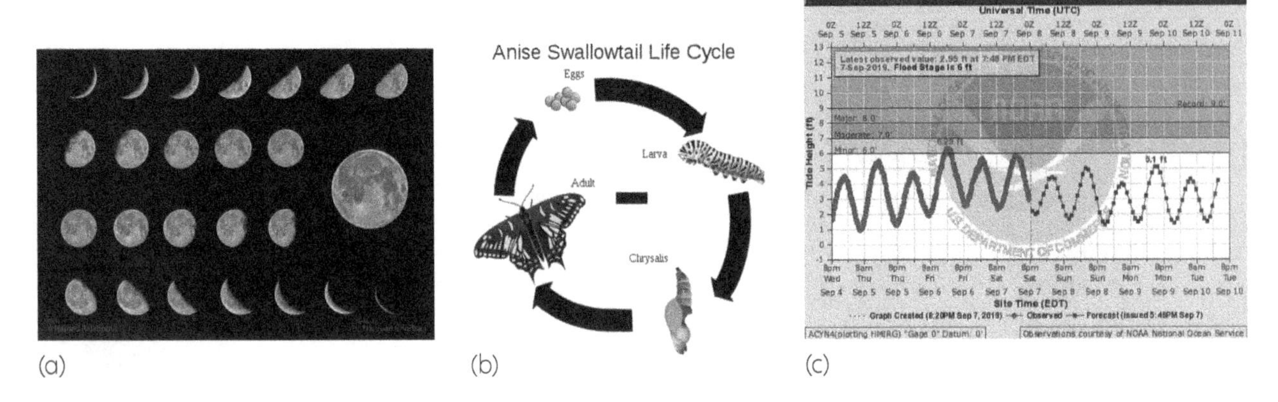

Photographs of the Moon phases (a), a diagram of the anise swallowtail's life cycle (b), and a graph of the daily tidal cycle (c)

Relationships Among Variables

Identifying how one variable changes in relation to another variable is important for explaining many phenomena, especially when searching for potential cause-and-effect relationships. Graphs, mathematical equations, and statistics are useful for analyzing these relationships, determining causality, and describing change. The graph of the tide cycle in Figure 4.8 represents the relationship between tide height and time of day. The ideal gas laws are classic examples of relationships among variables, represented with the mathematical equation $PV=nRT$ or with graphs that visually illustrate the relationships among pressure, volume, temperature, and moles of particles of a gas. Other examples include global temperature and the amount of carbon dioxide in the atmosphere and the relationship between dissolved oxygen and water temperature.

Successfully communicating relationships between variables relies primarily on a successful interpretation of how each variable is represented. Interpretation is enabled by appropriate labels and scales, as well as standard conventions (such as using the color red to represent heat or T for temperature). Figure 4.9 shows examples of graphs representing relationships between variables.

How Students Build Understanding of Patterns

We now shift to what and how students learn about patterns. We begin with some considerations for how students learn to recognize patterns and then outline the progression of goals in the *Next Generation Science Standards* (*NGSS*; NGSS Lead States 2013) for learning about patterns.

Figure 4.9. Examples of patterns in relationship between variables

Charles's Law (Volume vs. Temperature)
Volume (m³)
Temperature (K)

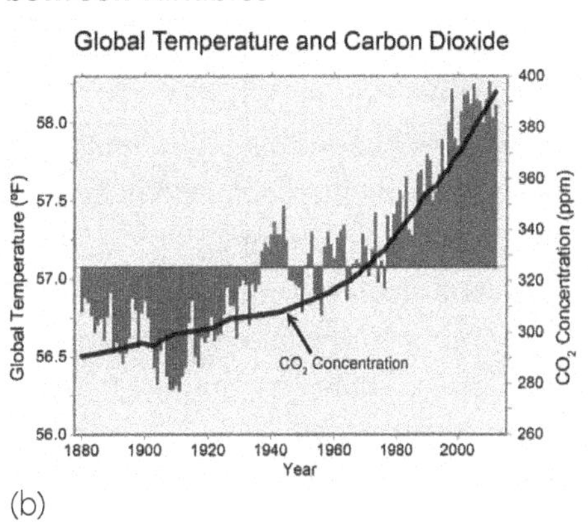

(a) (b)

A graph of Charles's law (a) and a graph of global temperature and carbon dioxide over time (b)

Challenges in Learning to Recognize Scientifically Important Patterns

A common notion is that children are naturally adept at recognizing patterns. To some extent, this is true, as children seek patterns in their daily experiences to make sense of their everyday world. However, not all patterns gleaned from everyday experiences build scientific understanding. Experts learn to recognize complex patterns that may not be obvious. Students may need guidance to notice important and relevant patterns. Instructional Application 4.1 illustrates this idea.

INSTRUCTIONAL APPLICATION 4.1
Patterns of Evolution

Think of the trees that dot a neighborhood street. To a botanist, these street trees display easy-to-spot evidence for the patterns of evolution. To a lay person, however, street trees are a pleasant source of shade, a splash of vibrant color in the fall and spring, or for much of the year, simply a giant mass of green. The evolutionary patterns that trees display are hidden in plain sight, often obscured by their most obvious characteristic—their enormous size. For students to notice evolutionary patterns, they must closely examine tree structures they would otherwise ignore, such as the shape of individual leaves, leaf arrangements, tree fruit, and flower structures. How do students begin to observe trees as evolutionary experts do? How do we support students in moving from novice to expert observation?

Continued

Instructional Application 4.1 *(continued)*

The first step in connecting tree features to their evolutionary history is for students to categorize these features by type, as the features trees exhibit are indicative of their evolutionary history. For example, in her sixth-grade classroom, Ms. Tappa brought in parts from five different trees (Figure 4.10). To help her students notice evolutionarily important tree features, she asked them to make groups of the different tree parts. Even with parts from just five trees, the students made many groups. The first pair of students focused on the tree feature that jumped out at them: the leaves. They grouped the leaves of plants C and D together because their leaves were simple and lobed. Then they made a bigger category to which they added plant E because its leaves were also simple. Another set of students started examining the other parts and grouped plants A and E together because both had pod-shaped fruits. Still other students became interested in how leaves were arranged along the branches. After looking closely, they decided to make a special group for plants B and D because their leaves were opposite from each other. They then grouped together the remaining plants, A, C, and E, because their leaves were arranged in an alternating pattern.

Figure 4.10. Botanical drawings of trees Ms. Tappa brought into her class

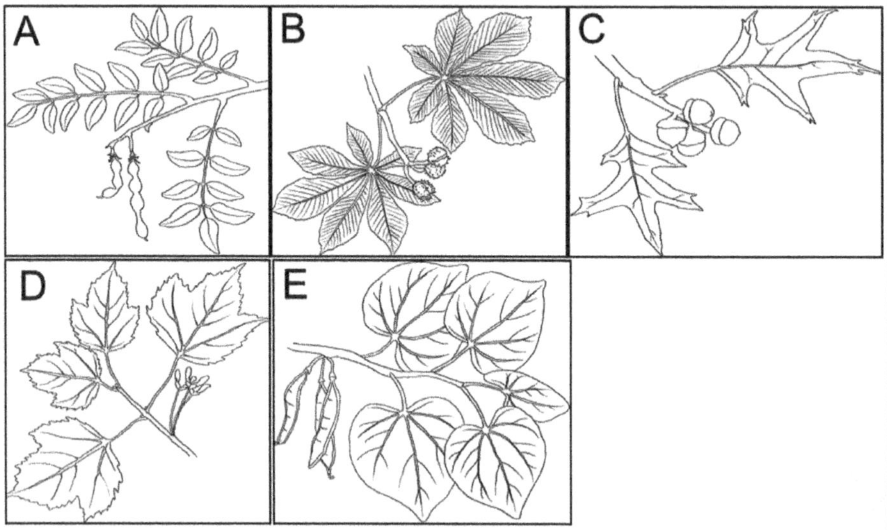

Chinese scholar (A), horse chestnut (B), pin oak (C), red maple (D), and redbud trees (E)

Once students finished making groups, they shared their findings and learned the proper vocabulary to describe what they noticed (e.g., terms like *lobed, simple, leaf arrangement*). With this basic knowledge of leaf- and fruit-type patterns, the students

Continued

Instructional Application 4.1 *(continued)*

were now equipped with the observation skills to notice evolutionary patterns in the trees outside their classroom. Students would now be able to collect observational data to support or reject tree identification hypotheses, and the ability to recognize leaf and fruit types would help students sort neighborhood trees into groups that represent common descent. Framing the tree features they noticed in the patterns of evolution helped students focus on the scientifically important patterns necessary to build models for how the common ancestor to present-day trees looked and diversified. Providing students with the observation skills to notice key tree features made it possible for them to see the evolutionary patterns in the biodiversity that they passed daily.

Progression of Goals for Learning About Patterns in School

At the youngest grades, students should learn to recognize and describe patterns and use patterns as evidence to make predictions or construct explanations. For example, the first-grade performance expectation (PE) 1-LS3-1 asks students to "make observations to construct an evidence-based account that young plants and animals are like, but not exactly like, their parents." Here, the DCI is about inheritance and variation in traits. To learn this core idea, the performance expectation requires students to recognize categories of patterns as similarities and differences in features between parents and offspring. For instance, students may observe that puppies share similar body shapes and hair types with their parents, even if their colors and sizes are different. Students learn to expect similar patterns in other animals. They also may notice that in some types of animals, such as frogs and toads, the young look different from their parents, but as they grow into adults, they look more like their parents. Students observe similar patterns in plants, with young and older plants of the same species having the same leaf shape, even if the size is different.

Young children should also begin to use graphs, charts, and maps to identify, represent, and communicate about patterns. In the previous example about inheritance and variation, tools such as Venn diagrams or other sorting and classifying diagrams can help students see, represent, and talk about patterns they notice between parents and offspring. In other performance expectations, the use of such tools is more explicit. For example, PE 3-ESS2-1 has students "represent data in tables and graphical displays to describe typical weather conditions expected during a particular season." In this performance expectation, the CCC of patterns is deeply intertwined with the SEP of Analyzing and Interpreting Data and the DCI that there are patterns in the types of weather expected during a particular season for a particular place. Engaging students in making bar charts or pictographs of data such as daily average temperatures, precipitation, or wind direction can help children recognize the patterns in the weather and use them to predict, in general terms, what type of weather they might expect in summer or winter.

In middle school, students should be becoming more sophisticated in the types of patterns they identify, the tools they use to identify those patterns, and the ways they use those patterns to explain and predict phenomena. Students can also begin to see how the CCC of patterns connects to other CCCs. For example, finding and identifying patterns is foundational to recognizing and explaining cause-and-effect relationships; and scale, proportion, and quantity is useful for identifying patterns (Rehmat et al. 2018). Consider MS-ESS1-1, which has students "develop and use a model of the Earth-Sun-Moon system to describe the cyclic patterns of the lunar phases, eclipses of the Sun and Moon, and seasons." The DCIs about lunar phases, eclipses, and seasons are grounded in recognizing and representing patterns, which are skills and concepts students have been working on since elementary school. Now, however, students are expected to use those patterns to recognize and explain cause and effect (another CCC; see Chapter 5), an essential part of modeling the Earth-Sun-Moon system. In middle school, students should also be extending what they have learned about finding and using patterns to recognize patterns across scales. In MS-PS1-2, students "analyze and interpret data on the properties of substances before and after the substances interact to determine if a chemical reaction has occurred." In order to understand the DCI that substances have physical and chemical characteristics that change during chemical reactions, students must understand that macroscopic patterns are related to microscopic and atomic-level structures, which is an idea that links to both the CCC of scale, proportion, and quantity (Chapter 6) and the CCC of structure and function (Chapter 9). Tools such as the periodic table are useful for making these patterns explicit.

Most of the high school performance expectations that incorporate patterns require students to integrate several patterns observed across scales to use as evidence for causality in explanations of phenomena. For example, HS-PS2-4 has students "use mathematical representations of Newton's law of gravitation and Coulomb's law to describe and predict the gravitational and electrostatic forces between objects." Both of these laws use mathematics ($1/r^2$) to express the squared inverse relationship, an important pattern observed in physics whereby the magnitude of a physical quantity (such as a force) decreases with the square of the distance from the source. This performance expectation requires students to use mathematical and computational thinking to identify this pattern at multiple scales and to think across these scales to explain and predict forces at a distance.

Although the *NGSS* weaves patterns as a CCC into specific performance expectations, students should use patterns to address any PE, not just those that explicitly name it as a CCC. Furthermore, students should learn to critically evaluate the evidence for patterns. Take, for example, a unit addressing HS-ESS3-6, which has students "use a computational representation to illustrate the relationships among Earth systems and how those relationships are being modified due to human activity." Here, students could generate computer models of groundwater contamination. To engage in the practice of mathematical and computational thinking, students could evaluate the distribution patterns of the sampling wells and the plume of contaminants (Figure 4.11). They might get further insight by adjusting assumptions used in the computer model and comparing plume solutions with different boundary conditions. Students might decide, for example, that the map is more reliable in certain regions near its center, recognizing the limits of extrapolation and observing regions with fewer wells around the edges. This type of critical thinking is important for analysis and interpretation of all patterns, regardless of whether patterns are explicitly mentioned in the relevant performance expectations.

Figure 4.11. Computer model of a groundwater contamination plume

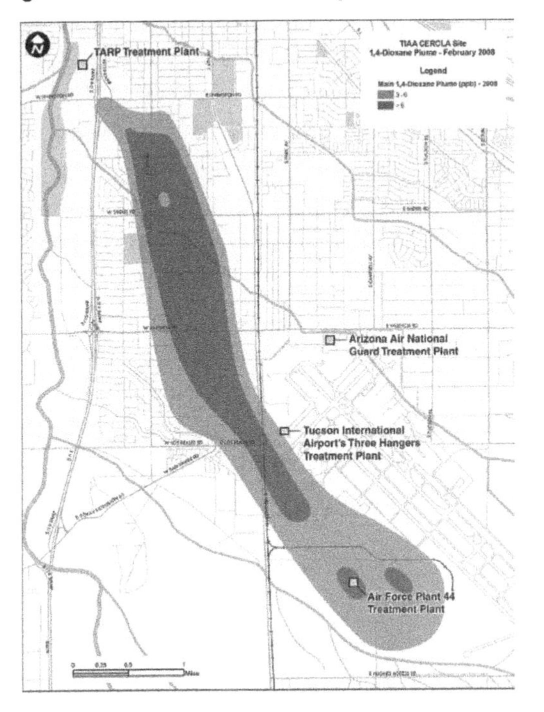

Teaching Science Using Patterns

In three-dimensional learning and teaching, where CCCs, DCIs, and SEPs are intertwined in order to make sense of phenomena and solve problems (NRC 2012), engaging students in doing science is the goal. Teaching that addresses three-dimensional learning must make the role of patterns in developing and using scientific knowledge explicit to students. One way to do this is to use the Experiences-Patterns-Explanations (EPE) tool.

The EPE tool translates the Observations-Patterns-Models (OPM) triangle shown in Figure 4.2 (p. 65) into a representation that is more manageable for the classroom (Figure 4.12). As we explained in the section on the role of patterns in science, developing and using models

Figure 4.12. The OPM triangle for science translated into the EPE triangle for school science

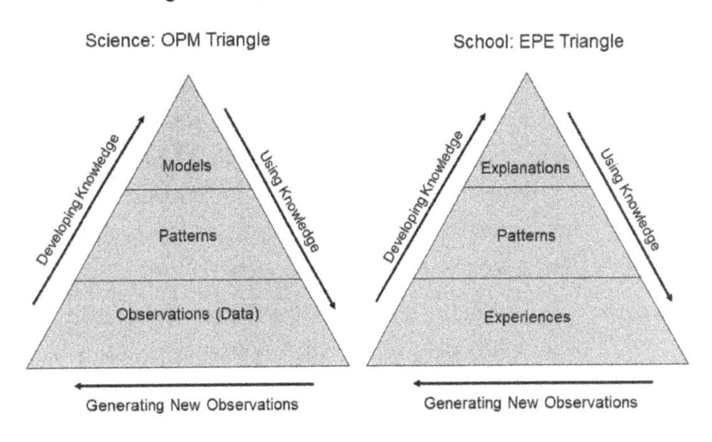

is long-term work that builds on millions of observations of numerous people over many years. However, the classroom context limits the number and types of observations that can be made, as well as the time period across which to make them. Nevertheless, school science lessons that align with the *NGSS* should engage students in the same practices associated with developing and using knowledge, as shown in the Observations-Patterns-Models triangle. In Figure 4.12 (p. 79), observations are translated into experiences because teaching science is more involved than just providing students with data. Teaching also involves drawing on students' multiple and diverse informal experiences, prior knowledge, and cultural funds of knowledge as sensemaking resources (Michaels, Shouse, and Schweingruber 2007). Experiences capture this broader category of resources, which teachers use to support students in making sense of phenomena. Models are translated into explanations because, depending on the performance expectation, the ultimate learning goal may not be the development of a model. In most cases, however, the disciplinary core idea is grounded in an explanatory model and students are expected to be able to articulate model-based explanations for the phenomenon they are investigating (NRC 2012).

"In three-dimensional science teaching, patterns are pedagogically important because they link experiences to explanations."

In the EPE triangle, as with the original OPM triangle, patterns play an important linking role. In typical school science, traditional instruction often moves directly from observations to explanations, or it simply presents explanations, without making patterns explicit to the students. Without visible patterns in experiences, students have nothing to explain, which can make understanding models more difficult for students and reduce science learning to memorizing facts. In three-dimensional science teaching, patterns are pedagogically important because they link experiences to explanations. Figure 4.12 emphasizes this important pedagogical, as well as scientific, role of patterns.

The Experiences-Patterns-Explanations table (Figure 4.13) turns the EPE triangle into a tool for planning and assessing three-dimensional science lessons that make patterns explicit (Gunckel 2010). The tool highlights the connecting role between patterns and experiences. It highlights what patterns to make visible during instruction and what patterns to feature to probe student sensemaking during assessment. (See Chapter 15 for more on assessment.)

To use this tool, begin at the top by identifying the phenomenon of interest, the driving question that will guide a series of lessons, and the relevant performance expectation. Then, in the right-hand Explanations column, write the ideal answer to the driving question, which students should understand and be able to use by the end of instruction. This answer should be a model-based response that aligns with the DCI from the performance expectation. In the middle Patterns column, list the patterns that support this explanation. In this column, make notes about the types of patterns employed (e.g., classifications) and the tools that students might use to see these patterns, such as graphs, maps, diagrams, and mathematical equations. The left-hand Experiences column includes all of the experiences that students will use to see the patterns. This column includes prior school experiences, out-of-school experiences, and funds of knowledge, as well as the experiences planned for the classroom. Reading the chart from left to right shows how patterns link experiences to explanations while developing knowledge (i.e., going up the left side of the EPE triangle); reading the chart from right to left shows how patterns link explanations to experiences while using knowledge (i.e., going down the right side of the EPE triangle). In practice, students will move back and forth between developing and using knowledge.

Figure 4.13. Experiences-Patterns-Explanations (EPE) table template

Phenomenon: ______________________________
Driving Question: ______________________________
Performance Expectation: ______________________________

Experiences	Patterns	Explanations
1. Experiences planned for the lesson 2. Prior school experiences 3. Cultural experiences and family funds of knowledge		

Developing Knowledge →

← Using Knowledge

Examples of Teaching About Patterns in the Classroom

This section presents two examples of teachers supporting students in using patterns in three-dimensional science learning. The first example (Instructional Application 4.2, p. 82) focuses on using patterns to develop knowledge and make predictions; the second (Instructional Application 4.3, p. 85) illustrates using patterns in an engineering design process.

INSTRUCTIONAL APPLICATION 4.2

Using Patterns to Develop Knowledge and Make Predictions in Science

MS-LS2-2: Construct an explanation that predicts patterns of interactions among organisms across multiple ecosystems.

In a busy eighth-grade classroom, Ms. Green's students are working in pairs to plot data on graphs. Some pairs of students are plotting the number of wolves on Isle Royale (the biggest island in Lake Superior) from 1980 to 2018. Other pairs of students are plotting the number of moose on Isle Royale during the same years. The class is familiar with Isle Royale from their social studies lessons on the Great Lakes. As the students are finishing their graphs, Ms. Green walks around the classroom, asking each pair what patterns they notice on their graphs. Emily and Deja are working together on a moose graph. Emily says that the number of moose was different each year and Deja adds that it seems like the number generally increased over the period from 1980 to 1996, then decreased abruptly. Darnell and Jacob notice that the number of moose was down for a while in the early 2000s but that it had been rising steadily since about 2012. Jamal and Kaitlin are working on a wolf graph. Kaitlin notices that in 1980 there were 50 wolves but that there are almost no wolves now. With this lesson, Ms. Green's students are engaging in the scientific practice of analyzing and interpreting data. By having them create graphs, Ms. Green is helping them organize data in order to see patterns in information that might have initially seemed to be a random set of numbers of wolves and moose (Figure 4.14).

Figure 4.14. Graph of wolves and moose on Isle Royale

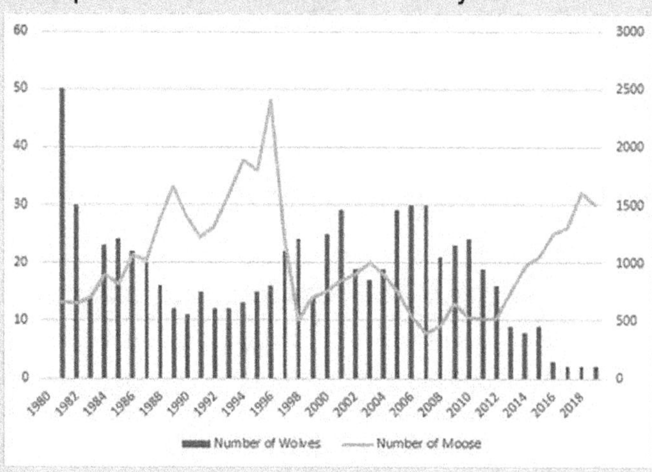

Continued

Instructional Application 4.2 *(continued)*

The next day, Ms. Green asks the class what questions they are prompted to ask based on the patterns they notice in their graphs. This activity engages the students in the SEP of Asking Questions and Defining Problems. Several students want to know why the number of moose seemed to go up and down. Lizabeth asks why the moose population dropped off so fast in 1996, and Julio wonders why the moose population is now rising. Many students wonder if the wolves are dying off. Ms. Green points out the patterns that the class is noticing and writes them in a list on the board. This strategy helps make the patterns explicit to students and also helps Ms. Green assess the students' sensemaking of the graph. Next to each pattern, Ms. Green writes the corresponding question from the students. In this way, she is helping students see that patterns can be a source of questions for further investigation.

Ms. Green then has students who had drawn graphs of the moose population get together with students who had drawn graphs of the wolf population. When they compare their graphs, the students notice some new patterns. Ms. Green makes a new list of patterns on the board, including that when the population of the moose were up, the wolf numbers were down; and when the population of the wolves were up, the moose numbers were generally down. She uses her document camera to highlight these patterns for everyone to see. Some students, who know that wolves are predators of moose, argue that when there were a lot of wolves, they kept the moose population in check. This prompts other students to suggest that if there were too many wolves, there wouldn't be enough moose for them to eat, and so some wolves might starve. Using their graphs as evidence, these students claim that the moose population would then recover. Matt even suggests that maybe when there weren't many wolves, the moose population would rise. Then there wouldn't be enough food for all the moose, and they could die off too. Gradually, the class begins to recognize a general cycling pattern. Ms. Green explains that this relationship is a predator-prey relationship, a key interaction among organisms included in the DCI.

However, there is still that strange trend since 2012 of a decreasing wolf population. With the moose population rising, some students argue, the wolf population should be recovering and beginning to go up instead of down. Even in the mid-1990s, Monica claims, pointing to the graph, the wolf population never dropped below 10 wolves. Ms. Green lets the class ponder this question over the weekend.

On Monday, Ms. Green introduces some new information. She tells the class that in 1985, visiting tourists had brought a dog onto the island. The dog was a carrier of the parvovirus that can be fatal to wolves. Several students have heard of parvovirus because they have pet dogs that received vaccinations against the disease. Ms. Green asks the students to consider this new information and use it to explain the troubling pattern they had noticed. Now the pattern seems to make sense. Maybe the

Continued

Instructional Application 4.2 *(continued)*

wolf population was being killed off by the virus, the class hypothesized. Ms. Green confirms that biologists on Isle Royale had identified parvovirus in the wolves and that they hypothesize that this may be one reason for the wolf population crash. She asks the students to discuss what they thought would happen to the moose population with so few wolves on the island. James and Tristan both argue that the moose population would also crash, like it had in 1996, because there were no predators to keep their population from eating all their food on the island. Here, James and Tristan are demonstrating their understanding of predator-prey relationships, using a DCI to identify a mechanism for the pattern they are seeing and using the pattern as evidence to predict a future event.

Figure 4.15 shows the Experiences-Patterns-Explanations table for Ms. Green's lesson. Moving from left to right, the students used the patterns to develop new knowledge about predator-prey relationships. Moving right to left, they used this knowledge to make predictions about future events.

Figure 4.15. Experiences-Patterns-Explanations table for lesson addressing MS-LS2-2

Phenomenon: Crash of the wolf population on Isle Royale
Driving Question: What happened to the wolves on Isle Royale?
Performance Expectation: MS-LS2-2 Ecosystems: Interactions, Energy, and Dynamics: Construct an explanation that predicts patterns of interactions among organisms across multiple ecosystems.

Experiences	Patterns	Explanations
Experiences planned for the lesson Construct graphs of wolf and moose populations. Each data point on the graph is an observation. Prior school experiences Prior knowledge of predator and prey animals. Cultural experiences and family funds of knowledge Possible familiarity with parvovirus as a canine disease from having a family pet dog.	1. Populations of moose and wolves cycle. 2. Cycles of wolves follow cycles of moose. 3. Wolf population crashed after 2012 while the moose population soared.	Wolves and moose had a typical predator-prey relationship until parvovirus was introduced to Isle Royale in 1985. The parasite killed most of the wolves. Without predators, the moose population increased and may have outstripped its food supply, eventually leading it to crash again in the near future.

Developing Knowledge →

← Using Knowledge

INSTRUCTIONAL APPLICATION 4.3
Using Patterns to Apply Knowledge to Solve Problems

4-PS4-1: Develop a model of waves to describe patterns in terms of amplitude and wavelength and that waves can cause objects to move.

4-PS4-3: Generate and compare multiple solutions that use patterns to transfer information.

Ms. Barnett presents her fourth-grade class with a new lesson. Students are grouped in pairs and then each pair is given a metal Slinky toy and a scenario describing two friends who live in adjacent apartment buildings. These friends suspend the Slinky between their buildings to communicate with each other. The students are directed to set up a similar situation across the spaces between their desks and consider how they could communicate with each other without talking and using only the Slinky.

The students are first tasked with creating a system of "Slinky codes" to say "Hi, I'm here!" and "Bye, I'm leaving." Experimenting with the Slinky gives them ample opportunity to generate a variety of (potentially new) experiences with waveforms, even though they currently know little about the appropriate terminology. Some students send *longitudinal* "push-pull" pulses, and others send "side-to-side" *transverse* waves. One group experiments with *amplitude*, varying between big and small motions while another takes advantage of *frequency* by using "fast wiggles" and "slow wiggles" to communicate. A couple of groups swing the Slinky like a jump rope, and, when encouraged to "try it faster," they "discover" the double and triple nodes of a *standing wave*. Some students even consider using the metallic sounds generated by the Slinky as elements in their code, especially when one student compares the Slinky to her violin strings. The students record their observations and ideas in a design notebook.

The next day, Ms. Barnett asks each student group to prepare a description of one of their "Slinky communication methods," using a few sentences and including a sketch. Groups are then asked to replicate their sketches on the board and to present their ideas, and each sketch is labeled with a letter. When the collection is complete, Ms. Barnett challenges each student pair to consider the entire collection of methods and group them into three to five short lists of methods "that are similar." (Students list the sketches using their label names.) After a short time, she adds two additional sketches—one of a waving hand and another of ocean waves—and asks students to classify these sketches within their groupings, as well.

Continued

Instructional Application 4.3 *(continued)*

Ms. Barnett starts the next discussion by pointing out the collective observation that motion at one end of the Slinky is repeated in different places throughout the entire length of the Slinky. She notes that this is also true of the physical examples of a hand waving and ocean waves and thus introduces the term *wave;* Ms. Barnett goes on to build a short vocabulary of associated terminology, annotate the class sketches (e.g., saying "Here is where Luisa and Elton demonstrated a large *amplitude.*"), and provide the "standard" up-and-down sine curve as another visualization.

She asks students to use past observations and provide examples of other physical systems that might behave similarly to a Slinky when "perturbed." Student ideas include ripples in water, earthquakes, and plucked guitar strings. Ms. Barnett then summarizes with an explanation that motion, in the form of waves, can be transmitted over distances, and she points out that the Slinky is a model for how this can occur in other materials. At its conceptual level, this approach sets the stage for lessons some years later regarding the transmission of sound and light waves.

The students are then given a set of prompts for research on coded messages (e.g., smoke signals, semaphore, Morse, binary) and a design challenge. The teacher groups together students and explains that half of each group will be given a short, spoken message that they will need to transmit to the other half of their group using the Slinky and a code of their own design. Ms. Barnett wants this process to simulate an engineering design problem with competing optimization goals; she tells the students that their solutions must transmit the message both accurately and quickly. She gives the students time to experiment and design while she reinforces the waveform vocabulary by using it in discussions with each student group as they describe their ideas. The students complete the design challenge with varying degrees of success, but most important, with an additional set of concrete experiences and associated patterns that they can draw on later to reinforce abstract concepts encountered in future physics or computer science lessons.

The use of patterns can be obscured in lessons similar to this one because waves themselves are a cyclical pattern, and waveforms are essentially the phenomenon that is being investigated. Indeed, cyclical patterns in nature are so pervasive that there is a well-defined mathematical model, the sine curve, to describe them. A primary objective of the lesson is to establish appropriate terminology for cyclical patterns. It is important, however, to also note the sequence of the lesson: Students are first led "up" the EPE triangle, discovering and developing familiarity with a system (i.e., having *experiences* with the Slinky), finding similarities and commonalities (i.e., examining *patterns* found within the variety of approaches), and summarizing (i.e., contemplating an *explanation* of how the system is a representative model) as they

Continued

Instructional Application 4.3 *(continued)*

develop new knowledge. Then student understanding is reinforced as they come "back down the other side" of the triangle, applying their knowledge in a challenging design context, thus creating new experiences for future discovery (Figure 4.16).

Figure 4.16. Experiences-Patterns-Explanations table for 4-PS4-1 and 4-PS4-3

Phenomenon: Interaction over distance (via waves)
Driving Question: How can a message be sent through a Slinky?
Performance Expectations: 4-PS4-1 Develop a model of waves to describe patterns in terms of amplitude and wavelength and that waves can cause objects to move. 4-PS4-3 Generate and compare multiple solutions that use patterns to transfer information.

Experiences	Patterns	Explanations
Experiences planned for the lesson Playing with Slinky toys Research on codes Challenges with transmission Prior school experiences Design Process Cultural experiences and family funds of knowledge Musical instruments Nature vacations (echoes, water) Use of codes	1. Classification of wave types (standing/traveling, longitudinal/transverse) 2. Describing waves as a cycle of motion 3. Sine wave with frequency, wavelength, and amplitude	Motion is replicated in time and space, creating a wave. A Slinky is a model for how waves travel in other media. Waves can be used to transmit information

Developing Knowledge →

← Using Knowledge

Summary

This chapter shows how patterns are fundamental to scientific discovery, engineering design, and science learning in the classroom. Scientists use patterns to connect the observations they make about the world to the models they construct to understand the world. Students in the classroom use patterns to connect their experiences with the natural world to the explanations they make to scientifically understand the world. In three-dimensional learning, tools such as graphs, charts, maps, and mathematical equations support students in finding, analyzing, and communicating about patterns as they engage in science and engineering practices to develop and use their understandings of disciplinary core ideas.

References

Covitt, B. A., J. M. Dauer, and C. W. Anderson. 2017. The role of practices in scientific inquiry. In *Helping students make sense of the world using next generation science and engineering practices*, ed. C. V. Schwarz, C. Passmore, and B. J. Reiser, 59–83. Arlington, VA: NSTA Press.

Gunckel, K. L. 2010. Experiences, patterns, and explanations to make school science more like scientists' science. *Science and Children* 48 (1): 46–49.

Michaels, S., A. W. Shouse, and H. A. Schweingruber. 2007. *Ready, set, science: Putting research to work in K–8 science classrooms*. Washington, DC: National Academies Press.

National Research Council (NRC). 2012. *A framework for K–12 science education: Practices, crosscutting concepts, and core ideas*. Washington, DC: National Academies Press.

NGSS Lead States. 2013. *Next Generation Science Standards: For states, by states*. Washington, DC: National Academies Press. *www.nextgenscience.org*.

Rehmat, A. P., O. Lee, J. Nordine, A. M. Novak, J. Osborne, and T. Willard. 2018. Modeling the role of crosscutting concepts for strengthening science learning for all students. In *Proceedings Summit for Examining the Potential for Crosscutting Concepts to Support Three-Dimensional Learning*, ed. S. J. Fick, J. Nordine, and K. W. McElhaney, 51–65: Charlottesville, VA.

Chapter 5

Cause and Effect: Mechanism and Explanation

Tina Grotzer, Emily Gonzalez, and Elizabeth Schibuk

While sitting in a restaurant, my (Tina's) six-year-old twins noticed a painting of colonial times above our table and wondered out loud about it. They asked, "Why are all those people under the tree?" and "Why is the sky so blue?" The conversation turned to what season it was. My daughter said it had to be summer because the people had no shoes. Her brother quickly responded that the leaves on the tree were very light green and that only would happen in the spring. As they disagreed, I asked what else they noticed that might help them decide. Listening to their animated conversation about what else they saw, I reflected, admittedly with pride, on how they were reasoning about claims and evidence—a pillar of scientific explanation. Then the conversation took an unexpected turn. One of them hiccupped, and they both simultaneously declared, "A magic crow flew by!" with accompanying giggles. "WHAT?" I asked, my sense of pride turning into surprise. They were both very happy to explain that when they hiccupped, it always happened when a crow was flying by!

It is common to notice patterns such as these in our everyday world. It is also human to wonder about why patterns occur—to want to know what happened to cause the patterns. Parents of toddlers can likely relate to hearing unending "why?" questions, but as the above story illustrates, it is also very human to jump to conclusions about the causes of the patterns around us. When we don't have sufficient background information, we may fill it in with what we imagine could be so, simpler causal stories, and even magic. We also don't track patterns between events very well because our attention to everything going on around us at once is necessarily limited. So we may not notice that we hiccup when we are inside with no crows around or that crows almost always gather in the tree over the yard whether we are aware of them or not.

Chapter 5

This chapter addresses the *Next Generation Science Standards* (*NGSS*) crosscutting concept (CCC) of cause and effect: mechanism and explanation. It begins by explaining what the CCC of cause and effect is and why it is so important to address. It describes how the CCC builds from the CCC of patterns and how it supports the CCC of systems and system models. It looks at how the understandings that are part of causal reasoning are important to how scientists think about causality. It shows how the CCC of cause and effect provides a productive thinking tool in making predictions and in testing for cause and effect in and across the scientific disciplines. It then considers how students' understanding of cause and effect grows over time, the challenges in learning about it, and what teachers can do to help their students build the ability to reason about cause and effect. Moreover, it offers promising instructional strategies for helping students achieve a deep understanding of the CCC of cause and effect.

What Is the CCC of Cause and Effect and Why Is It Important?

The crosscutting concept of cause and effect focuses on answering the question of why things happen. It looks at patterns, for instance, in the way two variables change, and identifies what makes them change (i.e., the mechanisms responsible for the change). As discussed in Chapter 4, noticing patterns or regularities in our world enables us to develop models and make predictions. Understanding what makes the patterns happen allows us to make better predictions about what might happen given certain conditions and to understand how to replicate the patterns. The human tendencies revealed in the story about hiccupping and crows illustrate why the CCC of cause and effect is such an important part of the *NGSS*, helping students learn to reason well about causality. The story illustrates that even the youngest learners bring a lot of causal reasoning skills to the classroom despite not necessarily having the background knowledge to support it. However, learning to reason well about cause and effect is a challenging endeavor and requires that we teach it in terms of its nuance and complexity from K–12 and beyond. The CCC of cause and effect is designed to support learners on this path to understanding what good causal reasoning looks like and what some of its pitfalls can be, which helps in our quest to understand why our world works the way that it does.

The CCC of cause and effect builds on the CCC of patterns in that noticing and reasoning about patterns is an essential step, and often an initial step, in causal reasoning. The CCC of cause and effect goes further than the CCC of patterns by addressing the "why" behind the patterns. The CCC of cause and effect also contributes to being able to understand and reason about the CCC of systems and system models. While systems thinking involves understanding a system's boundaries and developing models for how systems work, the types of relationships embedded within those system models involve cause-and-effect relationships.

How Is Cause and Effect Used in Various Scientific Disciplines?

Understanding cause and effect in the sciences involves reasoning about covariation patterns and mechanisms as described in the sections that follow. Both play an important role in being able to engage in causal explanation.

Covariation Patterns

Covariation patterns are the regularities between the occurrences of two or more events, actions, or entities—for instance, the flipping of a light switch and a light coming on or oil floating on water. These regularities make it possible to make some predictions even without understanding the underlying causes. For instance, in his theory of natural selection, Darwin addressed the regularity that certain species have traits that make them more successful in reproducing and thriving than other species. His theory came along before the discovery of the gene or DNA as mechanisms that enable the passing on of traits.

When noticing covariation patterns, scientists look beyond simple, linear, and direct relationships. Science involves varied possible relationships—for instance, the cyclic patterns integral to simple circuits and the relational patterns inherent to density when analyzing sinking and floating. Two-way or bidirectional patterns characterize such things as magnetism in which both entities repel or attract and the mutual causation in symbiosis wherein both organisms influence each other typically toward enhanced survival rates.

Mechanism

Mechanism refers to what makes the cause-and-effect pattern happen. Identifying the mechanism is important because it allows scientists to move from seeing that a correlation exists (i.e., observing two things happening with some regularity) to knowing how they are connected. Once scientists understand the underlying mechanism, they can make causal claims (i.e., positing that one thing affects the other). This is important because correlations can be spurious, as in the example in the opening story about the crows appearing during hiccups. A correlation can also be the result of a hidden cause. In other words, there might be a pattern of change in two (or more) variables, but neither is the cause of the change in the other; instead a hidden variable causes the change in both. Take, for example, the strong correlation between ice cream consumption and drowning, which can be explained by a third variable—hot weather. The weather causes more people to eat ice cream *and* more people to go into the water. Detecting patterns is an important step in reasoning about causality; however, knowing the mechanism offers justifiable evidence for making a causal claim.

A look at what happened with the COVID-19 outbreak provides insight into how scientists use covariation patterns and mechanisms to develop scientific explanations for phenomena. In the earliest days of the COVID-19 pandemic, noticing that people were getting sick with specific and severe symptoms and seeing that it was possible to trace their contacts to certain events, places, and/or sick persons revealed a pattern of contagion to scientists and medical professionals. This covariation information helped them develop models to learn about patterns of transmission. However, noticing these patterns of transmission was complicated by a host of factors—including that there were differences in how long it took some people to get sick, that some people were exposed and did not get sick, that others got sick but had no symptoms, and that some people seemed to spread the virus to a greater extent than others. With the level of complexity that characterizes COVID-19, models of patterns were not enough to predict the transmission behavior of the disease. Figuring out what the actual mechanism of the virus was and learning how it behaves offered crucial information for containing the transmission. For instance, the virus's characteristic spikes revealed information about how it attaches and persists in the nasal passages. Scientists continue to study the virus to learn more about this mechanism and its behaviors in an effort to explain the confusing patterns that it presents.

While scientists often find patterns and then seek mechanisms, their general knowledge of mechanisms often has a powerful influence on their investigations. Knowledge of how certain mechanisms, such as germs, viruses, and toxins, behave in general leads them to look for certain, expected effects. In the case of COVID-19, scientists were familiar with other forms of the virus that offered important information. Investigating the variations of the coronavirus responsible for COVID-19 led to an understanding of the specific mechanistic details of how these versions of the virus work, allowing scientists to move beyond their understanding of more general mechanisms that characterize the workings of viruses as a broader category.

The level of explanation—whether it's a broad or more fundamental level—affects how scientists describe causal mechanisms. For instance, the mechanism of an electrical switch can be described at a broad level as flipping a switch to make current flow. At a deeper level, it can be described in terms of voltage as a push that in relation to resistance in the circuit causes current. At an *even deeper* level, it can be described as density differentials in electrons that contribute to repulsion and attraction, resulting in flow across the circuit. In biology, scientists refer to infection and inflammation as a cause at a broad level and to bacteria as a cause at a deeper level. In physics, scientists talk about mechanical devices at a broad level and vectors and forces at a more fundamental level. Deep understanding involves knowing how to talk about causes at these different levels.

Scientific explanation sometimes allows for more than one type of accepted mechanism. For instance, buoyancy and density mechanisms are both used to explain sinking and floating. Buoyancy is described as the upward thrust exerted by a fluid that opposes the weight of the object resulting from gravity. Density refers to the amount of matter in a given space and influences whether one substance (or object) sinks or floats in another substance.

Scientists use their knowledge of covariation patterns and possible mechanisms to investigate cause-and-effect relationships. The CCC of cause and effect states, "A major activity of science is investigating and explaining causal relationships and the mechanisms by which they are mediated. Such mechanisms can then be tested across given contexts and used to predict and explain events in new contexts" (NRC 2012, p. 84). In order to test for a possible causal relationship, scientists often try to disrupt it or intervene on it in some way. Being able to predict, intervene, and assess outcomes hinges largely on the subdiscipline of science that one is investigating. This fits directly with the *NGSS* science and engineering practice (SEP) of Planning and Carrying Out Investigations (Windschitl 2017).

How Does the CCC of Cause and Effect Provide a Productive Tool Within and Across Disciplines?

When reasoning about cause and effect, scientists need to consider the covariation patterns that occur and think about possible mechanisms that give rise to them. How they do so differs in particular disciplines and depends on the information that is available to them. Some common approaches that scientists use to test for possible causal relationships and produce new scientific knowledge include (1) the isolation and control-of-variables approach, (2) the body-of-evidence approach, and (3) the approach of measuring the causal contributions statistically.

Isolation and Control-of-Variables Approach

A common and useful approach to understanding cause and effect in the sciences involves experimentation in a lab where an attempt is made to isolate and control for variables to examine their causal contributions to an outcome. This is known as a control-of-variables (COV) approach. Scientists might test individual variables or a set of variables to consider the interactions between them. By isolating each variable, scientists are able to gain a sense of the behavior of individual variables. Caveats to the COV approach include that variables may behave very differently in isolation than they do in interaction, and they may behave differently out in the world than they do in lab contexts. For these reasons, scientists may progress to testing variables in interaction and in real-world contexts. The COV approach is more common to the physical and engineering sciences than the biological sciences, where there may be no meaningful control to test.

A Body-of-Evidence Approach

Some disciplines are much less amenable to control or isolation, and important outcomes are typically at the systems level, making such an approach less meaningful. For instance, ecosystems scientists tend to use a body-of-evidence (BOE) approach in which they collect a variety of evidence to construct a causal analysis (Kamarainen and Grotzer 2019) and seek out corroborating or supporting evidence. By using many sources of information, they are able to gain a greater understanding of what is going on. These sources include opportunistic interventions or contrasts that naturally occur and are then studied. (Such methods are particularly important in contexts that require a "do no harm" stance so that intervention is limited, such as in the biological, medical, and environmental sciences.)

Measuring the Causal Contributions Statistically

Across the scientific disciplines, cause and effect is often analyzed statistically in terms of the causal contributions that particular variables make to particular outcomes. For instance, in looking at the onset of certain cancers, smoking would be factored into the equation, even though one would not set up a controlled experiment in which people are randomly assigned to a smoking or nonsmoking group, as that would be unethical. A statistical analysis focuses on the correlational patterns with attention to suspected mechanisms. This approach can reveal probabilistic causal patterns—ones that do not deterministically lead to the same outcome in every single instance. Smoking increases the likelihood of certain cancers but does not lead to cancer in every case. Understanding these relationships involves having large data sets. By using large data sets, scientists are able to see regular patterns that occur and how they co-occur with other possible contributing factors. This helps them know what variables to examine more closely.

How Does Student Understanding of the CCC of Cause and Effect Build Over Time?

How might students build understanding of the CCC of cause and effect over time? The *NGSS* grade bands recognize how students' abilities grow across developmental levels. Consider the following examples for how the focus of the CCC of cause and effect becomes more abstract and more sophisticated across the grades.

At the K–2 level, the CCCs of patterns and cause and effect focus on phenomena that can be observed. For instance, students are learning that "patterns in the natural and human-designed world can be observed" (2-PS1-1). They're also learning that "events have causes that generate observable patterns" (2-PS1-4). A focus on observable patterns enables primary students to develop the skills of noticing and thinking about covariation patterns. Given that they have relatively less background knowledge, having them focus on more familiar, observable changes enables them to connect possible mechanisms to the outcome. For example, in performance expectation (PE) K-ESS2-2, students are expected to show their understanding by "construct[ing] an argument supported by evidence for how plants and animals (including humans) can change the environment to meet their needs." To address this PE, students may observe a picture or watch a video of a beaver chewing a tree to build a dam. This enables them to see the mechanism and the pattern. Young students can easily see the animal taking actions to change its surrounding environment; the information that they need to hold in mind is not terribly demanding because they are simply gathering observational evidence and are not intervening on the pattern in any way. The evidence that students collect supports their developing explanation about the beaver and its relationship to the environment in a developmentally appropriate way.

At grade levels 3–5, the CCC of cause and effect focuses on understanding that "cause-and-effect relationships are routinely identified" and that they can be "tested and used to explain change." The shift is from observing patterns to comprehending the role of cause and effect in understanding how the world works and for providing evidence in support of explaining changes. Consider the following example of a performance expectation within this grade band: "Use evidence to construct an explanation for how variations in characteristics among individuals of the same species may provide advantages in surviving, finding mates, and reproducing" (3-LS4-2). The PE asks for examples of cause-and-effect relationships, such as this one: "Plants that have larger thorns than other plants may be less likely to be eaten by predators, and animals that have better camouflage coloration than other animals may be more likely to survive and therefore more likely to leave offspring" (3-LS4-2). At this level, students are introduced to the idea that, in addition to observations, evidence from testing cause-and-effect relationships can help them develop a scientific explanation and that changes can be tested to reveal the causal mechanism.

At the middle school level, the CCC of cause and effect includes the recognition that knowing cause-and-effect relationships—the typical patterns and the underlying mechanisms—enables prediction. It also includes understanding that phenomena can have more than one cause and that some causes can only be predicted statistically rather than tracked deterministically between cause and effect. Students apply these understandings to challenges such as to "use a model to support an explanation of the effect of resource availability on organisms and populations of organisms in an ecosystem." Middle school students have more background knowledge, are able to hold more information in memory, and are beginning to reason about more complex patterns. Instructional supports can help them bring these skills together in support of strong causal reasoning. For instance, as students consider the effects of resource availability on populations over long time spans, they use supporting skills such as graphing to visualize population data over time to make the covariation patterns—the way two variables change together—more visible to them. Instead of observing one animal engaged in concrete behaviors, as is the case in kindergarten, middle school students are reasoning about populations of animals and developing models to explain what might be going on.

At the high school level, students are learning that "empirical evidence is required to differentiate between cause and correlation and to make claims about specific causes and effects" (HS-PS4-1). Empirical evidence goes beyond simple tests and often comes from intervention or simulation in some way. The disciplinary core ideas at this level often include multiple facets of understanding related to causality. For example, consider the following performance expectation: "Evaluate the evidence for the role of group behavior on individual and species' chances to survive and reproduce" (HS-LS2-8). There are a number of conceptual parts in this standard. Students must identify a group behavior in an organism or in a group of organisms. They must find evidence of how this behavior could have a causal impact on both an individual organism's survival and reproduction rate, as well as on the whole group's survival and reproduction rate. Students must then evaluate the evidence in terms of "merits" of the identified behavior. They are handling more information and both obvious *and* nonobvious factors. They are modeling probabilistic causal outcomes. They can use a scientific simulation or model to see what happens at the species level when an individual organism's behavior is changed. Because there are clear inputs and outputs within simulations, using such a model can aid in illustrating the causal relationship between individual and species behaviors and the outcome on survival and reproduction.

Challenges Students Face When Learning About the CCC of Cause and Effect

The performance expectations set forth a developmentally appropriate path for learning the CCC of cause and effect. What challenges do students face when learning about and using this CCC? As the examples and discussion above suggest, children bring significant understanding related to causality to the classroom. Children as young as preschool-age hold understandings about the nature of causality that can support later classroom learning. For example, they do not allow for causeless effects, they will look for an obvious cause, and if they don't find one, they will seek one out or simply posit a nonobvious one—even if they have to resort to magic crows! However, there are also many challenges in learning to reason about the CCC of cause and effect.

When students are learning about covariation patterns, they often focus on simple, linear, and direct relationships. They tend to distort scientific explanations to fit with these simpler patterns instead of the more complex patterns within the concepts (See Grotzer 2010). When causes are spread out in space and time, it is harder for young students to see the patterns at all. Learning to reason about mechanisms can also be challenging because they are not always obvious to students. Consider the microbes responsible for decomposition. Students can see the pattern of an organic material decomposing but either attribute the process to large visible decomposers such as earthworms or think that things just fall apart. Also consider climate change: Students can witness effects of climate change but cannot see carbon in the atmosphere.

These challenges often lead students to make simplifying assumptions about the nature of causality. See Table 5.1 (p. 99) for descriptions of how students tend to think about causal patterns contrasted with how the patterns are structured in scientific explanations. In order to help students learn to reason about cause and effect, it is important to help them recognize the features of causal relationships, also shown in Table 5.1. This includes understanding covariation patterns that are nonlinear, bidirectional, and simultaneous. It includes realizing that causality can have a passive nature, is not always event-based as in steady states that play out over time, and that it does not necessarily involve intentionality. It can include distant and delayed causes and effects. The patterns can be probabilistic regularities requiring students to reason about them statistically instead of deterministically. Effects can emerge from distributed causes arising out of many microinteractions.

Unless teachers are aware of students' simplifying tendencies, instruction can reinforce them and make learning the more complex structures difficult. For instance, it is common for teachers in early elementary to have students make a list of objects that sink and objects that float. Unfortunately, this reinforces a simple linear cause and effect in which students envision that the features of the object alone—such as the weight—account for whether something sinks or floats instead of the relationship between the density of the object and the density of the liquid. A "discrepant events" approach can help students revise their underlying causal models. Video examples of this approach can be found on the Causal Patterns in Science website (Understandings of Consequence Project 2008):

- Seesaw example: *www.causalpatterns.org/causal/causal_examples.php*
- Sinking and floating with candles example: *www.causalpatterns.org/recast/density_examples.php*

Classroom learning can also unintentionally pull against building understanding of how scientists analyze cause and effect. For instance, realizing that scientists assess causality through statistical patterns is an important lesson, and it is something addressed in MS-LS4-6: "Some cause-and-effect relationships in systems can only be described using probability." It is supported by the SEP of Analyzing and Interpreting Data in the *NGSS* (Rivet and Ingber 2017) in that students are engaged in looking at data sets for general trends as opposed to considering the outcome of one experiment. However, in the classroom, students often interpret outcomes from one experiment. They also learn about the importance of reliability in trusting their results. This is an important scientific concept but one that can make it hard to understand how scientists use statistical patterns (such that the outcome may not occur every single time) in reasoning about cause and effect. A study of sixth graders showed that they applied the criterion of reliability so rigidly that they thought there should be 100% reliability if a causal relationship exists. Younger students who knew less about the concept of reliability were less rigid (Grotzer et al. 2011). Teachers can help students learn the nuanced message that reliability is a goal of experimentation but that cause and effect often act probabilistically in ways that we can only see across broader data sets.

Table 5.1. Curriculum concepts for teaching causality

Simplifying Assumption	How Does It Work?	Reframing for Causal Complexity	How the Reframed Version Works ...	How We Tend to Get Stuck ...
Simple Linear ("This Makes That Happen")	One thing makes another thing happen.	Extended or Nonlinear (Domino, Cyclic, Escalating, Mutual, or Relational Causality)	Causal patterns can be extended or nonlinear. They can include indirect or bidirectional effects.	We often adopt simple storylike patterns where one thing happens and causes something else. We miss patterns that are like dominos, cyclic as in a feedback loop, mutual as in symbiotic relationships, or relational where a relationship between two variables results in an outcome.
Event-Based ("What Happened?")	Something has to happen in order for us to think about causality.	Steady State ("What's Going On?")	Systems in balance entail causal relationships even if nothing is happening at the moment.	Unless something is happening, we often don't think about the causal relationships in play. We tend to focus on the balance of forces that made a bridge stand only if it falls down. Similarly, we tend to realize what forces sustained a government once it collapses.
Sequential ("Step by Step")	Causes always come before effects in a step-by-step pattern.	Simultaneous ("All at Once")	Causes and effects can co-occur in time and still have a causal relationship.	We expect that causes have to occur before effects in our explanations even when it doesn't fit the phenomenon. Gears all turn at once even though one causes another to turn.
Obvious Variables or Mechanisms ("Easy-to-Notice Causes and Effects")	Some causes can be directly perceived.	Nonobvious Variables or Mechanisms ("Hard-to-Notice Causes and Effects")	Some causes are nonobvious because they are microscopic, imperceptible, or inferred.	We focus on obvious instead of nonobvious causes unless there is no obvious possible cause. We are more likely to report worms as decomposers than microbes. We don't think about air pressure as the cause of earaches on planes unless we have been told to. We do reason about inferred causes such as intentions in social situations, but we reason in more complex ways about ourselves than others.

Continued

Table 5.1. *(continued)*

Simplifying Assumption	How Does It Work?	Reframing for Causal Complexity	How the Reframed Version Works ...	How We Tend to Get Stuck ...
Active or Intentional Agents ("Someone Did It" and "On Purpose")	Many cause-and-effect relationships involve an actor who intends a certain outcome. Many actions have a purpose or intent.	Passive or Unintentional Agents ("No One Did It" and "By Accident" or "Didn't Mean To")	Some causal relationships don't involve action or intentionality to work. Alternatively, actors and their intentions may not correspond with effects arising at another level.	We tend to look for who made something happen and assume active and intentional agents. Seatbelts, a passive restraint system, keep us in the car when it stops without actively doing anything. Electrons are active and protons are passive, yet outcomes are caused by their attraction, not one or the other. We drive our cars intending to get somewhere, not intending to contribute to global warming. We refer to "Mother Nature" to assign agency to natural occurrences.
Deterministic ("It's Always Supposed to Work")	An effect always follows a given cause.	Probabilistic ("It Usually Works")	An effect sometimes follows a given cause.	In part, we judge whether a causal relationship exists by how reliably the effect follows the cause. Although we sum across instances, allowing for some unreliability, probabilistic causes make it harder to notice and attend to causal patterns. It also can lead to risk-taking as in "I did it before and I didn't get sick, so I'm not going to get sick now" instead of "Even if I didn't get sick before, I can still get sick now." Over-reliance on correlation can lead to assuming that relationships are causal so we don't seek out other possible causes.

Continued

Table 5.1. *(continued)*

Simplifying Assumption	How Does It Work?	Reframing for Causal Complexity	How the Reframed Version Works ...	How We Tend to Get Stuck ...
Spatially and Temporally Close ("Local Causes and Effects" and "Immediate Causes and Effects")	Causes and effects physically touch each other or are close to each other in space and time.	Spatially and Temporally Distant or Delayed ("Distant Causes and Effects" and "Slow" or "Delayed Causes and Effects")	Causes can act at a distance and there can be delays between causes and effects. Effects may need to accumulate to a certain level to be noticeable or may need to reach a trigger point before which there is no effect.	We often limit our search for causes and effects to those that are close together in space and time. We miss more distant causes of events. We also don't attend to effects that are at an attentional distance—for example, we don't think about affecting polar bears when we drive to work. We misunderstand events in the Middle East because we view them in too brief a time span. Teens think, "I can't see any bad effects of getting a suntan right now," instead of "The hurtful effects of getting a suntan accumulate and show up after a long delay between cause and effect."
Centralized and Direct ("Someone's in Charge")	A central figure or leader causes (and typically intends) the outcome.	Distributed and Emergent ("No One's in Charge")	Individuals interacting give rise to an emergent effect where the intent of the individuals may not align with the emergent outcome.	We tend to focus on centralized causes and don't attend enough to the power of distributed causality. Therefore, we think that government institutions and leaders rather than our individual, daily civilized actions give rise to a civilization. Or we wait for our leaders to enact legislation to combat global warming rather than changing our individual actions to contribute to the emergent solution.

Source: Grotzer 2012, reprinted with permission.

Visit *www.causalpatterns.org* to learn more about the details of these challenges and access curriculum materials designed to support students' learning. To adapt existing curriculum to address underlying causal patterns, read a curriculum module called *Reasoning About Causal Complexity in Science and Beyond* (Grotzer 2010) at *www.causalpatterns.org/resources/natureofscience/resources_overview2.php*. Finally, see the book *Learning Causality in a Complex World* (Grotzer 2012), written for educators about the significant body of research that contributes to our knowledge of students' default assumptions.

What Big Understandings Do Students Need in Order to Reason Well About the CCC of Cause and Effect?

An understanding of the nature of causality such that it supports the disciplinary core ideas and science and engineering practices boils down to five big ideas:

1. Patterns can be correlational or causal. It is important to understand the differences. *Covariation relationships suggest the possibility of causal relationships but may just be correlations.*
2. Making a causal claim requires evidence of mechanism(s). *Intervening on a pattern or collecting corroborating evidence can help us understand the mechanisms behind the pattern in order to make a causal claim. Knowledge about different kinds of mechanisms can lead us to look for patterns that might exist based on how the mechanism behaves.*
3. Causal patterns can be complex and have features that make it hard to notice them. *Instead of just assuming that patterns are simple, direct, and/or linear relationships, it is important to consider more complex patterns.*
4. Vocabulary can unintentionally simplify cause-and-effect concepts, so terminology matters. *Everyday descriptions of causal relationships are not very sophisticated (and often depend on the use of metaphors, such as* dominos *or* snowball effect*). Developing vocabulary related to cause and effect can be very helpful in understanding science and beyond.*
5. The disciplines solve the problems of analyzing cause and effect differently. *Experiments that isolate and control variables are not meaningful in all disciplines; some disciplines have adopted other approaches that depend on corroborating evidence.*

What Instructional Strategies Are Promising for Helping Students Learn the CCC of Cause and Effect?

This section covers the five following instructional approaches to supporting students in learning about the CCC of cause and effect: (1) engaging students in opportunities to learn the difference between correlation and causality, (2) using a corroborating-evidence or BOE approach for supporting causal claims, (3) restructuring or ReCASTing students' default assumptions about the underlying causality as they learn new concepts,

(4) developing a more extensive and more nuanced vocabulary for talking about causality, and (5) using thinking moves related to reasoning about cause and effect. Each of these resonates with a three-dimensional approach to learning such that the crosscutting concepts, disciplinary core ideas, and science and engineering practices work in concert with one another.

Engaging Students in Opportunities to Learn the Difference Between Correlational and Causal Claims

It is common for students to confuse correlation and causation. Focusing on the difference can help them learn when the patterns they are considering are only correlations and when it is warranted to make a causal claim. Many teachers use a Claims, Evidence, Reasoning (CER) framework (McNeill and Krajcik 2012). When doing so, it is important to help students realize that a claim can be a correlational claim, where one is stating that two variables have a pattern of change between them, or a causal claim, where one is stating that one variable causes a change in another (or others).

In order to make a causal claim, students need to have evidence of impact or knowledge of the behavior of a mechanism. One way to gather this evidence is by intervening on the covariation pattern and seeing what happens. Classroom Snapshot 5.1 provides an example of what this might look and sound like in a middle school science class.

CLASSROOM SNAPSHOT 5.1
Intervening on Covariation Patterns

Serena is using EcoXPT, a multiuser virtual environment, to learn about a pond ecosystem. The seventh grader has explored the virtual ecosystem and its surroundings, including the avatars in the program, the human-made structures, and the organisms that inhabit the pond. After discovering that some of the avatars have placed fertilizer around the virtual world, which subsequently got into the pond during a rainy day, Serena wonders if the fertilizer is what caused some of the fish populations to die.

To see if fertilizer causes harm to fish, Serena moves her avatar into the lab building. After talking to an ecosystems scientist character in the virtual world about the different kinds of experimentation they employ, she chooses to use the lab's tolerance tanks because she has learned that this experimental tool can isolate the effects of one factor on the fish. The scientist character explains that sometimes scientists test effects of different variables on a few fish in these tanks to figure out how to save

Continued

Classroom Snapshot 5.1 *(continued)*

a population of fish. Serena increasingly adds more fertilizer to the tolerance tanks, only to find out that the level of fertilizer does not affect the fish. While this finding doesn't tell her what directly killed the fish in the pond, Serena now develops a new question from the evidence she has gathered: Did the fertilizer affect something else in the pond that, in turn, caused the bluegill and largemouth bass to die off?

Another way to gather experimental evidence is by using naturally occurring opportunities, or "natural experiments." This is common in the biosciences. Classroom Snapshot 5.2 gives an example of what this looks like in practice for learners.

CLASSROOM SNAPSHOT 5.2

Using Naturally Occurring Opportunities to Learn About Causes and Effects

Billy, a fifth grader, learned that animals exhibit various behavioral adaptations that help them survive in their surrounding habitat. He wants to gather evidence from his local ecosystem to support this claim. To do so, Billy has decided to sit in his backyard and observe his surroundings. Billy notices a squirrel meandering about underneath a nearby tree. At first glance, Billy thinks nothing of this instance because he always sees squirrels in his backyard. As he starts to look at the squirrel more closely, Billy notices that it's actually digging a hole. After the squirrel scurries away, Billy walks up to the place where the squirrel was digging and notices that there are a few similar looking holes close by. Billy recalls learning that squirrels like to eat acorns, so he goes back to his post and waits to see if the squirrel will come back to dig some more. Shortly after, Billy sees the squirrel return with an acorn, dig a new hole, and place the acorn inside. He writes these observations in his science notebook.

Using naturally occurring opportunities allows students to observe natural phenomena in the environment around them and to collect evidence for developing claims and causal reasoning. Both of the Classroom Snapshot examples directly connect to the science and engineering practices of Developing and Using Models and Planning and Carrying Out Investigations, specifically within the life sciences disciplinary core ideas (LS2 and LS4).

Using a Body-of-Evidence Approach

Helping students see how cause and effect applies in the real world often involves using a body-of-evidence (BOE) approach instead of engaging in experimental interventions.

The CCC of cause and effect approach of collecting corroborating evidence and the science and engineering practice of Developing and Using Models supports this thinking. In Classroom Snapshot 5.3, consider how Ms. Schibuk uses a teachable opportunity that arises from students' questions to launch from noticing patterns to analyzing for causality. She uses language related to pattern and mechanism and helps students see how the theoretical explanation is based on corroborating evidence from numerous sources.

CLASSROOM SNAPSHOT 5.3

Using a Body-of-Evidence Approach to Analyze a Theory

Ms. Schibuk's seventh-grade students are questioning the theory of tectonic plates. "Why is it called the theory of tectonic plates? Are they real? How do we even know?" they ask. In response, she decides to work with students on seeing patterns and constructing explanations of cause and effect from data. Leveraging an earlier assignment in which students were given large world maps and asked to plot live earthquake data from across the world for the past 30 days, she invites students to consider the emerging patterns. It appears to be a tedious task initially. However, as students add more and more data points to their maps, they begin to see patterns unfold as the data points begin tracing out, with increasing precision, large sections of Earth. Ms. Schibuk acknowledges the importance of the students' questions that recognize how difficult it would be to construct a controlled experiment to test the theory of plate tectonics. She asks the class to think deeply about the following question: "In a case like this, how can one move from noticing patterns to finding out about an underlying causal mechanism?" She asks them to consider how, in many instances in science, we cannot explicitly view a causal mechanism, and so we move from patterns to theoretical explanations, often termed *conjectures*. Over time, we continue to build our body of evidence that will either support or refute the working theory. With time, if scientists are unable to refute a claim, the working theory becomes accepted scientific knowledge. She shares that scientists use corroborating evidence from seismic surveys, radioactive dating, and other technologies to support the idea that the mechanism of moving plates over time accounts for the emerging patterns. For instance, scientists have found fossils of animals on different continents that would have been on the same landmass when they were alive. Scientists have yet to find evidence to refute the theory of plate tectonics, so it holds.

ReCASTing Simple Causal Structures

As outlined in Table 5.1 (p. 99), scientifically accepted explanations often include complex forms of causality, and students tend to reduce them to simpler forms. Teachers can help students revise their simpler causal structures to develop more sophisticated structures that do not distort the science concepts.

ReCAST activities *reveal the causal structure* of a phenomenon and invite students to revise or "recast" it. Classroom Snapshot 5.4 is related to understanding the role of density in sinking and floating and occurred in an eighth-grade classroom participating in a study of students' thinking about the nature of causality. As previously discussed, students often construct a simple linear explanation for why an object either sinks or floats. For instance, a student may say, "Weight makes things sink, so that means a heavier object will sink and a lighter one will float." ReCAST activities can help students reconceptualize sinking and floating as a relational causality—usually one of balance or imbalance between the density of the object and that of the liquid or between the densities of the two liquids—that causes an outcome.

CLASSROOM SNAPSHOT 5.4
ReCASTing the Causal Structure of Sinking and Floating

Ms. Sayed shows her eighth-grade class two beakers filled with liquid. Then she holds up two pieces of wax, one large and one small, cut from the same candle. She asks, "What do you think will happen when I drop each of these into the liquid?" Students predict that the smaller one will float and the larger one will sink. She drops the smaller piece into beaker A and the larger piece into beaker B. The students let out a cheer as they see that their prediction was correct; the small piece of candle is floating in beaker A and the big piece has sunk to the bottom of beaker B. They argue that the weight of the larger candle piece made it sink.

Then Ms. Sayed asks them to keep watching as she takes both pieces of candle out and drops the larger piece into beaker A and the smaller piece into beaker B. Now the big piece is floating and the smaller piece sinks to the bottom! (See Figure 5.1.) "What?!" the students exclaim. "Wait, you tricked us," says Ayanna. "What did you do to the water?" Now they are attending to the broader causal system, not just the object itself. Ms. Sayed explains that one container is water (beaker A) but the other is isopropyl alcohol (beaker B). Whether something sinks or floats depends on the relationship between the density of the object and that of the liquid or between the densities of the two liquids. Together, the class brainstorms a list of examples that

Continued

Classroom Snapshot 5.4 *(continued)*

illustrate this concept, such as oil and water, the layers of the atmosphere, and so on. This phenomenon highlights the relational causality between the liquids and pieces of candle. Surfacing the causal preconceptions to students firsthand allows them to think deeply about the inherent causality and to ReCAST their ideas about sinking and floating.

Figure 5.1. Demonstration of candles floating or sinking in different liquids

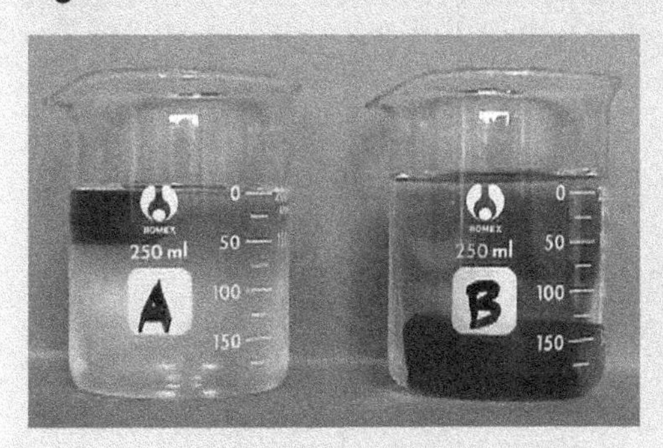

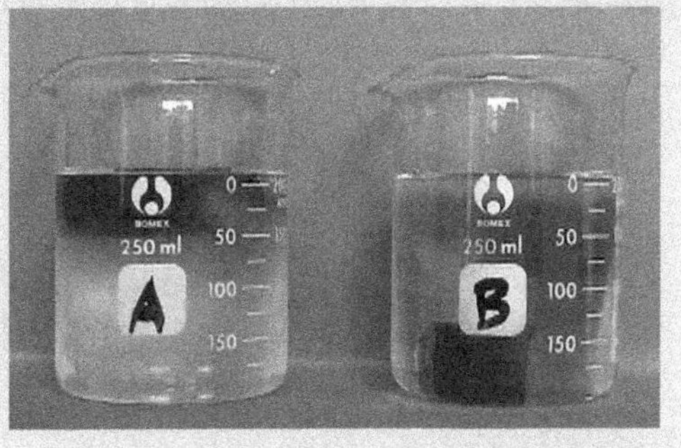

Ms. Sayed helped students ReCAST their causal understanding of sinking and floating by showing them that candles, both big and small, float in water (beaker A) but sink in isopropyl alcohol (beaker B).

Figure 5.2. Activity to demonstrate how air pressure differentials work

Source: Basca and Grotzer 2003, reprinted with permission.

Figure 5.3. Student explanation of the phenomenon using a relational causal model

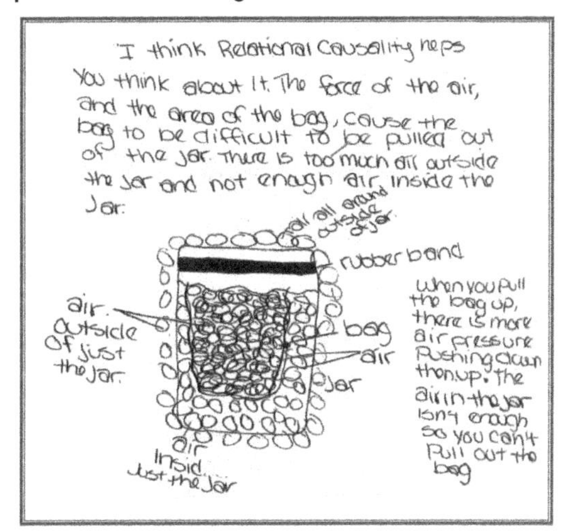

Source: Basca and Grotzer 2003, reprinted with permission.

Figures 5.2 and 5.3 illustrate another example of using a ReCAST activity to revise students' default assumptions. The students in this instance were attempting to model what was going on when they tried to pull a plastic bag sealed around the edge of a glass jar out of the jar (see Figure 5.2). Figure 5.3 features a student response to the activity. This activity helped students realize that air pressure differentials are better explained using relational causality than linear causality. (To view more examples of ReCAST activities, visit *www.causalpatterns.org*.)

Introducing a Vocabulary for Talking About Causality

Even though children engage with causality every day, they seldom have a chance to develop a vocabulary to talk about its structure and features. Helping students develop such a vocabulary offers them a way to reflect on causality, discuss it with others, and develop more sophisticated understandings of its nature. In Classroom Snapshots 5.5 and 5.6, the teacher introduces vocabulary and builds on vocabulary the students use.

CLASSROOM SNAPSHOT 5.5

Using Causal Vocabulary While Collecting Evidence About Cause and Effect With Simple Tests

The second graders in Ms. Rodriguez's class are experimenting to see what happens when they heat or cool different substances. They put cups of water in the freezer and later find that the water is frozen. Ms. Rodriguez invites them to use the language of causality as they talk about what happened. "What was the effect of putting the water into the freezer?" she asks. "The freezer turned the water into ice," they respond. "Can you describe what caused the water to turn to ice?" she inquires.

Then Ms. Rodriguez poses a question related to the performance expectation 2-PS1-4, "Construct an argument with evidence that some changes caused by heating and cooling can be reversed and some cannot." She asks, "How can we show whether the changes can be reversed or undone? Did freezing the water cause permanent changes?" The second graders work in small groups to think about the questions. They agree on the expected outcomes with the frozen water—that leaving it out of the freezer would allow the ice to melt back to its liquid state. Kiyanna asserts, "It is a back-and-forth kind of cause." They are much less sure about what will happen when you heat water, however. They agree that when water is heated in a pot, it leaves the pot as steam, but they don't agree on what happens after the water leaves the pot. Troy argues, "It disappears as it leaves the pot." Mia shakes her head, "I think that it is still there. It is just like clouds," she argues. "These are interesting ideas," says Ms. Rodriguez. "How might you investigate the claims that you are making? Is there an experiment or test that you can do to gather evidence about what happens?" "You would have to capture all of the steam as it was leaving, collect the clouds," Meher suggests. "Then when it cools it will become water again." Jake says, "You can't see all the water anymore, so you don't know if it is really gone

Continued

Classroom Snapshot 5.5 *(continued)*

or still there—so how can you catch it all?" Talia has a different idea. She suggests making a little world inside a globe—like a terrarium—and letting the steam escape into the globe. That way, the steam can't get away and then they would be able to see if it disappears or turns back into liquid.

When they are older, the students will learn more about what happens at the molecular level when something freezes or turns to a gaseous state. For now, testing their ideas will help them understand the effect that heat or cold have on different substances and whether those effects are reversible or not.

CLASSROOM SNAPSHOT 5.6
Talking About the Causality of Drinking From a Straw

The students in Mr. Mduba's sixth-grade class are thinking about what happens when you drink from a straw. Their first idea is that they suck really hard on the straw and make the liquid in the straw come up. Their teacher asks them to think about what happens when they are drinking from a juice box and reach a point where they cannot get any more liquid out unless they let air into the juice box.

Mr. Mduba: If you are the cause of the liquid moving up the straw, why do you have to stop to let air in?

Dinesh: Well the air pressure inside the box must be doing something, too. Maybe it is not just a linear causality where we make the juice come up. Is it a relational causality?

Mr. Mduba: What does it mean to have a relational causality?

Ian: It's not enough just to have two things in a relational causality. There has to be a relationship between them.

Mr. Mduba: What is the relationship between, in this case?

Stephanie: The air pressure in the straw and in the container. For it to be a relational causality, you have to be able to compare them—like one has to be higher and one lower, or they have to be equal.

Keisha: Wait yeah, actually, this liquid going up happens because there is a lower pressure here. This happens because of the low pressure. If the pressure were equal, nothing would happen because this [ambient air pressure in the container] would be pushing down as much as this would be pushing up [air pressure in the straw]. So it is a relational causality.

Teaching Thinking Moves Related to Cause and Effect

The teaching of thinking moves in support of good scientific reasoning invites explicit discussion of the CCCs. Thinking moves can highlight what good thinking about a specific CCC looks like. An example is the set of thinking moves (and related posters and introductory videos) developed as part of the EcoXPT curriculum (EcoLEARN Project 2020), which is available at *https://ecolearn.gse.harvard.edu/projects/ecoxpt*. Although these thinking moves were developed in service of ecosystems science learning and the EcoXPT curriculum, they can easily be used in other areas of science instruction. The thinking moves include the following:

- Deep Seeing (focused on observation)
- Evidence Seeking (focused on gathering corroborating evidence and evidence from multiple sources and evaluating these sources)
- Pattern Seeking (focused on noticing patterns and how certain variables change together or not)
- Analyzing Causality (focused on using experimental evidence and intervention to try to influence change in the patterns in an effort to discern the underlying causal mechanisms)
- Constructing Explanations (encouraging students to develop the best "story" or explanation they can from the available evidence)
- Building a Body of Evidence (focused on helping students assess whether their overall collection of evidence is strong enough to support their claims)

As a set, the thinking moves support scientific explanation; each is used in support of causal reasoning. As mentioned in this chapter's discussion on grade bands, engaging in observation (Deep Seeing) and collecting evidence (Evidence Seeking) are important skills for being able to notice possible patterns and to consider whether one can make causal claims or not. Noticing patterns (Pattern Seeking) often precedes determining a causal relationship (although knowledge of the behavior of causal mechanisms can also lead us to look for expected effects). The Pattern Seeking thinking move involves looking over extended time periods, using numerical or visual data as evidence, and noticing if there are changes before and after an event. These scaffolds for noticing patterns support students as they begin to look for possible causal relationships. The Analyzing Causality thinking move highlights intervention and experimentation as effective ways of distinguishing causality from correlation. Conducting experiments enables scientists to isolate a factor in a system to see its place in the bigger picture and also to observe its role(s) in the system. Experiments can be used in conjunction with observations, data, and other forms of evidence to support the existence of a causal relationship. The Constructing Explanations thinking move focuses on putting together a coherent explanation without gaps or leaps in the causal logic. It focuses on bringing together bodies of evidence to assess what causal story they collectively tell and whether there are competing explanations to be considered. Finally, the Building a Body of Evidence thinking move focuses on the overall strength of the collection of evidence in supporting the claims.

Figure 5.4 features the posters that correspond to each thinking move and include important prompts to facilitate students' use of the move. The accompanying introductory videos can be downloaded from here: *www.ecolearn.projectzerohgse.com/ecoXPT/resources.php.*

Figure 5.4. Posters describing thinking moves for supporting causal reasoning

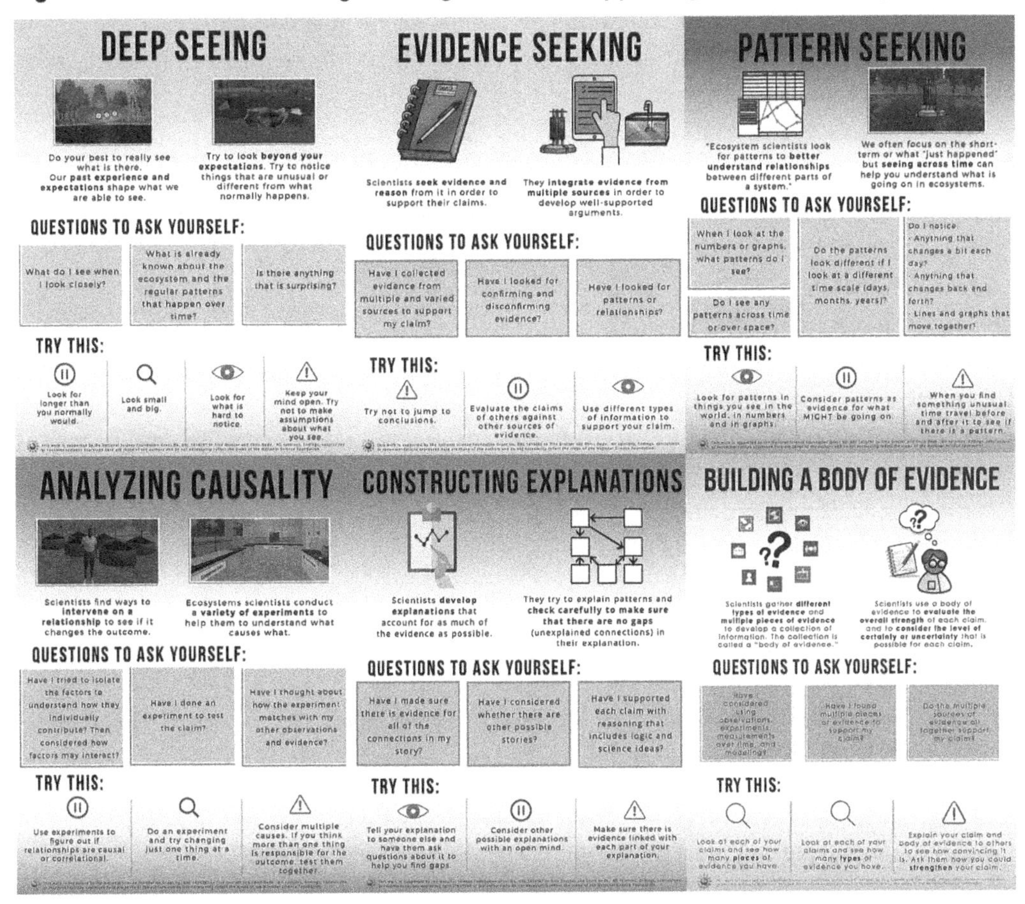

Summary

Effective instructional strategies need to focus on the features of causal reasoning, the types of challenges that students have in learning to reason about cause and effect, and the ways in which cause and effect are conceptualized and used in the scientific disciplines. Supporting students through thinking moves helps them know what good causal reasoning looks like. It also provides a window into how scientists use the CCC of cause and effect in their respective disciplines. Developing a vocabulary of causality opens the door to those conversations. This is all essential to enabling students to understand how scientific explanation is achieved. Without an understanding of how science builds causal connections between patterns and mechanisms, the scientifically accepted explanations for many different phenomena can indeed seem magical. Understanding how causal claims are built is what enables scientists to justify the explanations they stand behind and is what keeps magic crows at bay!

Note

Tina Grotzer's research was supported by the National Science Foundation Grant No. DRL-1416781, 0845632, 0455664, and 9725502. All opinions, findings, conclusions, or recommendations expressed here are those of the authors and do not necessarily reflect the views of the National Science Foundation.

References

Basca, B. and T. A. Grotzer. 2003. Causal patterns in air pressure phenomena: Lessons to infuse into pressure units to enable deeper understanding. Understandings of Consequence Project, President and Fellows of Harvard College.

EcoLEARN Project 2020. EcoXPT. President and Fellows of Harvard College. *https://ecolearn.gse.harvard.edu/projects/ecoxpt.*

Grotzer, T. A. 2010. Reasoning about causal complexity in science and beyond. President and Fellows of Harvard College. *www.causalpatterns.org/resources/natureofscience/resources_overview2.php.*

Grotzer, T. A. 2012. *Learning causality in a complex world: Understandings of consequence.* Lanham, MD: Rowman & Littlefield.

Grotzer, T. A., M. S. Tutwiler, L. S. Solis, and L. Duhaylongsod. 2011. Interpreting probabilistic causal outcomes in science: A microgenetic study of sixth graders' patterns of reasoning. Paper presented at the Annual International Meeting of NARST. Orlando, FL.

Kamarainen, A. M., and T. A. Grotzer. 2019. Constructing causal understanding in complex systems: Epistemic strategies used by ecosystem scientists *BioScience* 69 (7): 533–543.

McNeill, K. L., and J. Krajcik. 2012. Supporting grade 5–8 students in constructing explanations in science: The claim, evidence and reasoning framework for talk and writing. New York: Pearson Allyn & Bacon.

National Research Council (NRC). 2012. *A framework for K–12 science education: Practices, crosscutting concepts, and core ideas.* Washington, DC: National Academies Press.

Rivet, A., and J. Ingber. 2017. Analyzing and interpreting data. In *Helping students make sense of the world using next generation science and engineering practices*, eds. C. V. Schwarz, C. Passmore, and B. J. Reiser, 159–180. Arlington, VA: NSTA Press.

Understandings of Consequence Project. 2008. Causal Patterns in Science website: President and Fellows of Harvard College. *https://ecolearn.gse.harvard.edu/projects/ecoxpt.*

Windschitl, M. 2017. Planning and carrying out investigations. In *Helping students make sense of the world using next generation science and engineering practices*, eds. C. V. Schwarz, C. Passmore, and B. J. Reiser, 135–157. Arlington, VA: NSTA Press.

Chapter 6

Scale, Proportion, and Quantity

Cesar Delgado, Gail Jones, and David Parker

Scale as a Thinking Tool: Discovering New Worlds

Jake is a seventh-grade student at a public school in the Midwest. He was asked to order some objects by size (from smallest to largest) and came up with this answer:

head of a pin - virus - ant - molecule - mitochondrion - atom - cell - human

Jake was then asked, "So you think mitochondria and molecules are bigger than an ant?" He responded, "Yeah, 'cause if you look at an ant, ants are very small."

Like Jake, many students initially struggle to conceive of objects that are smaller than a tiny visible one, such as the head of a pin. In fact, Jake said the smallest thing he could think of was a baby ant. For these students, the crosscutting concept (CCC) of scale, proportion, and quantity is a stairway to entirely new worlds: the realms of objects too small to see with the naked eye. These microscopic objects are at the heart of life and physical science mechanisms and entities, which underlie the macroscopic phenomena we experience every day.

"For these students, the crosscutting concept (CCC) of scale, proportion, and quantity is a stairway to entirely new worlds: the realms of objects too small to see with the naked eye."

As *A Framework for K–12 Science Education* (the *Framework*; NRC 2012) explains, students' proficiency with this concept develops over a span of years, following a coherent pathway. Students in grades K–2 investigate tangible phenomena they can experience directly. Students in grades 3–5 begin to study objects that are not directly accessible. These objects are inside the human body or Earth but are mainly still macroscopic (i.e., large enough to be seen with the naked eye). At grades 3–5, "When microscopic entities are introduced, no stress is placed on understanding their size—just that they are too small to see directly" (NRC 2012, p. 34). However, experiences with Jake and many other students show us that the very *existence* of objects that are too small to be seen directly is a major conceptual leap. Teachers must recognize this leap as a potential challenge that should be explicitly addressed and carefully scaffolded. In middle school, phenomena at the cellular and atomic levels are introduced, and in high school, students learn subatomic and subcellular explanations.

Per the *Next Generation Science Standards* (*NGSS*; NGSS Lead States 2013), this very progression of the phenomena to be studied is organized in part by scale, showing how this CCC interacts with and provides a logical organization to sequencing of the disciplinary core ideas (DCIs). As the focus moves to objects too small to experience directly without aid for our senses, instructional activities begin to rely on the practice of modeling; moreover, the practice of explaining incorporates atomic or cellular processes for macroscopic phenomena. As subcellular and subatomic phenomena are studied, students use mathematical tools such as those featured in the science and engineering practice (SEP) of Using Mathematics and Computational Thinking. This sequence illustrates the three-dimensional nature of student learning in science, incorporating a CCC (scale, proportion, and quantity), SEPs (modeling, scientific explanations, mathematical thinking), and DCIs (LS1: From Molecules to Organisms: Structures and Processes).

Read Classroom Snapshot 6.1 for an example of a middle school teacher instructing students using the CCC of scale, proportion, and quantity.

CLASSROOM SNAPSHOT 6.1
There's a Whole Other World Out There

Ms. Perez is a sixth-grade science teacher in an urban public school. She teaches units on life science, physical science, and Earth and space science. Today, she is starting her life science unit. Since the *NGSS* recommend stressing macroscale systems in elementary life sciences, she is excited to help students delve into the microscopic realm of cells now that they are in middle school. Her students are intrigued and excited that she has optical microscopes set up around the room on the counters. She introduces the day's driving question, "What does a cell look like?" Then she leads her students through a realistic online simulation of an optical microscope so students learn how to place slides, turn on the light, focus, and safely switch from lens to lens.

Next, students prepare a slide with one of their own hairs. Ms. Perez intends to leverage the hair as an anchor object that is at the boundary of macroscopic and

Continued

Classroom Snapshot 6.1 *(continued)*

microscopic. Students work in pairs to view the hair, place a thin plastic metric ruler next to the hair on top of the slide, and sketch their observations. Some students are surprised at the amount of distance between millimeter markings, saying "What a big millimeter!" They are asked to estimate the thickness of a single hair; most estimates are around one-tenth to one-twentieth of a millimeter. Students then swab their cheeks and transfer cells to the same slide with the hair. After staining, they again use the microscope and sketch their observations. Some students are amazed to see that the cells are even smaller than the hair. As they grapple with estimating the diameter of the cell, Ms. Perez circulates around the room, monitoring their thoughts. One group suggests "one-half of one-tenth of a millimeter." Several others are puzzled and ask, "What's smaller than a millimeter?" Ms. Perez calls on a couple groups to report their ideas, then she introduces the micrometer as a thousandth of a millimeter. Her students then volunteer that a hair's thickness might be around 100 micrometers (mm) and a cheek cell around 50 mm. (For the activity that inspired this snapshot, see Delgado et al. 2015.)

Why Are Scale, Proportion, and Quantity Important?

Scale, proportion, and quantity (SPQ) is present and important in all science disciplines. An understanding of chemistry depends on conceptualizing the relationships across the macro, micro, and symbolic levels (Gabel, Samuel, and Hunn 1987). As part of the *NGSS* (NRC 2012; NGSS Lead States 2013) DCI PS1, students are asked to recognize the micro-macro relationship found in atoms and molecules: "How do atomic and molecular interactions explain the properties of matter that we see and feel?" Micro-macro relationships are also important in physics, where thermal energy (macro) is conceptualized as resulting from the movement of particles (micro) (see MS-PS1-4). Biologists use SPQ to examine life, from molecular components to entire ecosystems, as recognized in LS1: From Molecules to Organisms and LS2: Ecosystems. Earth and space science also "involve[s] phenomena that range in scale from the unimaginably large to the invisibly small," and seek to place Earth relative to the universe in both spatial and temporal scale.

Scale, Proportion, and Quantity as a Thinking Tool

Within a Discipline

Using SPQ as a thinking tool—a way of understanding multifaceted phenomena—enables us to generate explanations and mechanisms for important scientific phenomena. In many cases, smaller-scale processes underlie observable macroscopic phenomena. For example, the configuration and interactions among molecules or atoms determine physical processes such as boiling, freezing, and other phase changes, as well as chemical

> Using SPQ as a thinking tool—a way of understanding multifaceted phenomena—enables us to generate explanations and mechanisms for important scientific phenomena.

processes such as the combustion of fossil fuels. However, students often attribute macroscopic properties to atoms and molecules (e.g., molecules burn or melt), hindering their understanding of chemical reactions and physical changes (Driver 1985; Kind 2004). Once students begin to model phenomena such as phase changes as being caused by the aggregate behavior of myriad atoms or molecules, they can then begin to predict and explain macroscopic phenomena. For instance, water's high boiling point is unusual for such a small molecule. It's explained by the hydrogen bonds among neighboring molecules, which require additional energy to overcome in order to produce a gas of freely moving individual molecules. Similarly, Coulomb's law explains how the charged components of atoms (protons and electrons) interact in chemical bonding and the formation of new compounds from elements. Knowing how distance and charges determine attraction (as set forth in Coulomb's law) provides insight into periodic trends such as atomic radius or ionization energy.

In other cases, it is the large-scale perspective that sheds light on phenomena of interest. Understanding the vast temporal scale of life on Earth has been found to correlate with understanding and acceptance of evolution (Cotner, Brooks, and Moore 2010), which explains the diversity of life-forms on Earth. Similarly, the theory of plate tectonics is fundamental in Earth science and involves both immense geological times (e.g., continental drift) and sudden, short events (e.g., earthquakes), which underlie understanding of volcanism, rift valleys, and many mountain ranges. In life science, the levels of organization—from cells and organelles to tissues to organs to organisms to populations and webs—are directly related to SPQ. These examples demonstrate how SPQ is used *within* a science discipline to make sense of fundamental phenomena and processes.

Connections Across Disciplines

SPQ is also an intellectual tool for making connections and distinctions across disciplines. For example, students commonly confuse cells and atoms (Flores, Tovar, and Gallegos 2003; Harrison and Treagust 1996). This confusion may occur because both atoms and cells are described as "building blocks," are made of components students must learn about, and are too small to see. The relationship between cells and atoms may be neglected in both chemistry and biology classes or units, leading to a lack of distinction between cells and atoms and a disconnect between chemistry and biology. This confusion might be avoided if students had an idea of the size of these objects. Understanding that cells are 10,000–100,000 times larger in a linear dimension than an atom helps

> "SPQ is also an intellectual tool for making connections and distinctions across disciplines."

differentiate the two. Understanding scaling from one dimension (length) to three dimensions (volume) leads to the recognition that a cell is composed of billions or trillions of atoms, which could help students appreciate the complexity of the cell and of life itself. Furthermore, understanding that all the disciplines of science converge at the smallest of scales can help students understand the foundations of science. The knowledge of SPQ need not be quantitative, as the example of cells versus atoms illustrates. Knowing that a cell can be seen using an optical microscope but that an atom cannot can be a useful initial distinction between cells and atoms or molecules. Research shows that practicing scientists tend to operate within scale "worlds" that are characterized by tools, measurement units, and characteristic objects. Understanding relative size (i.e., which objects are larger than others) is an important part of conceptualizing science and understanding phenomena at different scales (Tretter, Jones, and Minogue 2006).

Other science disciplines can be connected through scale as well. Examples of the connections across biology and physics that involve SPQ include (1) the limit to the size of organisms that rely on diffusion to transport nutrients and waste products and (2) the increasing role of gravity in dictating the structure of land organisms of larger mass (leading to progressively thicker legs relative to body size) (Stevens, Sutherland, and Krajcik 2009; Tretter and Jones 2003). Understanding the particle nature of matter—a key in chemistry—requires comprehension of the physical forces that dominate at the atomic-molecular scale; the effect of gravity is almost negligible, whereas electrostatic interactions are relatively much stronger at that scale. These properties allow students to think about how a gecko can cling to the ceiling or why Styrofoam pieces adhere to a cardboard box. Helping students think about and visualize how the size of the material influences the behavior of the material is useful in understanding science phenomena. The study of the water cycle in Earth and space science relies on an understanding of the molecular structure of water and how this structure determines water's properties. Scientific practices such as Developing and Using Models, Analyzing and Interpreting Data, and Engaging in Argument From Evidence are also deployed while making sense of these phenomena and their explanations or mechanisms.

Read Classroom Snapshot 6.2 (p. 120) to see how one teacher uses the CCC of scale, proportion, and quantity to link various science units over the course of the year.

CLASSROOM SNAPSHOT 6.2
Promoting Connected Knowledge

Mr. Lieu teaches middle school science and for the last several years has taught units for life science, physical science, and Earth and space science. Although he feels that students learn effectively during each unit, he has noticed that by the end of the year, many students seem to have forgotten what they have learned in the previous units. This year, he is hoping to leverage SPQ to help students build connections across units and to keep their knowledge fresh and relevant throughout the whole year.

Since Mr. Lieu wants to use SPQ to help his students make connections, he has prepared two paper scales on butcher paper that extend across a wall of his classroom. One scale is marked 0–0.1 mm. The other, which is placed below the first, is marked 0–0.001 mm and the equivalent in micrometers, 0–1 µm. The scales are color-coded: The smallest 1% of the 0.1 mm scale, corresponding to 0–0.001 mm, is color-coded red; and the 0–0.001 mm scale is color-coded red, as well. Mr. Lieu will keep the scales posted throughout the year, using them to indicate the size of the objects discussed in the science classroom. This will spur the students to develop an overall model of SPQ. Mr. Lieu points out the scales to his class and asks, "Where would you place the size of a hair? A cheek cell? Discuss with a neighbor for two minutes." As the students discuss, Mr. Lieu listens in. Most groups are using inches, but Rosa and Kwang are instead using meters and millimeters. Both are recent immigrants from countries that use the International System of Units (SI). Although they are both still developing their English, Rosa and Kwang are advantaged by their cultural background on scale. Mr. Lieu asks Kwang and Rosa to collaborate on locating the size of a hair and a cheek cell on the appropriate scale. They paste images of hair at the right end of the larger scale and label its size in millimeters and micrometers. Then they paste images of the cheek cell in the appropriate place near the middle of the larger scale and label its size, as well. Mr. Lieu informs students that this year they will track objects on these scales, which he calls "sizelines." He notes that they will ultimately need even smaller scales to locate things such as atoms and viruses. When he later covers macroscopic objects in Earth science, he will add larger scales, too.

In addition to encouraging students to make connections across entities and phenomena included in the DCIs, using SPQ in this way will prompt students to employ practices such as using models, analyzing data, using mathematics, and communicating information. Other crosscutting concepts also play a role. For instance, patterns in the colors of the sizelines will indicate to students how neighboring sizelines relate to one another.

Development of Understanding of Scale, Proportion, and Quantity Over Time

Attributes, Quantification, and Scales

Concepts of scale, proportion, and quantity developed as scientists have worked to provide better and more detailed descriptions of phenomena in order to increase understanding and prediction of phenomena. Phenomena of interest can be observed holistically, without any attempt to quantify or even define parts or aspects of the phenomena. A more analytical view can result in defining characteristics or attributes of the phenomena. By defining these characteristics, we are then able to categorize and distinguish among phenomena.

> "Concepts of scale, proportion, and quantity developed as scientists have worked to provide better and more detailed descriptions of phenomena in order to increase understanding and prediction of phenomena."

Historically, Galileo distinguished between *natural motion* (e.g., water flowing downhill) and *violent motion* (e.g., a rock propelled upward by a catapult). Students might have experiences with continued effects for an initial cause (e.g., striking a bell) or with cases where a continuing force is required for a continued effect (e.g., dragging an object on a surface with friction), and distinguish between these two types of phenomena prior to learning physics formally (Hammer and Elby 2002). These attributes are not yet quantitative but can be captured with qualitative terms such as *smaller/bigger*, *faster/slower*, or *easier/harder to lift or drag*. Objects can be ordered by an attribute (e.g., ordering animals by size). Once attributes are defined, learners can first seek to quantify and measure, using body-based measures like hand spans or common objects like paper clips. Then they can use more formal scales—"the spatial, temporal, quantitative, or analytical dimensions used by scientists to measure and study objects and processes" (Gibson, Ostrom, and Ahn 2000, p. 219).

The *NGSS* proposes that young children begin their school science experience with observations without measurement. (For example, the assessment boundary of K-PS3-1 states, "Assessment of temperature is limited to relative [terms] such as *warmer/cooler*.") Instead, they use holistic observation and qualitative attributes. Students then progress to measuring at grade 2, starting with length (2-PS1-2). Length is one of the fundamental base units in the SI (metric) system, and it is likely the most intuitive and thus the first measurement students should engage with. Later, students measure mass and weight at grade 5 (5-PS1-2) and density, melting point, and solubility in middle school (MS-PS1-2). The *NGSS* proposes starting with everyday, macroscopic objects, then introducing microscopic entities in grades 3–5 but without emphasizing their actual size; by middle school, students learn about cells, molecules, and atoms.

One of the ways people learn how to estimate length and distance is through the development of landmarks (Tretter et al. 2006). Studies of scientists and engineers have found that experts develop a mental set of anchor points that they use to navigate from one scale "world" to the next (Jones and Taylor 2009). For example, scientists refer to the size of red blood cells, an embryo, or an atom as mental landmarks that can be used when reasoning about scale (Jones and Taylor 2009). As people experience different sizes and scales through everyday life, they can compare new objects to known objects and develop mental categories (e.g., smaller than a human, roughly human-size, room-size, football field–size, larger). These categories are indexed by landmarks (Tretter et al. 2006). More expert learners tend to have more mental categories and thus more differentiated and developed conceptions of size, whereas novices may not distinguish between small macroscopic objects and objects too small to see. Helping students develop a set of meaningful landmarks for sizes can enable them to apply scientific concepts in multiple domains. This can be done by pointing out the relative size of one object to another and helping students develop their own landmarks for sizes and distances. Making a set of objects and their sizes available for students will likely spur their construction of a set of coordinated landmarks for size.

There is evidence that having kinesthetic experiences in real-world contexts contributes to the development of a mental map of spatial scale (Tretter et al. 2006). Students report learning about size and scale from measuring objects and using microscopes and telescopes and from experiences like running track, studying maps, watching television, looking at mountains, and traveling. Scientists and engineers, like students, report learning scale through physical experiences such as walking, riding a bicycle, or traveling in an airplane (Jones and Taylor 2009). Students can relate to a meter by thinking about a track or swim meet that was measured in meters. This builds a conceptual landmark for distances that can be used when students encounter new measurements.

Experiencing science in a human-size world makes it easier for students to understand scientific concepts than experiencing science with phenomena that are too large or too small to see (Tretter, Jones, and Minogue 2006). The use of physical models can provide physical experiences that help students build conceptual bridges to science that exist outside the human scale.

Surface Area to Volume

The CCC of scale, proportion, and quantity is a valuable thinking tool that can help students connect knowledge across the disciplines. However, students are unlikely to make connections between the domains of science without help from their teacher. The relationships between surface area and volume can provide a scaffold for students to understand phenomena in physics, as well as in biology, Earth and space science, and chemistry (Taylor and Jones 2008, 2012).

The effects of the relationship between surface area and volume can be found in a variety of contexts in each of the science domains. At the macroscale in physics, for example, properties such as conductivity and melting point are independent of the amount of material that is present. When these same phenomena are examined at a nanoscale, though, surface area increases and surface-related properties have different effects; at the nanoscale, melting point actually varies depending on the size of the sample. An example of this would be the difference in how water sticks to a tiny capillary tube (adhesion, cohesion, and capillary action) but drains out of a larger-size drinking straw. Moreover, different forces are dominant at different scales, with gravity more important for larger objects than small ones, for example.

Table 6.1 shows that as a cube gets larger in size, both the surface area and volume increase. What is really important to notice is that volume increases more rapidly, and the ratio between the surface area and the volume of the cube decreases (Taylor and Jones 2008, 2012).

Table 6.1. Surface-area-to-volume ratios with increasing cube size

Size of Cube (length of each side)	Total Surface Area (cm^2) L × W × 6	Volume of Cube (cm^3) L × W × H	Total Surface Area Divided by Volume SA (cm^2)/Vol (cm^3)	Surface Area (cm^2) to Volume (cm^3) Ratio
1	1 × 1 × 6 = 6	1 × 1 × 1 = 1	6 / 1	6 : 1
2	2 × 2 × 6 = 24	2 × 2 × 2 = 8	24 / 8	3 : 1
3	3 × 3 × 6 = 54	3 × 3 × 3 = 27	54 / 27	2 : 1
4	4 × 4 × 6 = 96	4 × 4 × 4 = 64	96 / 64	3 : 2
5	5 × 5 × 6 = 150	5 × 5 × 5 = 125	150 / 125	6 : 5
6	6 × 6 × 6 = 216	6 × 6 × 6 = 216	216 / 216	1 : 1

The relationship of surface area to volume can be experienced in everyday life with different sizes of sugar. Even young students can see and feel the effects of different surface-area-to-volume relationships using this item. Very fine powdered sugar sticks to their hands, but as the sugar particle gets larger (table sugar and raw sugar), less and less of it sticks. The powdered sugar has more surface area (to adhere to skin) for a given volume. For instance, if powdered sugar were to come in grains just three times smaller than the table sugar grains, the powdered sugar would have three times the surface-area-to-volume (SA/V) ratio. (See Table 6.1). Getting students to think about why the sugar sticks more when it is smaller can help them reason about other phenomena such as gecko feet, which have hairlike structures called setae that increase the surface area of the toes, allowing them to stick to the ceiling.

Within living organisms, surface-area-to-volume relationships are critically related to organ, tissue, and cell functions. For example, for digestion to take place efficiently, a large surface area is needed in the intestines for digested nutrients to be absorbed through diffusion. Villi (small protrusions from the intestinal wall that are long and thin) greatly increase the surface area to allow for nutrient exchange in the tissues. The same high surface area design can be found in our kidneys where the tiny nephrons come in contact with the capillaries, allowing for the transfer of waste products. (Nephrons are the functional unit of the kidney, and a human kidney is made up of around one million nephrons.)

Cell size is another biological example that can help students visualize the importance of surface-area-to-volume relationships. Once students have considered how cells absorb nutrients and remove waste products through the cell membrane, they can be challenged to think about how size influences this process. As cells get larger, they have relatively less exposed surface for effective gas exchange to take place. Surface-area-to-volume relationships are predictive of limits to the size of cells. A huge cell simply cannot exist with the typical organelles present in average-size cells because the large cell does not have enough gas and nutrient exchange to survive. This limitation on the size of cells is likely one of the reasons that specialized tissues have evolved with multicellular organisms.

In addition to biology-related examples, there are a number of physical science phenomena that illustrate the SA/V ratio effects. Elementary students can experience how a candy mint dissolves faster in their mouths when broken into little pieces than when it is whole. They can see how granulated sugar dissolves faster than rock candy in a beaker of water. Seeing how larger particles dissolve more slowly than smaller particles is a visual way to illustrate processes such as weathering on rocks. By designing an experiment exploring the factors involved in how fast sugar dissolves in water and explaining their results with a poster or video, students can engage in argumentation, planning and carrying out experiments, analyzing and interpreting data, and communicating information. This activity can be tweaked for different age groups. For example, K–2 students could focus on the mass of sugar and water before and after the sugar dissolves, helping them develop the essential concept of conservation of mass. Many may think that sugar simply "disappears" based on their visual observation, and they may expect the mass to decrease. High school students can develop molecular-level mechanisms for the dissolution of ionic solids like table salt and for molecular substances like sugar based on Coulomb's law and their knowledge of dipoles and bonding.

Thermal cooling is another process influenced by surface-area-to-volume ratios. A tiny mouse and a large elephant have heating and cooling challenges that are directly related to their size. The mouse has a high SA/V ratio that leaves it susceptible to being overcome by cold temperatures. Likewise, the elephant, with its large body mass and limited corresponding surface area, can experience challenges to cooling the body when the temperature is hot. Elephants are able to compensate for their body's low SA/V ratio by having very thin and large ears with high SA/V ratios that serve as radiators, releasing excess heat. The differences in heating and cooling exhibited by the mouse and elephant can also be seen in lakes that differ in depth. Consider, for example, two lakes that have the same surface area but different volumes due to different depths—one is shallow, the other is deep. The shallow lake will heat and cool much more quickly than the deep lake. In order to learn about thermal cooling and SA/V ratios in middle and high school, students can develop models to explain how heating and cooling involve particle collisions and movement. Additionally, the PhET simulation "States of Matter: Basics" can be used for students to learn about how water acts at the molecular level, when heated or cooled. This simulation can be found here: *https://phet.colorado.edu/sims/html/states-of-matter-basics/latest/states-of-matter-basics_en.html.*

Plants provide us with yet another example of the influence of SA/V ratios. More specifically, plants have evolved with high and low SA/V ratios respective to their leaves that allow them to survive in different environments. Desert succulent plants often have fleshy, plump, and smooth leaves that have limited surface area, and cacti may have no leaves at all. In more tropical and wet environments, plants often have thin, flat leaves that have much more surface area in relation to the volume of the leaves. The desert plant has less moisture loss than the tropical plant due to the shape and size of its leaves.

Developing an understanding of the relationships between surface area and volume in science can begin in grades K–2 with observations of macroscale phenomena such as the dissolution of particles of different sizes, the melting of ice cubes that are large and small, and the evaporation rates of shallow and deep containers of water. (See Table 6.2 on page 126 for examples of surface-area-to-volume applications by grade level.) These applications of surface area to volume allow students to investigate and experience at a macroscale how changing either the volume or the surface area influences the behavior of the materials.

Table 6.2. Examples of surface-area-to-volume applications by grade level

Level	Science Applications of Surface Area to Volume
K-2	Comparing and modeling for the stickiness of sugar of different sizes (i.e., powdered, granular, and cubed)
	Testing the dissolution of salt particles of different sizes (e.g., table salt and rock salt) and generating a scientific explanation
3-5	Examining and explaining electrostatic attraction of materials of different sizes (e.g., Styrofoam particles of different sizes in a plastic bottle)
	Obtaining and graphing the time of reactions for materials of different sizes (e.g., whole and crushed denture cleaning tablets placed in water in a film canister)
Middle School	Predicting and testing the cooling rates for containers of water that are different sizes
	Calculating surface-area-to-volume ratios for cubes of different sizes to visualize how the ratio decreases as the cube size increases and graphing the relationship
	Modeling weathering of rocks of different sizes
	Graphing heat loss of animals with different body sizes and shapes to examine the relationship of surface area to volume and the impact on cooling
	Modeling respiration rates for eggs of different sizes
High School	Measuring and modeling diffusion rates for cells of different sizes
	Analyzing respiration rates of large and small animals

In the middle grades, students typically develop their abilities in conceptualizing ratios and can apply proportional reasoning to their investigations. Proportional reasoning starts with understanding how a whole object is composed of repeated parts. For measurement, proportional reasoning is applied as students consider units of measurement and how these units are distributed. The *Common Core State Standards for Mathematics* (NGAC and CCSSO 2010) recommend that, as proportional reasoning develops, students can learn how to calculate surface area and volume for various shapes (6.G.A.4). They also learn how to make measurements of such things as the temperature and volume of materials. At this point, they can begin to identify and combine variables, collect data, and see patterns in the data that help them model the phenomena.

In high school, students can learn how SA/V ratios can predict the behavior of materials at different scales. For example, gravity is not as relevant at small scales where other forces such as intermolecular forces dominate. Students can apply the idea that surface area increases relative to volume as particles get smaller to predict that wood shavings, steel wool, or flour will burn more easily than a log, an iron bar, or a loaf of bread; they can then design an investigation and model and explain their experimental findings. This understanding of the relative importance of forces at different scales also allows students to think about how that gecko really does stick to the ceiling.

Teaching students the scaling effects of surface area to volume can help them develop the ability to conceptualize and predict phenomena, processes, and behaviors of materials across different science domains. Developing mental models of how surface area and volume ratios influence processes and systems takes time to develop; once mastered, students can use this scaling ratio as a thinking tool to understand and predict phenomena and the behavior of materials in different domains.

"Teaching students the scaling effects of surface area to volume can help them develop the ability to conceptualize and predict phenomena, processes, and behaviors of materials across different science domains."

Challenges to Learning and Using Scale, Proportion, and Quantity

Thinking and learning about objects and phenomena outside human scale is difficult for learners of all ages, and accuracy of their knowledge of the scale of objects falls quickly at the small range (Tretter, Jones, and Minogue 2006). Middle school students often mention a small macroscopic object (e.g., grain of salt) when asked for the smallest thing they can think of (Waldron, Spencer, and Batt 2006). Learners of all ages in the United States may be unfamiliar with units suitable for microscopic objects, and close to half cannot order millimeter, micrometer, and nanometer by size (Waldron, Spencer, and Batt 2006).

Students may have difficulty with the symbolic aspects of scientific representations and intuitively assume constant scale across images. This is illustrated in a comment made by a middle school student: "A quarter and a blood cell are the same size because I don't know the size of a blood cell, but I have seen pictures in my book of cells, and they were a little bit bigger than a quarter" (Jones 2013, p. 147). Students may also assume there can be no object that is too small to see and have inaccurate landmarks for size—for instance, believing that a small macroscopic object is smaller than atoms or cells (Waldron et al. 2006). The classic video "Powers of 10" has proved to be an effective tool at helping students conceptualize the relative sizes of objects (Jones et al. 2007).

Scientific phenomena vary across an enormous range of magnitudes, and this poses difficulties when seeking to relate, connect, or compare phenomena. Familiar and easy-to-understand linear scales do not adequately represent more than three or four orders of magnitude, but logarithmic or "powers of 10" scales—which can effectively portray ranges of data across any number of orders of magnitude—are counterintuitive even for college undergraduates. This is shown in the statement of one university student: "So you're saying a logarithmic scale doesn't even take into account whether or not it's, like, spatially accurate? ...Yeah, that's how the physics graph worked 'cause we got rid of the 10 and just used the exponents. But then you get a really distorted graph!" (Delgado and Lucero 2015, p. 646). The conceptual difficulties must be taken into account when designing instructional activities that include SPQ, and students must be scaffolded in building their understanding.

Knowledge of unfamiliar measurement units can be developed based on familiar objects (e.g., a four-year-old child is about 1 m in height; a football field is about 100 m long), and knowledge of unfamiliar objects can be developed in part through measurement units (e.g., a red blood cell is about 5–7 micrometers in diameter). A major challenge for SPQ is when both the units *and* the objects are unfamiliar.

Promising Instructional Strategies for Scale, Proportion, and Quantity

Teachers can make SPQ an intrinsic part of instruction about scientific entities and phenomena. Many objects displayed in diagrams or photographs in curriculum materials lack any indication of scale, and part of learning about these objects could involve (1) adding a scale bar or a better-known object at the same scale to provide context or (2) calculating the magnification or reduction factor. In Classroom Snapshot 6.1 (p. 116), cells were the object of study, and their size was contextualized by placing them (and having students sketch them) next to a human hair on a single microscope slide. In addition, a scale bar was included by placing a translucent metric ruler with millimeter markings next to the hair and cells. Students could calculate the magnification by measuring the apparent size of the millimeter on the ruler under the microscope or could use the magnification factor directly from the microscope lenses. By integrating instructional activities about SPQ into the explorations of the novel concept of cells, students can use each concept to make sense of the other, as shown in the Classroom Snapshot. Recall that some students in this vignette spontaneously asked for units smaller than a millimeter when studying their cheek cells next to the hair under the microscope.

Generating a coordinated set of mental landmarks for size is difficult, but external representations of the size of scientific objects can reduce the cognitive load students encounter. Sizelines, as suggested in Classroom Snapshot 6.2 (p. 120), can help leverage

SPQ to make connections across the topics and disciplines of the science classroom. These external representations can help students internalize landmarks for size in relation to one another, which can start with familiar, human-scale objects in elementary school and move into microscopic objects in middle school and subatomic particles and organelles in high school. Coordinated sizelines can overcome the limitations of a single linear scale to represent sizes, and they might be used to scaffold an understanding of the powers of 10 scale.

Making scale drawings of objects or places—for instance, using one inch to represent one yard—could scaffold students' understanding of scale factors and proportional reasoning, starting in elementary school, in coordination with activities involving measurement.

Relationships between two objects of scientific interest can be better understood if SPQ is addressed explicitly. For instance, some students believe viruses are larger than cells because viruses attack our bodies. Furthermore, in unpublished research, certain students believe viruses "eat" our cells, rather than inject their genetic materials into cells. Some, having heard that viruses "spread" and "invade an organ," believe that an individual virus can grow as large as an organ. Teachers should follow good constructivist practice and bring student ideas out into the open through such things as class discussions, drawings or models, and small-group discussion because many ideas about scientific phenomena may include non-normative ideas, often related to SPQ or other CCCs.

Following the insights of project-based learning and the sequencing of the popular 5E model (Bybee et al. 2006), it is important to *engage* students in instructional activities, including those that feature SPQ. For instance, the activity described in Classroom Snapshot 6.1 (p. 116) was part of a unit with the following driving question: "How can nanotechnology keep me from getting sick?" The anchoring activity for the unit was a news article (revised for vocabulary and ease of comprehension) about a middle school student who suffered from an antibiotic-resistant bacterial infection that was in the news at that time.

There are a variety of free resources on SPQ that can be used in lessons. Some online simulations and visualizations for scale are listed in Table 6.3 (p. 130). Students can use these simulations in investigations to learn about scientific objects, to calculate how many times one object is bigger than another, or to build a set of landmarks or their understanding of temporal scale.

Table 6.3. Online simulations on scale

Simulation (Author)	URL and Range of Sizes
"Universcale" (Nikon 2019)	*www.nikon.com/about/sp/universcale* Proton through known universe
"The Scale of the Universe" (Huang 2010)	*http://htwins.net/scale* Planck length (smallest length theoretically possible) through known universe
"The Scale of the Universe 2" (Huang 2012)	*http://htwins.net/scale2* Planck length through known universe
"Secret Worlds" (Hahn, Burdett, and Davidson 2017)	*http://micro.magnet.fsu.edu/primer/java/scienceopticsu/powersof10* Proton through Milky Way
"Cell Size and Scale" (Genetic Science Learning Center 2020)	*https://learn.genetics.utah.edu/content/cells/scale* Carbon atom through coffee bean
"Depth" (Munroe n.d.)	*https://xkcd.com/485* Proton through human
"If the Moon Were Only 1 Pixel" (Worth n.d.)	*http://joshworth.com/dev/pixelspace/pixelspace_solarsystem.html* 3474.8 km through solar system
"How Big Is a ... ?" (Sullivan 2018)	*www.cellsalive.com/howbig_js.htm* Rhinovirus through pinhead

Classroom Snapshot 6.3 provides an example of a teacher who made SPQ an integral part of a lesson on population.

CLASSROOM SNAPSHOT 6.3

Quadrats and Biodiversity

Mr. Parker teaches environmental science to eleventh and twelfth graders at an independent school. Each year, Mr. Parker utilizes SPQ as a tool in units dealing with population. Since understanding scale and proportion of a population is a ubiquitous concept in environmental science, Mr. Parker chooses to utilize a lab on quadrats (small areas selected at random as samples for estimating the variety and density of plants or animals) to help students understand proportions. The *NGSS*

Continued

Classroom Snapshot 6.3 *(continued)*

expects students to be able to integrate their understanding of mathematics with carrying capacities of a population (HS-LS2-1).

For the quadrat lab, Mr. Parker takes his students outside to a forested area located on the school property. Here, he instructs a student to toss a marked object randomly into the woods. The object is used as the center of a one-meter-by-one-meter plot the students create. The plot is turned into a grid with subsequently smaller squares, and a tally is made of the various plant types found there. Students then use these data to extrapolate to the rest of the forest, employing the practice of mathematical thinking and concepts of SPQ to estimate the abundance of particular species in the grounds being studied.

The activity also leads to discussions about error. One group proposes that lady ferns are the most common species, but other groups have different findings. Mr. Parker orchestrates a discussion about sampling and extrapolating from a sample to the population. This helps students realize that the forest is so diverse that one single quadrat will not give an accurate understanding of the full ecosystem. However, the students also realize that if multiple quadrats are completed, these data could be used to estimate the population density of a specific species in the area. Students suggest pooling their data to obtain a more accurate representation. The larger proportion that is sampled, the better the measurement, which is an important understanding for modeling.

Each time population comes up in class, Mr. Parker directs the students back to the quadrat lab, reminding them how understanding frequency and density of a specific species can help in interpreting human impact on the environment (LS2.A). The scaffolding this activity offers leads students to a better understanding of ecosystem dynamics, functioning, and resilience (LS2.C).

Supporting Diverse Student Learners

Research on learning size and scale has shown that experiences exploring the world through activities such as riding a bicycle or flying in an airplane, as well as experiences using measurement tools like a measuring tape or a meterstick, are fundamental in developing accurate concepts of size and scale. However, not all students have the same opportunities to travel or use tools at home. Furthermore, many tools have roots in culturally linked activities. While some students' parents or relatives have involved them in using rulers to measure wood for a construction project or birdhouse, other students may not have had these experiences or may use different tools such as metersticks, tape measures, or ultrasonic measurement tools. Chesnutt et al. (2018, p. 895) describe this type of access to cultural tools as "science scale capital." These researchers measured students' access to scale experiences and tools (scale capital) and found that reported science scale capital significantly predicted students' accuracy of size and scale. Furthermore, this study found significant differences in accuracy of size and scale concepts by race and ethnicity. The researchers argue that these differences are most likely related to scale capital tied to socioeconomic status. Such differences should be taken into account when designing assessments.

SPQ offers the opportunity for diverse learners, including English language learners and students with visual impairments, to adopt roles as experts and help their peers. As noted in Classroom Snapshot 6.2 (p. 120), students who grow up abroad and are familiar with the metric system may have some advantages (Delgado 2013). Such students have a more developed conceptual understanding of size and scale than students who have grown up using U.S. customary ("Imperial") units. Jones et al. (2013) also found cross-cultural differences in size and scale across Austria, Taiwan, and the United States. Research on visually impaired students shows that they have difficulty with linear estimations, particularly when using the SI ("metric") system (Jones, Taylor, and Broadwell 2009), but they have more accurate estimations at very large and small scales than their peers with unimpaired vision.

Opportunities to travel and engage in creative building projects that are part of a students' science scale capital are often linked to family financial resources, and not all students have access to these opportunities. If we are to help all students develop accurate concepts of quantity, size, and scale, educators must make sure students have access to and opportunity for rich experiences to learn and apply their knowledge related to size and scale. We must also recognize and leverage students' home experiences that may have built informal knowledge of SPQ, including cooking (measures and proportions), construction (measurement), and more.

Summary

SPQ is a thinking tool that allows students to conceptualize and reason across the science disciplines and across very large and small scales. Furthermore, by understanding SPQ, students can predict how the behavior of materials might vary across very large and very small scales.

Teaching SPQ begins with helping students understand units and measurement. Starting with simple comparisons (e.g., "larger than" or "smaller than") and visualizing parts of a whole (i.e., multiplicative reasoning) can set a foundation for more sophisticated concepts of size and scale. Providing students with physical experiences (e.g., showing that powdered sugar sticks more than granulated sugar) and then using physical models for small and large science phenomena can help students develop landmark concepts and reason with SPQ.

As students develop their understandings of SPQ, they can apply SPQ concepts to look at relationships that can predict and explain science phenomena, like heat loss for mice and elephants or rates of gas exchange in large and small cells.

References

Bybee, R. W., J. A. Taylor, A. Gardner, P. Van Scotter, J. Carlson Powell, A. Westbrook, and N. Landes. 2006. *The BSCS 5E instructional model: Origins and effectiveness.* Colorado Springs, CO: BSCS.

Chesnutt, K., M. G. Jones, R. Hite, E. Cayton, M. Ennes, E. N. Corin, and G. Childers. 2018. Next generation crosscutting themes: Factors that contribute to students' understandings of size and scale. *Journal of Research in Science Teaching* 55 (6): 876–900.

Cotner, S., D. C. Brooks, and R. Moore. 2010. Is the age of the Earth one of our "sorest troubles?" Students' perceptions about deep time affect their acceptance of evolutionary theory. *Evolution* 64 (3): 858–864.

Delgado, C. 2013. Cross-cultural study of understanding of scale and measurement: Does the everyday use of U.S. customary units disadvantage U.S. students? *International Journal of Science Education* 35 (8): 1277–1298.

Delgado, C., and M. Lucero. 2015. Scale construction for graphing: An investigation of students' resources. *Journal of Research in Science Teaching* 52 (5): 633–658.

Delgado, C., S. Y. Stevens, N. Shin, and J. S. Krajcik. 2015. A middle school instructional unit for size and scale contextualized in nanotechnology. *Nanotechnology Reviews* 4 (1): 51–69.

Driver, R. 1985. Beyond appearances: The conservation of matter under physical and chemical transformations. In *Children's ideas in science*, ed. R. Driver, E. Guesne, and A. Tiberghien, 145–169. Philadelphia: Open University Press.

Flores, F., M. E. Tovar, and L. Gallegos. 2003. Representation of the cell and its processes in high school students: An integrated view. *International Journal of Science Education* 25 (2): 269–286.

Gabel, D. L., K. V. Samuel, and D. Hunn. 1987. Understanding the particulate nature of matter. *Journal of Chemical Education* 64 (8): 695–697.

Genetic Science Learning Center. 2020. Cell size and scale. Learn.Genetics. *https://learn.genetics.utah.edu/content/cells/scale.*

Gibson, C. C., E. Ostrom, and T. K. Ahn. 2000. The concept of scale and the human dimensions of global change: A survey. *Ecological Economics* 32 (2): 217–239.

Hahn, D. A., C. A. Burdett, and M. W. Davidson. 2017. Secret worlds: The universe within. Florida State University. *http://micro.magnet.fsu.edu/primer/java/scienceopticsu/powersof10.*

Hammer, D., and A. Elby. 2002. On the form of a personal epistemology. In *Personal epistemology: The psychology of beliefs about knowledge and knowing*, ed. B. K. Hofer and P. R. Pintrich, 169–190. Mahwah, NJ: Erlbaum.

Harrison, A. G., and D. Treagust. 1996. Secondary students' mental models of atoms and molecules: Implications for teaching chemistry. *Science Education* 80 (5): 509–534.

Huang, C. 2010. The scale of the universe. *http://htwins.net/scale.*

Huang, C. 2012. The scale of the universe 2. *http://htwins.net/scale2.*

Jones, M. G. 2013. Conceptualizing size and scale. In *Quantitative reasoning in mathematics and science education, monograph no. 3*, ed. R. Mayes and L. Hatfield, 147–154. Laramie, WY: University of Wyoming.

Jones, M. G., and A. Taylor. 2009. Developing a sense of scale: Looking backward. *Journal of Research in Science Teaching* 46 (4): 460–475.

Jones, M. G., A. Taylor, and B. Broadwell. 2009. Concepts of scale held by students with visual impairment. *Journal of Research in Science Teaching* 46 (5): 506–519.

Jones, M. G., A. Taylor, J. Minogue, B. Broadwell, E. Wiebe, and G. Carter. 2007. Understanding scale: Powers of ten. *Journal of Science Education and Technology* 16 (2), 191–202.

Jones, M. G., M. Paechter, G. Gardner, C. F. Yen, A. R. Taylor, and T. R. Tretter. 2013. Teachers' concepts of spatial scale: An international comparison between Austrian, Taiwanese, and the United States. *International Journal of Science Education* 35 (14): 2462–2482.

Kind, V. 2004. Beyond appearances: Students' misconceptions about basic chemical ideas, 2nd ed. Report prepared for the Royal Society of Chemistry, London. *https://edu.rsc.org/resources/beyond-appearances/2202.article*.

Munroe, R. n.d. Depth. *https://xkcd.com/485*.

National Governors Association Center for Best Practices and Council of Chief State School Officers (NGAC and CCSSO). 2010. *Common core state standards*. Washington, DC: NGAC and CCSSO.

National Research Council (NRC). 2012. *A framework for K–12 science education: Practices, crosscutting concepts, and core ideas*. Washington, DC: National Academies Press.

NGSS Lead States. 2013. *Next Generation Science Standards: For states, by states*. Washington, DC: National Academies Press. *www.nextgenscience.org*.

Nikon. 2019. Universcale. *www.nikon.com/about/feelnikon/universcale/index.htm*.

Stevens, S. Y., L. M. Sutherland, and J. S. Krajcik. 2009. *The big ideas of nanoscale science and engineering: A guidebook for secondary teachers*. Arlington, VA: NSTA Press.

Sullivan, J. A. 2018. How big is a…? *www.cellsalive.com/howbig_js.htm*.

Taylor, A. R., and M. G. Jones. 2008. Proportional reasoning ability and concepts of scale: Surface area to volume relationships in science. *International Journal of Science Education* 31 (9): 1231–1247.

Taylor, A. R., and M. G. Jones. 2012. Students' and teachers' application of surface area to volume relationships. *Research in Science Education* 41 (3): 357–368.

Tretter, T. R., and M. G. Jones. 2003. A sense of scale. *Science Teacher* 70 (1): 22–25.

Tretter, T. R., M. G. Jones, and J. Minogue. 2006. Accuracy of scale conceptions in science: Mental maneuverings across many orders of spatial magnitude. *Journal of Research in Science Teaching* 43 (10): 1061–1085.

Tretter, T. R., M. G. Jones, T. Andre, A. Negishi, and J. Minogue. 2006. Conceptual boundaries and distances: Students' and adults' concepts of the scale of scientific phenomena. *Journal of Research in Science Teaching* 43 (3): 282–319.

Waldron, A. M., D. Spencer, and C. A. Batt. 2006. The current state of public understanding of nanotechnology. *Journal of Nanoparticle Research* 8 (5): 569–575.

Worth, J. n.d. If the Moon were only 1 pixel. *http://joshworth.com/dev/pixelspace/pixelspace_solarsystem.html*.

Chapter 7

Systems and System Models

Sarah J. Fick, Cindy E. Hmelo-Silver, Lauren Barth-Cohen, Susan A. Yoon, and Jonathan Baek

How does using the crosscutting concept (CCC) of systems and system models as a lens help us make sense of the world? Many of the confusions that students have about how phenomena work can either be complicated or simplified by the use of a systems lens. Many phenomena that we examine in science courses, particularly at the younger grades, are systems—where keeping track of what is coming into, happening within, and leaving the system would help students make sense of what is going on. For example, using a systems lens to support students in keeping track of the inputs and outputs of a plant cell during photosynthesis would help them identify where the carbon that makes up the plant structure and sugars is coming from (carbon dioxide); this would thereby support students in working toward overcoming a common difficulty—that the mass of a plant comes from the dirt. In another example, when trying to determine the source of a pollutant in a stream, students might use the system of a watershed to narrow down the location of the source, checking the mouth of each water input into the larger watershed (e.g., showing how subsystems are nested within systems) for traces of the pollutant, until they can figure out which subsystem is the source. Then students can check the source until they find where the pollutant enters the stream. Without using the systems lens, students' search for the pollutant might be much less organized and purposeful. In regard to the previous example, they might never confront common difficulties about photosynthesis. Using systems as a lens is a powerful approach that can help students, as it does scientists and engineers, make sense of how the natural and designed world works. With that in mind, we need to help students understand when and why to use the systems lens to support their understanding.

Why Are Systems and System Models Important?

Systems are an important tool for understanding the natural world that cuts across different science disciplines and serves as a connection between science and engineering (NRC 2013). As described by *A Framework for K–12 Science Education* (the *Framework*), a system is "an organized group of related objects or components that form a whole" (NRC 2012, p. 92) determined by the investigator (scientist or engineer) based on the question being asked or the problem being solved. The investigator then describes a subset of the interactions and processes that are going on within that space to either design solutions or better understand processes and relationships.

Furthermore, systems are important in professional scientific sensemaking. For example, we know that the world is constantly changing and environmental conditions are continuously evolving, and there is an increasing frequency of catastrophic events like hurricanes and drought, all exacerbated by human-generated activity. Scientists have focused recent efforts on setting research agendas to investigate and manage issues related to systems that affect our lives, such as the spread of disease, power grid robustness, and biosphere sustainability (NRC 2009). The goal of this work is to identify the processes, flows, and changes within systems so interventions can be applied to restore stability in the face of perturbations. For example, during the COVID-19 pandemic, understanding the conditions within social and biological systems that promote the spread of the virus and how to mitigate that spread within population systems has been critical in making decisions about prevention and treatment.

Defining Systems and System Models in the Context of the *Next Generation Science Standards*

As we have illustrated, there are different ways that systems can be defined. Within science education, there are policy documents that have historically provided instruction on how to define them. For example, the *National Science Education Standards* defined systems as "the units of investigations … [and] an organized group of related objects or components that form a whole. Systems can consist, for example, of organisms, machines, fundamental particles, galaxies, ideas, and numbers. Systems have boundaries, components, resources, flow, and feedback" (NRC 1996, p. 116). These components also make up what the *Next Generation Science Standards* (*NGSS;* NGSS Lead States 2013) describes as an understanding that students should have when leaving high school, with students building increasing complexity to their understanding as they progress from elementary school. The following systems aspects and their descriptions are compiled from the *Framework* and Appendix G of the *NGSS*:

- **Boundary:** Using the lens of systems is a way to simplify the natural world by determining what parts will be in focus. One step in this simplification process is to determine boundaries between what is under examination and everything else. These boundaries are sometimes artificial but are often based on physical structures (e.g., cell walls or membranes, the presence of elevation in the landscape). The boundaries allow the scientist or engineer to consider how things outside the boundary affect things inside the boundary.

- **Inputs and Outputs:** Every system can be characterized by what goes into it and what comes out of it. Scientists and engineers ask questions about the effect of objects outside the system on the things inside the system. The effects of things coming from outside the system—flows of matter or energy or forces acting on the system—serve as inputs into the system, whereas the energy, matter, or forces leaving the system serve as outputs.
- **Processes and Interactions:** Within a system, there are interactions among the parts that cause observable changes. These interactions can vary across large and small scales. In an ecosystem, the interactions might be between different organisms or between organisms and their environment. In our solar system, these interactions involve gravitational forces between different celestial bodies. All of these interactions include transfers of energy, matter, or, in some cases, information.
- **Flows Within the System:** The movement of energy, matter, and/or information into, out of, or within the system is often described as the behavior of the system. These movements and changes are often what characterize the system and how it works, and they are often the focus of study.
  - *Initial Conditions:* Defining the initial conditions for a system helps frame what you predict might happen and allows you to make predictions about how those conditions change within the system. The aspects of the system for which you define initial conditions shapes the focus of your predictions. Depending on the initial conditions and how they evolve over time, different patterns will emerge.
- **Systems at Different Scales:** Depending on the scale at which you are examining the system, a subsystem at one scale could be a whole system at another. For example, the area of land that drains water into a river is a watershed, but this watershed could be one tiny portion of a larger watershed—a subsystem within a system.
- **Subsystems (Nested Systems) and Between-System Interactions:** Though we define a boundary in one particular location, there are likely interactions between the system being examined and other systems that have an impact on what you are observing (whether neighboring systems or systems within the system). It is hard to truly isolate one system, so we need to consider how the interactions between systems may affect our results.
- **System Models:** System models are representations of a system in which the modeler simplifies the phenomenon by focusing attention on particular aspects or processes. These system models can be used to predict behaviors or failure points in the system, as well as help identify problems. The models can take a range of forms, from simple drawings or lists of the components and interactions to mathematical

models to complicated computer simulations or prototypes. Regardless of the complexity of the model, it is important to understand the limitations, approximations, and assumptions that affect the model's ability to make predictions about the system's behavior. As a part of investigating the system, students should define the boundaries of the system, initial conditions, and the inputs and outputs to the system.

The systems aspects mentioned in this bulleted list can be the organizing principles that teachers use for supporting students to learn about how and why to use systems to make sense of science concepts. They also can be the transferable nuggets of systems that students bring across science ideas and disciplines to make use of this lens to understand new concepts (Goldstone and Wilensky 2008). When learning about a new concept, these aspects can be used to ask questions that guide investigations and inquiries into deeper understandings in making sense of phenomena (Fick 2018). (See Classroom Snapshot 7.1.)

CLASSROOM SNAPSHOT 7.1

Grounding Learning in One Disciplinary Core Idea to Bridge to Another[1]

In an Earth science unit within Mr. Baek's middle school class, students were investigating watersheds as a tool to understand water flow on and through Earth's surface. As part of this unit, students conducted investigations to define aspects of the watershed using a systems perspective (Fick, Arias, and Baek 2017; Fick and Baek 2017). The students were active participants in defining the system components. For example, the unit began with an investigation of the slope of the landscape. Students developed clay mountains and sliced them horizontally in equal parts to understand how elevation is represented using topographic lines. Then they poured water on surfaces with different slopes to observe the impacts of slope on the flow of water. Finally, students used topographic maps to make predictions about the flow of water based on the slope of the landscape. At the end of the lesson, students drew lines on the maps defining the boundary of the watershed due to elevation. After doing this, they defined these lines as a boundary to water flow. Through this process, students developed an understanding of where the system boundary was located by defining its location based on its purpose and used the boundary to understand where water flowed.

1. *Note:* This Classroom Snapshot is inspired by a collaboration between chapter authors Sarah J. Fick and Jonathan Baek. It has been condensed for clarity.

Continued

Classroom Snapshot 7.1 *(continued)*

In a subsequent lesson, students were investigating where the water went within that boundary, showing that all of the water within a particular area flowed toward the same destination. They then named the watershed according to the destination of the flow. This raised an important question that the teacher then posed to the class: "Does the ocean have a watershed?" There were a lot of replies—many yeses, some nos. "Does the ocean have a watershed?" Mr. Baek repeated, and then he asked for some students to answer the question with justification. "Yes, I think the ocean has a watershed because all of the water flows to the ocean," Samantha answered confidently. "What do other people think? Do you agree with that reasoning?" Mr. Baek asked. Samara wasn't sure she agreed with Samantha, saying, "I agree that the ocean has a watershed, but I think it is because all of the water flows to bigger and bigger bodies of water. We started in a creek, then went to a stream, then a river, a great lake, and then it has to flow to the ocean." These points left Emily with a question: "But where does the water from the ocean go? What is its destination?" Mr. Baek nodded. "That is a great question, Emily. Where does the water from the ocean go? Does it just keep getting fuller and fuller?" asked Mr. Baek. "No, it evaporates," contributed Kobe. "That isn't what I meant," clarified Emily. "I thought that water in a watershed had to go somewhere; the water in the stream out back eventually goes to a river. If the ocean has a watershed, doesn't the water have to go somewhere?" Mr. Baek now saw where Emily was confused, and wondered if anyone else could help Emily clarify her confusion about the system output. "Can anyone answer Emily's question?" he asked the class. Samara responded, "You're right, we have been naming our watersheds based on where it goes. So the watershed for the ocean would be all of the land that has water that flows to the ocean." This explanation helped Samantha see where she had gotten confused, and she thanked Samara for the explanation.

In this example, Mr. Baek saw that there seemed to be some inconsistency in how the class was using their newly defined concepts, so he posed a less straightforward example for the students to explain, which allowed Samantha to ask a question about the outputs of the flow of water. This process of defining and then using the concept was repeated for the different aspects of the watershed, including the inputs, interactions, and outputs, in addition to the boundary examples provided here. By supporting students to develop knowledge of the system components through using them, students understood how they were applied. In the last lesson of the unit, students were given the concept of a system and the components that made up a system (inputs, outputs, boundary, subsystems, and interactions). Students were then asked to explain why a watershed was or was not a system.

Continued

Classroom Snapshot 7.1 *(continued)*

In the next unit, students were eager to apply their knowledge of systems to the new concepts and almost immediately started asking what in the classroom was a system. In response, Mr. Baek always asked them to explain what the different components of the system would be. When they began learning about human body systems, students were ready to apply the systems aspects to consider the parts and their relationships to one another.

Key Characteristics of Using Systems as a Lens

Many of the science concepts students learn about can be productively considered from the systems lens. We often use the term *system* in science instruction without really supporting students to know how and why something is classified as a system or how the idea of a system helps us understand how it works.

There are many ways that systems can be explicitly incorporated into science instruction. The important part is to be explicit with students about what the system is and why is it classified as such. The goal is for students to ultimately be able to use the lens of systems to solve problems or answer questions about how science ideas or phenomena work. Explicit inclusion of systems does not mean quizzing students on what the parts are, but it might mean asking them why a component plays a particular role in the system. For example, it is more important for students to be able to apply the concept of the system boundary in response to the question "How do we know where a new system starts?" than it is for students to be able to define the boundary in response to the question "What is the name of the edge of a system?" An important aspect of using the systems lens is being able to apply it—not just know the "terms." After students have learned about systems as a concept, it can be used to support them to ask questions of their own understanding. If students have trouble describing an element of a system, then they might have a question in need of additional investigation. We want students to be able to use the idea of a system to make sense of science.

In order for students to be able to use systems in new science contexts (both across disciplines and across phenomena), they need to first learn why it is important and useful in the context of one particular science idea. After that, they can begin to try and use the lens to explain another science idea or phenomenon (Goldstone and Wilensky 2008). In thinking about how this might work, one could support students to understand and use the aspects of the systems lens (inputs, outputs, boundary, subsystems, and interactions within systems including feedback loops) within the context of one science idea and then support students to see how these aspects might be useful to explain multiple science ideas or phenomena (Fick 2018). The way the aspects of the lens operate in different disciplines might change; for example, how scientists define the boundary of a system following the movement of energy and matter (a boundary defined by the investigator) might look different than the boundary of an aquarium (an actual physical boundary), but the aspects can help students ask questions about what they do not understand.

Promising Instructional Strategies for Systems and System Models

One challenge to supporting the use of the systems lens is that students are often not aware of how or why we use this lens to make sense of phenomena. The following instructional strategies have been shown to help students understand how and why to use the lens for making sense of science phenomena and designing solutions.

Instructional Strategy 1: Make Systems an Explicit Part of Instruction

Explicit discussion about systems means that its framing can become apparent to the students during instruction, and students are supported to see how and why systems help them make sense of the phenomenon. Recall the watershed example from Classroom Snapshot 7.1 (p. 138). Helping students understand how and why a boundary exists for a watershed can enable them to quickly narrow down the possible locations for a source of water pollution by first recognizing that the source must exist within the watershed or that the pollutant was contributed by an input (such as falling rain). By explicitly supporting a student to define the watershed boundary as an area of higher elevation that divides the flow of water, the student has set limits to where the water could have entered the system. This is also seen in Classroom Snapshot 7.3 (p. 148). Here, students define the boundaries of their system using rope on the floor and then must describe those boundaries to peers. Understanding systems involves two ways of knowing about their use. Students need to know (1) how and why the systems lens is important to learning about phenomena and (2) how and why we are defining elements of the system within that lens. Through these two ways of knowing about systems, students can begin to understand how they might apply the lens to other phenomena. There are many ways to make the systems lens explicit to students, and we describe a few of those ways in the next sections.

Instructional Strategy 2: Use Common Language to Refer to the Systems Aspects

During instruction, it is important to use consistent language around systems because it can help students in using the systems lens for understanding phenomena and designing solutions across content topics. Furthermore, using consistent language can help English language learners (ELLs) to engage in a sensemaking process in science by reducing their struggle with terminology. Consistent language that includes the previously mentioned aspects (e.g., boundary, inputs and outputs, and processes) can support students in having multiple opportunities to engage with the concepts and build understanding through use.

While this finding may seem obvious, research has shown that implicitly including the systems lens does not support most students to independently use those aspects to explain phenomena (Fick et al. 2019). Historically, students have not been taught to reference systems and their defining characteristics explicitly. As a result, students are unable to apply the systems lens to support their own sensemaking. For example, students studying how total rainfall affects absorption and runoff (a unit discussed in Classroom Snapshot 7.2 on p. 144) frequently do not include in their conceptual models the amount of rainfall in the area and its relationship to where the water is going (i.e., the inputs and outputs of the system). The primary goal of the learning unit is to attend to differences in the amount of rainfall (input) as related to the amount of runoff and water absorbed (outputs), which is determined by how absorbent the surface is (interactions); therefore, failing to attend to these relationships in the model makes it unclear what the students are learning. A small proportion of students do represent equal inputs and outputs in their representations. In this particular example, attending to this relationship is critical to the learning goal of understanding the relationships among the different components in order to be able to design solutions.

Providing all students with explicit instruction about how and when to use the systems aspects helps create more equitable opportunities for science learning. In Classroom Snapshot 7.1 (p. 138), we see an example of students being supported in learning about the systems aspects in the context of a particular science phenomenon. Then the students use those same terms in another unit. This illustrates the transportability of the systems lens terminology (i.e., the systems aspects) across science ideas, units, and disciplines. Using consistent common language to refer to the systems aspects can help students make connections across science ideas and disciplines for themselves.

Instructional Strategy 3: Have Students Define the Systems Aspects

As a part of learning how to use the CCC of systems and system models, it is important to (1) frame the learning within the context of a particular scientific phenomenon and (2) help students understand how and why the aspects of the system are being defined and how they are situated within the phenomenon being studied. For example, when learning about photosynthesis, you might ask students to think about the following: What is the boundary of the system we are studying? Why is that the boundary? Could we choose a different boundary? How would a different boundary change our understanding of the process? At the elementary level, photosynthesis might be considered in terms of what goes into and out of a leaf with a broad description of the interactions taking place within photosynthesis. In this case, the system boundary would be the leaf. As students add complexity to their understanding of photosynthesis, they might examine what goes on within a cell, considering the role of chloroplasts in this work. In this case, as students' understanding increases in complexity, the boundary might become smaller and the elements of the system being studied might become smaller, as well—first, leaf; then, cell; then, chloroplast. These decisions are important factors for students to understand as they learn how and why to use the aspects.

It is possible to have students define the aspects through identifying a need for them. In the curriculum for Classroom Snapshot 7.1 (p. 138), students spend a class period using topographic maps to predict where the water would flow based on the topography of the landscape (Fick and Baek 2017). They then draw lines in areas where the water flow faces two different directions. These lines form the basis for a watershed boundary, which is precisely how watershed boundaries are defined (areas of relatively higher elevation that divide the flow of water toward two different bodies of water). Engaging in the activity to develop the boundaries themselves, students are supported to see why the system boundary is an important and useful concept for sensemaking. This also allows students to understand which areas of land can be eliminated from consideration when attempting to identify sources of pollution. In Classroom Snapshot 7.3 (148), students define the boundaries of the system and the interactions that occur between the boundaries when they describe the energy transformations that take place. In Classroom Snapshot 7.4 (p. 153), students work to describe and explain the relationships between the components of the system. In these cases, students are being explicit about what the relationships are, and these interactions and processes serve as an aspect of the system.

Moving from concrete to abstract ideas can be difficult for students. At times, a boundary is being defined based on something under investigation that does not have a physical barrier, such as what happens within a specific property of land. There can be a political boundary serving as a separation between one area and another, but rarely is that separation marked by an impenetrable physical barrier, such as a wall. In other instances, the overall system is more concrete, such as with the aquarium example in Classroom Snapshot 7.4 (p. 153), where the boundary is a glass container and the components are more easily defined. In Classroom Snapshot 7.3 (p. 148), students model the process of energy flow, transfer, and transformation, and they must make decisions about where the boundaries in the system are placed by physically adding rope lines to the floor. This type of activity gives students opportunities to make clearer decisions about how the systems aspects relate to the phenomenon.

CLASSROOM SNAPSHOT 7.2

Using Systems to Bridge Science and Engineering Practices[2]

As part of an elementary Earth science and engineering unit, students in the class have been doing science investigations to understand how absorption and runoff are related to total rainfall in order to engineer solutions to a water runoff problem on their school campus. To do these investigations, the students have been pouring water on different surface materials held in a container and measuring the amount of water that stays on the surface after a set amount of time. They are then able to calculate how much water is absorbed by subtracting the amount of water on top of the surface from the total amount of water added. In order to engineer solutions to the water runoff problem, the students need to understand the relationship between the amount of rainfall, how much of that water is absorbed, and the amount of runoff. As they progress through the unit, students work on revising their conceptual models to show their changing understanding. During this process of revision, students are given the opportunity to reflect on how their understanding has changed and how they want to be able to represent that.

To support the students, the teacher, Ms. Small, is helping them attend to the parts of their model for which they have a changing understanding. Ms. Small says, "As you're working on your models, think about what we learned in this activity and in the last activities—how has your understanding changed? What questions do you still have?" DeRay responds, "I changed the amount of water that is being absorbed. Before I only had a little bit of water going into the ground, now I have a lot of water." Ms. Small follows up by asking, "How much is a lot of water? Do you have water going somewhere else, as well? How would you compare those amounts?" DeRay responds, "I have a lot of water going into the ground—more than is going to runoff." Ms. Small further prompts DeRay, saying, "And how does that compare to the amount of rainfall?" DeRay says, "The amount of water I have for runoff and absorption are equal to the amount of rainfall." Opening up the discussion to the broader class, Ms. Small asks, "That's an interesting point; why would that be important?" Shawna explains, "All of the water has to go somewhere, so the water that falls as rain either becomes runoff or is absorbed." Ms. Small clarifies, "So let's go back to our understanding of the system we're using here—how does what you learned relate to a system?" Djuna thinks for a moment and then responds, "I think

2. This Classroom Snapshot is inspired by work from the National Science Foundation–funded SPICE project (Award #1742195). It has been condensed for clarity.

Continued

Classroom Snapshot 7.2 *(continued)*

we learned about a couple parts. I learned about where the water goes, and that is an output from the system. But I also learned about how the surface material, whether it is grass or concrete or whatever, changes how much water goes there. That is how the water interacts with the surface."

Ms. Small knows that students are going to need to use the relationship between inputs and outputs as they move to the next part of the unit, which involves programming a computer model to test different design solutions. So she wants the students to explain this relationship mathematically. They have been using the concepts to do calculations over the last couple of days, but they have not yet used an equation to explain it. Therefore, she makes a move to support students to develop an equation. "I know that over the past couple days we have been doing some calculations in order to be able to find out how much water is absorbed. Can we express that relationship mathematically?" she asks. Michael says, "Can I write it on the board?" He then moves to the front of the room and writes out the following:

Amount of Water Absorbed = Total Rainfall – Amount of Water on Surface

Ms. Small asks, "How would you say that as a sentence?" Michael responds, "The amount of water absorbed is the total amount of water minus the amount of water still on the surface." Ms. Small asks, "Did anyone think of another way to say the same thing?" José says, "I would have said, 'Total rain is water absorbed plus water on top.'" Ms. Small responds, "Can you write that as an equation?" José moves to the board and writes:

Total Rain = Water Absorbed + Water on Surface

The students have not yet covered the use of variables with algebraic expressions in their math classes, but going over the different ways of thinking about how to answer the question here leads them to more advanced mathematical expressions than they would have gotten through the math alone. Both of these equations will be useful for the next part of the unit. When the students need to program the computer model, they will employ the equations to test different surface materials.

When Ms. Small goes back later and reviews the models that students created showing the absorption process, she finds that some students have included different surface materials in their models to show the different ways that water is absorbed and runs off as a result. She also finds that some students have labeled parts of their models as inputs and outputs to help represent that the water coming in as rain is equal to the amount of water leaving the system either through absorption or runoff. These ways of representing the system as both a conceptual model

Continued

Classroom Snapshot 7.2 *(continued)*

and a mathematical model will help the students as they transition from the scientific investigations to the computational modeling and, finally, to engineering their design solutions to reduce the runoff in their school yard.

Instructional Strategy 4: Use Systems to Make Connections Across Curricular Activities, Units, and Disciplines

One role that systems can play is to help students make connections. These connections can occur on multiple scales including across activities within the same unit, across units, and across disciplines. For example, as students move between different modeling activities in the classroom (e.g., computational, mathematical, physical, and conceptual models), the aspects of the system can help them keep track of important parts of the DCI, which helps students make connections across activities. In Classroom Snapshot 7.2 (p. 144), we see students taking their understanding from a conceptual model to a mathematical model in support of their transition from scientific investigations to computational modeling. Focusing on the inputs and outputs of the system helps students keep track of where the water is coming from and going to. In these examples, using systems as a lens to frame the DCI supports the students and the teacher in making connections across curricular activities, including shifts in the use of the science and engineering practices (SEPs). The systems lens highlights the important parts of how the phenomenon works so that those parts can remain the focus as students move through the unit. Similarly, the lens of a system can be applied to students' understanding to reveal aspects that are less clear and in need of additional information. When doing this, using the aspects of the systems lens can make the implicit connections and framing more explicit and available to students.

Research and observations in classrooms have shown evidence that students are able to apply the systems lens in new situations (Goldstone and Wilensky 2008). (This is illustrated in Classroom Snapshot 7.1 [p. 138] in which students make connections between watersheds and human body systems.) Researchers have shown that, in order to do this, students need to learn about the framing systems aspects (e.g., inputs, outputs, subsystems, boundary, interactions) within one context, being taught both what the concept is and how it is useful to make sense of the science. Then understanding can be taken to the next learning experience, whether it is a different concept within the same science discipline or a different science discipline. Once students understand the conceptual aspects of a system, they are capable of making connections to other topics, and this can be used to organize how to approach the material. When we teach students about using systems as a lens, it is possible that students can learn about systems in one particular context and understand how one phenomenon is a system without understanding how the broader principles are applicable across contexts and ideas. This is why students need explicit support to understand how systems help them explain a science idea (see Instructional Strategy 1: Make Systems an Explicit Part of Instruction on p. 141).

Supporting students to use the framing of systems to bridge across disciplines can occur at many grade levels with varying amounts of depth. In one middle school example, we created a frame for how to describe and define the components of the system through learning experiences designed to support each of the intersections between the DCI being addressed and the CCC of systems and system models (Fick, Arias, and Baek 2017). Classroom Snapshot 7.1 (p. 138) shows an example activity from this curricular unit that helps students build understanding of the systems concepts and apply them in the context of watersheds. Following this unit, the teacher, Mr. Baek, saw students independently use the systems lens when moving to their next unit on the human body, which led him to reconsider how he framed the unit and the way he talked about the body systems with students. The first time this happened, he did not expect them to make that connection, but the next time he started planning for it. Once students had gained an understanding of how to use systems as a lens on the science idea by defining the system components for that concept and had seen how the lens helped them ask questions and design investigations, they used that understanding in a new context. In their human body unit, Mr. Baek modified the lessons to include opportunities for students to construct models of the systems that make up the human body, thereby supporting students' understanding of subsystems and how they work together. For example, through identifying the inputs, outputs, boundaries, nested systems, and interactions within the system of digestive processes, students could explain the function of the system and how the components of the system work together to complete that process. Once students moved to another body subsystem, they could make connections between the subsystems, which supported the students in seeing where an output of one becomes the input to another.

Instructional Strategy 5: Use Systems to Ask Questions

In addition to the frequently used modeling approach, systems can also support the use of SEPs by introducing an opportunity to ask questions. The systems aspects can be a lens for examining students' ideas and representations. For example, when learning about a new system, a teacher might have students write or draw their current understanding about how the concept works or make a prediction about what is going to happen. These initial ideas can be used to identify what aspects of the system students are struggling to understand. A teacher might directly ask students questions about their work, or the teacher may have students ask it of themselves to frame their thinking. Such questions might include, "What is the boundary for this system? Why is it there?" Alternatively, students might realize they were told what the boundary is but do not understand why it is the boundary. They might learn about the inputs and outputs but have questions about how the inputs are converted into the new materials that become the outputs. Using systems in conjunction with asking questions might help students understand what aspects of the phenomenon or science idea are in need of investigation or additional research. This practice is highlighted in Classroom Snapshot 7.3, when the students are asking questions of each other based on aspects of the system's framing. In Classroom Snapshot 7.1 (p. 138), one student asks a question about the output of the system in order to clarify how the group is defining the system. Both students and teachers can use the aspects of systems to implicitly and explicitly ask about how a science concept is examined from a systems perspective.

CLASSROOM SNAPSHOT 7.3
Systems as a Framework for Asking Questions[3]

Partway through a unit on force and motion, Ms. Lopez begins her eleventh-grade class by explaining that today they are going to do an interactive modeling activity called "Energy Theater" (Daane, Wells, and Scherr 2014). She breaks the class in half and pushes all the tables and chairs to the side of the room. Then she writes the following rules of "Energy Theater" on the board:

1. Each person is only one unit of energy at a time.
2. Each person is only one form of energy at a time.
3. Each person indicates his or her form of energy with a hand signal.
4. Regions of the floor correspond to objects in the scenario.
5. People move from one region to another as energy is transferred, and they change hand signals.
6. The number of people in a region corresponds to the amount of energy.

Ms. Lopez assigns the students to use these rules and act out the following scenario in groups: Someone gives a block at the top of a ramp an initial push, and the block then slides down a ramp. Although the eleventh graders have previously discussed this and other similar scenarios, "Energy Theater" is a new activity. Ms. Lopez is curious how her students will think about the scenario and how "Energy Theater" will highlight nuances in their understanding.

3. *Note:* This Classroom Snapshot is a condensed depiction of classroom interactions observed during research, teaching, and professional-development activities that were conducted as part of the University of Maine Physical Sciences Curriculum Partnership (NSF #0962805).

Continued

Classroom Snapshot 7.3 *(continued)*

Ms. Lopez hands each group pieces of yellow rope that can be used to indicate regions of the floor that correspond to objects in the system (e.g., block, ramp, and hand), and the students begin engaging with the scenario. The classroom becomes very loud with everyone talking at once, and it is hard for Ms. Lopez to follow all the bubbling ideas. She steps back and watches the students as they begin placing the rope in circles on the floor and move in and out of the rope. After 15 minutes, she tells everyone that it is time to show their peers their scenario and after that there will be an opportunity to ask questions.

The first group to present places three circular ropes on the floor to represent the hand, block, and ramp. All of the students in the group start by squeezing into one of the circular ropes. Kim says, "We're starting in the hand that pushes the block and making a *C*-shaped hand signal because we are all chemical energy." Then, as two-thirds of the students move into the next rope that signifies the block on the ramp, she continues explaining, "The chemical energy turns into potential energy." The students in this rope now make a *P*-shaped hand signal with both hands. Continuing to narrate, Kim says, "As the block is pushed down the ramp, the potential energy turns into kinetic energy." As she says this, some of the *P*-shaped hand signals turn into *K*-shaped hand signals. Then, all at once, a few of the *K*-shaped hand signals turn into *T*-shaped hand signals, and those students walk together from the rope that signifies the block to the third rope that signifies the ramp. Meanwhile, Kim says, "Some of the kinetic energy turns into thermal energy that heats up the ramp."

At this point, Ms. Lopez says, "Thank you for presenting your scenario. Now it is time to ask questions. Who has a question?" José chimes in, "Why did the *K* become *T* at the end? Doesn't the ramp heat up slowly?" Ana, one of the presenting students, says, "Yeah, I guess you're right, we didn't think about kinetic energy becoming thermal energy throughout."

Then Kaylie raises her hand. "I have another question: When the *K*s became *T*s, then the block would be slowing down because it has less kinetic energy. But last week in class we did a problem where the kinetic energy is maximum at the bottom of a ramp." Ana again jumps to answer, "Um. Oops. Yeah, that makes sense. It's all *K* at the end. All of the chemical energy has turned into kinetic energy." Then Ms. Lopez says, "Thank you for those important questions. Now let's switch and have the other group present."

The second group stands up and adds a fourth rope to the three circular ropes on the floor. José explains that the fourth rope represents the surrounding air. Then he says, "We're beginning with half of us in the hand signifying chemical energy and the other half in the block signifying potential energy because there is potential

Continued

Classroom Snapshot 7.3 *(continued)*

energy in the block to begin with." Then the students start moving, with all of the students who signify chemical energy walking into the rope that signifies the block, as they change their hand signals to a *K*. Meanwhile, some of the other students who originally signified the block's potential energy start changing their hand signals to a *K* while remaining in the block. At the same time, other students in the block change their hand signals from a *P* to a *T*. Some of the *T*s then move to the rope that signifies the ramp while other *T*s move to the rope that signifies the air.

After they conclude the presentation, Kim, one of the observing students asks, "How come there was no chemical energy after the block started moving? I thought the block gets pushed all the way down the ramp." One of the students who just presented answers her by saying, "We think that after it gets moving, kinetic energy is more important. The block starts sliding down the ramp." Kim responds, "But the hand is still there?" The presenting student explains, "Yeah, but the hand is only part of the setup. Once the block is moving, it's not part of our scenario."

Then Ms. Lopez asks, "Does anyone have a different question?" In response, Jayden asks, "How come some *T*s go into the air? I might feel the ramp heating up, but the room temperature wouldn't change, right? I mean the thermostat wouldn't change." José then says, "Yeah, the room is too big. But I think if you were able to measure the temperature of the air right next to the block, it would increase."

Now Jules raises his hand and says, "In your scenario, you began with half of the people being chemical energy in the hand and half being potential energy in the block. But remember my group was different—we all started as chemical. Which one is right? And does it matter if we both ended with kinetic and thermal energy?"

At this point, Ms. Lopez stops the class and says, "Jules, we'll answer your question in the following class. But first I want everyone to take a blank piece of paper and draw out a model of this scenario." As students are walking out the door, Ms. Lopez collects the models. (A sample of these models appears in Figure 7.1.)

Figure 7.1. Sample student models for the "Energy Theater" activity

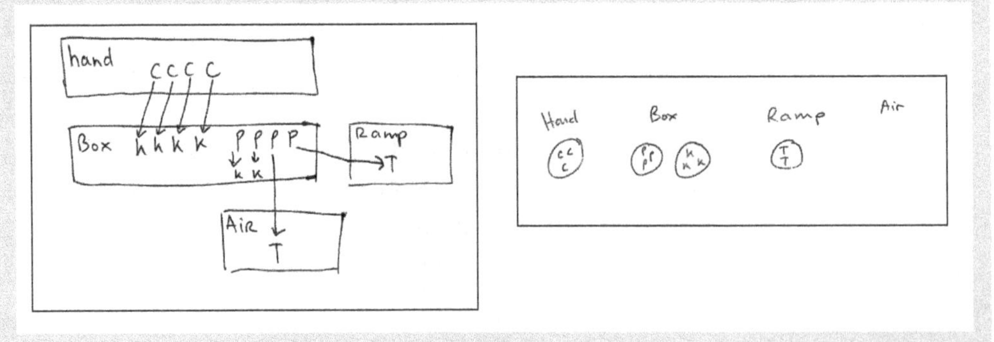

Continued

Classroom Snapshot 7.3 *(continued)*

From looking at all the drawn models, Ms. Lopez is pleased to find that many students included in their illustrations potential energy when the block is at the top of the ramp and kinetic energy when it is at the bottom, which is an important idea to cover in this unit. However, she wonders how the students are thinking about the role of thermal energy. Including thermal energy in their scenarios suggests friction, but previously in class they had focused on ideal scenarios with no friction. She decides that next week in class she will spend more time comparing ideal versus real-world scenarios.

Instructional Strategy 6: Use Systems to Support Modeling

Modeling systems is one way to help students be explicit about what is in the system and the relationships within that system that are important. Models can be used to make predictions about the abundance of inputs, outputs, and components or about interactions that might occur between components within a system. Models take many forms, including drawings showing processes and relationships, computer simulations, mathematical models, or physical replicas (Passmore, Schwarz, and Mankowski 2017). As models are created, decisions need to be made about what should and should not be included. Since it is impossible to include all aspects of a system, models are a simplification of a phenomenon, as shown in Classroom Snapshots 7.1 (p. 138) and 7.2 (p. 144). When creating models, it is important to be explicit about how the components you are including relate to the systems aspects (e.g., labeling inputs and outputs within a conceptual model) and how you are defining the boundary of the system. Using systems aspects to explain what is being modeled will support students in seeing how systems help to develop scientific understanding.

How Is the CCC of Systems and System Models Different From the SEP of Developing and Using Models?

One might say, "I have my students develop and use models all the time. That means they are learning about systems and system models, right?" The science and engineering practice of modeling involves representing an understanding of a system. Making decisions about what is important to represent and what is hidden in the representation is inherent within the practice of modeling. Because making such decisions will likely draw on discussion of systems aspects, covering this SEP involves the use of the CCC of systems and system models.

Because students are in general using the lens of a system *implicitly* when modeling, the fact that they are representing a system is not always apparent to them. Decisions about the boundary of the system, for example, are often implicitly made, meaning that a student is representing a process and not actively using the systems lens. Using the systems lens is an important tool to help students understand what they are modeling and why—and how the model can be used to better understand the system. By having the system remain implicit, several challenges might occur. For instance, some students might become aware of the idea of systems, whereas others do not. Students might be talking about different systems and are unable to reconcile the different frames for their discussion. Additionally, for teachers, it could be difficult to understand students' thinking if the ideas are not directly comparable. Using the systems lens is not always a part of modeling, but they make a great combination, allowing for clarity of ideas and access to one another's thinking within the classroom.

Using Systems With Conceptual Models to Highlight Student Thinking

Taking a systems lens is particularly important in classrooms where conceptual models—like the one shown in Figure 7.2—can be a tool for understanding students' thinking.

Figure 7.2. An example of a conceptual model

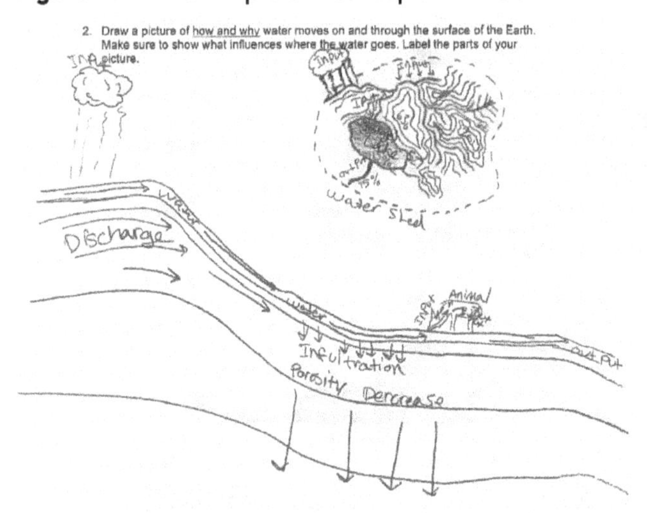

This conceptual model was developed by a middle school student during the curricular unit described in Classroom Snapshot 7.1 (p. 138).

Conceptual models, which are representations of a student's current understanding, provide the student with opportunities to show this understanding without requiring complex writing. In particular, this strategy supports students who are just beginning to write or are reluctant writers to share their thinking through another mode.

Conceptual modeling tasks given to students at the beginning of a lesson or unit before they have learned about a concept can help make students' prior knowledge visible and can be used to help students make predictions about what might happen in a given scenario. A systems-focused addition to this practice helps make the components that are present and gaps in student understanding apparent. One way to add the systems lens would be to ask

- "What is going in?"
- "Where is it going?"
- "What is coming out?"

These questions about the inputs and outputs of what is happening in the students' model can help students realize that matter and energy do not disappear; they have to go somewhere, so the students need to specify where they think they are going. Similarly, we could ask, "What happens to the matter to cause it to change forms in your system?" to prompt students to attend to the processes that are going on in their system. In Figure 7.2, we might notice that the upper part of the drawing shows elevation but that the elevation does not have an apparent relationship to the location of the watershed boundary. We could ask the student who made the model, "Why is the boundary where it is? What decides where the boundary is located?" These questions prompt students to consider components of their model that they might not have considered, but it is also important to help the students connect these questions back to the systems framing so they can begin to think about these aspects of a system for themselves in the future. These kinds of prompting questions have always been a part of instruction, but we often overlook the part of the practice that helps students connect it to why the systems framing is important and how it helps them make sense of the phenomenon.

Some popular computational modeling programs (e.g., C2STEM, NetLogo, StarLogo Nova) help students both use models to define the system and make predictions about what will happen within the system by enabling the students to modify the code. Such programs allow students to observe how changes to the system can influence or alter system states. This process of programming how components of a system interact could, with explicit support, allow students the opportunity to define the system under study, giving them a clearer understanding of how and why to use the systems lens. As students build up to these activities within the class lessons, it is also helpful to describe these components and their interactions from a mathematical perspective so students can more easily translate their understanding to the code. Using the systems lens helps students transition from one representation of a science concept to developing and using other representations.

CLASSROOM SNAPSHOT 7.4
Using Systems to Connect the Micro and the Macro[4]

In the Systems and Cycles project, middle school students learn about aquaria as model systems during a life sciences unit (Hmelo-Silver et al. 2017). As Ms. Seashore starts the unit, she asks her students to draw what they think is in an aquatic system. They have just gotten a 20-gallon aquarium for the classroom, and she wants her students to think about how to set it up. In one small group, Talia says, "Well of course

4. *Note:* This Classroom Snapshot is based on classroom interactions observed as part of research and teaching in the IES-funded Systems and Cycles project (Hmelo-Silver et al. 2017).

Continued

Classroom Snapshot 7.4 *(continued)*

we need fish in there." Other students in the group chime in with suggestions to add rocks, fish of different sizes, plants, and sand.

The model the students initially draw is just a picture of macroscopic elements in a fish tank. Ms. Seashore then asks the students what else they would need and how many fish could even fit in the tank they are setting up. Max jumps in and says, "How would we know that?" Keisha says, "We could just look it up on the internet." Talia wonders if they can just put as many fish as they want in the aquarium. Ms. Seashore asks them what other ways they could think about what they need to maintain a healthy aquarium system.

After her students brainstorm for a while, Ms. Seashore says, "Sometimes scientists use computer programs to simulate what they are trying to study. It allows us to change the inputs to the system and quickly see how that will affect other components." She then shows the students a computer program called the Fishspawn simulation (Hmelo-Silver et al. 2015). Talia says, "I see—we can change the number of fish in the aquarium. It also shows a filter and food. I guess we didn't think of those things." Max asks, "Can we all try the simulation and see what happens?" Ms. Seashore is excited to see that the simulator has stimulated some student-generated questions and agrees, making sure that there is a computer for each group in her class. She reminds students to record the initial conditions for the simulation (i.e., number of fish, how likely the fish are to reproduce, amount of food, and filter flow) and their observations as the simulation is running.

One student group includes Keisha, Max, and Talia. Keisha asks Max and Talia how they think they should set up their simulation. Talia wants to see how many fish they can put in there. Keisha suggests only putting in a few. Max asks, "Why does it matter? If we give them enough food, it will all be fine." Talia says, "Let's start with a number in the middle." Max and Keisha agree. The slider for the number of fish ranges from 0 to 30, so they set it to 15. Keisha says, "Why don't we start with everything in the middle?" They agree and Max records the initial conditions. The first time they run the simulation, there is a lot to observe—fish swimming around, yellow squares, graphs with water quality and fish population, boxes with numbers in them and causes of death for fish. After a few minutes, Max pauses the simulation and says, "That is really cool to see all the babies when the fish reproduce." Keisha says, "On the graph, it looks like when the fish population goes up, the water quality goes down. I wonder why that is happening?" Talia then says, "Wow, after the water quality goes down, the fish population decreases a bit. That is interesting, but I am not sure I understand it." Max looks at the little boxes and says, "There are a few starvation deaths and a lot of deaths because of water quality. So, I guess they have enough food. Let's run it again with more fish to see what happens." Keisha

Continued

Classroom Snapshot 7.4 *(continued)*

responds, "First, let me write down what we saw in the first run. Then I'll move the slider to have the most fish. But I won't change anything else." Talia says, "I will do the recording for this run." Keisha makes the changes and runs the simulation. Max says, "This is happening just like before—whenever there are a lot of fish spawned, the water quality goes down and then fish die. I wonder why that is." Talia says, "Some of them are starving so it might be that they needed more food." Keisha responds, "But lots more of them died because of the change in water quality. I still don't understand why that happens."

Ms. Seashore starts a class discussion, and students from all the groups have similar observations about the patterns in relationships between the number of fish, the quality of the water, and the fish deaths. They draw a class model showing the relationship between the components in the system, including water quality and fish population, but they can't agree on why the fish population increase affects water quality. Ms. Seashore then says, "Maybe we need to use another simulation to observe what we can't see with our eyes." She has all the groups open up a micro-level simulation that shows how fish excretion affects water quality. As the groups start the micro-level simulation, they notice something: They initially just see red dots but no fish. After a class discussion, they realize that only microscopic aspects of the systems are shown in this simulation.

For the first run of the simulation, Talia says to her group, "Let's start with just a few fish in the system." Keisha and Max agree. This time, Keisha records the settings. At first, they just see the red dots—and lots of them. They also see a graph that shows a red line rising, and the legend on the graph says it is ammonia. Keisha asks, "Where is that ammonia coming from? How is it related to the fish?" Ms. Seashore suggests that they try looking it up on the internet or in some of the textbooks that are available as classroom resources. Max looks in some of the books in the class while the others look online. With some surprise, Talia says, "It is fish pee … this site from an aquarium society says that fish excrete ammonia." Keisha says, "I found the same thing at a Marine Science Research Center website!" As the simulation runs, they start to see patches of different colors. In the blue patches, the red dots turn to white, which the graph legend says is nitrite, and the amount of ammonia starts to drop. After a little while, the white dots turn yellow after they go through a purple patch. Keisha looks in one of the classroom books and says, "I think that the purple squares are bacteria, and they are turning ammonia to nitrate, and the nitrate doesn't really hurt the fish but the ammonia does." Max says, "The ammonia doesn't seem like a big deal, then, if the bacteria convert it to something harmless." Talia says, "It took some time for the bacteria to start doing that. It didn't happen right away." Keisha adds, "I wonder if it would be faster if there were more fish—let's try that."

Continued

Classroom Snapshot 7.4 *(continued)*

And so they do. Shortly after starting the simulation with the maximum number of fish, they see lots of red ammonia dots; and before the bacteria kick in, they get a message: "All the fish have died." Then the simulation stops.

As class is ending for the day, Ms. Seashore returns to the class model to see if they can add more information to explain why the water quality got worse when the fish population was bigger. Talia offers part of the explanation, saying, "More fish make more fish pee." Keisha jumps in saying, "That is ammonia, and it is poisonous to the fish. If there are only a certain number of fish, the system can convert the ammonia into nitrates, which I read were plant food." Max then notes, "Yes, the plant food is nitrate. Bacteria in the water transform the ammonia into nitrate. But it takes a while for the bacteria to grow and complete the transformation process. So when we make our fish tank, we need to add fish slowly." And with this, the group has connected what they saw at a macro-level to what was going on below the surface.

Using Simulations to Support Students to Observe System Interactions

Simulations are important for highlighting or flagging aspects of systems that are difficult to see in real time. These may involve scale, interactions and mechanisms, or processes that occur over time. Professionally developed computer simulations, and in particular ones that allow the students to manipulate inputs, outputs, and/or interactions (e.g., PhET Interactive Simulations, C2STEM, NetLogo and StarLogo, and Concord Consortium), provide students with opportunities to run virtual experiments focused on a system. Depending on the system under study, a frequent challenge for understanding systems is that much of what happens in many systems is not visible because it is either too large or too small to observe. For example, when studying chemical and thermal energy in systems, it would be challenging to observe the movement of atoms. Nevertheless, simulations can show students representations of what is happening on a microscale. Simulations help students keep track of the inputs and outputs of a system and the interactions that occur within the system.

Systems work in ways that may be at odds with students' understanding, and simulations provide ways to challenge this prior understanding. People interpret what they see through the context of what they already know (NRC 2000). Elementary students are often asked to observe interactions that make a system, despite challenges in making those observations. For example, in a lesson about aquatic ecosystems, students might be asked to observe interactions in an aquarium, which might result in the observation of two snails next to each other. A student may assume the snails are "friends" and "hanging out together" or possibly eating each other when what the student is actually observing is proximity (Arias and Davis 2016). In this case, there are two problems: first, that the student is making inferences rather than observations, and in this case, the inferences include an anthropomorphic intention; second, that the student is trying to observe an interaction (feeding) that is occurring on a difficult-to-observe scale. One way to help students make scientific observations of microscale interactions would be through the use of an aquarium simulation that allows students to observe patterns in data to draw conclusions based on evidence. In an aquarium simulation, with appropriate scaffolding, students can observe how different inputs and outputs affect the overall balance within the system (see Classroom Snapshot 7.4 on p. 153).

Simulations also allow for students to test their ideas and manipulate inputs and outputs that would take too much time to do as an experiment. With simulations, students have the ability to quickly change inputs to the system and observe the effect on outputs. The inputs of a simulated aquarium system might include the addition of fish food, the neutralization of the tank water, and the addition of oxygen to the water (via aeration)—all of which are important inputs that help make the system stable. By changing these inputs to manipulate the system, students might explore the effects of, for example, putting too much or too little food into the system over time. In this setting, teacher orchestration needs to support students as they reflect on what they observe, practice how to record it systematically, and reflect on what they learned about the system from using the simulation. Simulations in physics also allow students to test their ideas as they manipulate the simulations. For example, the PhET Battery-Resistor Circuit program highlights the flow of electrons (the systems aspect of energy flows) within a simulation. With scaffolding, students can test their ideas about how voltage or resistance affects other aspects of the system, including temperature, as they manipulate the simulation inputs.

Systems frequently operate at different levels, and it is challenging to understand the relationships between them. For example, the interactions between organisms in an ecosystem drive the patterns seen in the ecosystem overall. Much of what happens in a system is hard to understand because these interactions are dynamic. (This is shown, for example, in the relationship between the fish population and water quality in Classroom Snapshot 7.4.) This suggests that simulations help support learners connect the system boundaries, interactions, and processes; manipulate inputs; and better understand the mechanisms and their resulting effects on outputs. Simulations that capture the interactions in a system can also allow students to see interactions that occur at time and space scales too large for practical classroom use. The PhET simulation My Solar System is an orbital simulator where students set inputs such as initial positions, velocities, and masses of several bodies, and then observe the resulting outputs of orbital paths and interactions. In this simulation, students are scaffolded to make predictions and only change one variable at a time while trying to achieve stability.

In addition, simulations of systems can be used to support the acquisition of important scientific practices, such as aggregating data, controlling variables, changing variables to test hypotheses and to make claims, arguing from evidence, and making sense of patterns through representations like graphs that run in real time alongside the simulation (Yoon et al. 2016). For example, in Classroom Snapshot 7.4, students move the slider representing the number of fish and keep the other inputs the same (i.e., controlling variables) to observe the effect on the output. They then find that increasing the number of fish decreases the water quality (i.e., planning and carrying out investigations and analyzing and interpreting data). Not being able to describe the process that causes this pattern, students have more questions to investigate and use a microscale simulation to observe the changes in chemical concentration (i.e., making sense of patterns).

Simulations are a tool that can provide students with the opportunity to run controlled experiments focused on particular aspects of a system. Simulations are specific to a particular process or interaction, providing students with opportunities to examine the outputs associated with changes in the student-controlled inputs, and therefore cannot be used for every process.

How Might Students Build Understanding of Systems and System Models Over Time?

In this section, we present the *NGSS* description of what students should know about systems across the grade bands (Table 7.1). We also offer an illustrative example with one particular DCI in order to show how the sophistication of students' understanding about the phenomenon builds in relation to the increasingly more sophisticated systems lens.

Illustrating Students' Developing Understanding Through Concrete Examples

Here we present an example from the life sciences—students learning about forest ecosystems across grade levels—to illustrate how the DCI of Interdependent Relationships in Ecosystems (LS2.A) increases in sophistication and uses progressively more complex understandings of the aspects within systems as the grade level increases in the *NGSS*.

Elementary School

NGSS Appendix G specifies that by the end of second grade, students should be expected to "understand that objects and organisms can be described in terms of their parts and that systems in the natural and designed world have parts that work together" (NGSS Lead States 2013, p. 85).

The performance expectation for LS2.A does not include reference to systems, but the text implies relationships between components that exist in a common area. By the end of second grade, students should know that a system contains components that exist within a defined boundary, which all work together to form the system. For example, a forest ecosystem contains plants and animals that work together.

Table 7.1. How the CCC of systems and system models changes across grade levels

Level	Grade Band	Systems and System Models
Elementary	K-2	"In grades K-2, students understand that objects and organisms can be described in terms of their parts and that systems in the natural and designed world have parts that work together."
	3-5	"In grades 3-5, students understand that a system is a group of related parts that make up a whole and can carry out functions its individual parts cannot. They can also describe a system in terms of its components and their interactions."
Middle School	6-8	"In grades 6-8, students understand that systems may interact with other systems; they may have subsystems and be a part of larger complex systems. They can use models to represent systems and their interactions—such as inputs, processes, and outputs—and energy, matter, and information flows within systems. They also learn that models are limited in that they only represent certain aspects of the system under study."
High School	9-12	"In grades 9-12, students investigate or analyze a system by defining its boundaries and initial conditions, as well as its inputs and outputs. They use models (e.g., physical, mathematical, computer models) to simulate the flow of energy, matter, and interactions within and between systems at different scales. They also use models and simulations to predict the behavior of a system and recognize that these predictions have limited precision and reliability due to the assumptions and approximations inherent in the models. They also design systems to do specific tasks."

Source: The *NGSS* Appendix G description of Systems and System Models (NGSS Lead States 2013, p. 85).

NGSS Appendix G specifies that by the end of fifth grade, students should be expected to "understand that a system is a group of related parts that make up a whole and can carry out functions its individual parts cannot. They can also describe a system in terms of its components and their interactions" (NGSS Lead States 2013, p. 85).

Following with the example of the forest ecosystem, by the end of fifth grade, students should be able to develop a model of a food web made up of plants, animals, and decomposers in which the individual components are interdependent. The food web should illustrate the movement of matter within the ecosystem, representing the interdependent relationships between the elements of the system.

Middle School

NGSS Appendix G specifies that by the end of eigth grade, students should be expected to "understand that systems may interact with other systems; they may have subsystems and be a part of larger complex systems. They can use models to represent systems and their interactions—such as inputs, processes, and outputs—and energy, matter, and information flows within systems. They also learn that models are limited in that they only represent certain aspects of the system under study" (NGSS Lead States 2013, p. 85).

By the end of eighth grade, students should know that a forest ecosystem is composed of subsystems that are nested within each other. For example, they should comprehend that trophic level producers and consumers operate within their own microsystem, but how energy flows between them will dictate the overall stability of the ecosystem (e.g., a disease that sweeps through the vegetation [producer] would severely affect the population of animals [consumers] that feed on it). Students should be able to model this interaction and make predictions in terms of how changes to one system will affect the other over time. Those predictions should be based on informed decisions made from data that have been collected by running the model. For example, students could look at population graphs and food webs to identify evidence to support predictions of periods of resource scarcity.

High School

NGSS Appendix G specifies that by the end of 12th grade, students should be expected to "investigate or analyze a system by defining its boundaries and initial conditions, as well as its inputs and outputs. They use models (e.g., physical, mathematical, computer models) to simulate the flow of energy, matter, and interactions within and between systems at different scales. They also use models and simulations to predict the behavior of a system and recognize that these predictions have limited precision and reliability due to the assumptions and approximations inherent in the models. They also design systems to do specific tests" (NGSS Lead States 2013b, p. 85).

By the end of high school, students should be able to use mathematical and computational models with attention to all the systems aspects, including the following: the boundary, initial conditions, inputs and outputs, flows within the system, interactions and processes that make up the system, and the way interactions and processes at different scales explain the patterns seen at the larger scale. For example, students can examine the molecular changes that occur with photosynthesis and cellular respiration in trees and how the cycling of carbon affects the forest ecosystem. At the high school level, students should also learn that models and simulations are not an exact representation of the real world but can be used to help describe and make predictions about the system being investigated.

Funds of Knowledge

The systems lens can also be used to incorporate students' funds of knowledge into classroom learning experiences. This is historically accumulated and culturally developed knowledge and skills that come from lived experiences in the world (Gonzáles, Moll, and Amanti 2005). For instance, in terms of science, a student's and their family's funds of knowledge might include information about farming and animal husbandry. In one example, as described in Gonzáles, Moll, and Amanti (2005), an elementary teacher in Arizona visited students' homes and learned about their funds of knowledge about horses. The students often had chores caring for the horses. One father participated in rodeos, and the family would visit horseracing events and watch videos of such events. They often visited other family members' ranches where there was a variety of other animals. Following that, the teacher designed a cross-disciplinary unit on horses in which students conducted research to investigate questions about horse care, health, behavior, and reproduction. From a systems lens, one can imagine that the students would possibly build on their knowledge about the relationships between different animals on a ranch and between the ways in which horses are cared for and the horseracing events that bring families together.

In one class where the curriculum from Classroom Snapshot 7.2 (p. 144) was taught, many of the students had parents who worked in landscaping, and the students themselves had helped their parents with some of the work informally. In conversations with their parents, the students realized that the way in which surface materials absorb water was missing from their representations of the process and from their computational models. This led to a conversation with the teacher in which the students discussed what was and what was not being included as components of the models. For example, although many of these surface materials in reality required subsurface drainage, this was not an aspect of the computational models. The students wanted to include it but realized they did not know how to represent it using code. Ultimately, the group realized this was a decision they had to make about what parts of the process would be modeled, and drainage was not included. The students' knowledge and experience with this aspect of the phenomenon was valued and encouraged; ultimately, though, it was something that students could only represent in their conceptual models—not in their computational models.

What students might bring to a science learning experience related to systems can change according to who your students are and what experiences they have had; this knowledge may look very different depending on the context. What is important is that we notice what knowledge and skills students have when they begin these learning experiences and value the expertise that our students bring, supporting and building on it with the learning experiences.

Summary

The CCC of systems and system models is an important aspect of making sense of phenomena across science disciplines. Students need explicit support to be able to apply this systems lens as a tool for thinking. There are limitless possibilities for learning activities to support students to engage in using systems as a lens. Here, we have presented a selection of activities that have been found to support students' learning, but it is important to remember that supporting students to use the practice of modeling alone will not engage the lens of systems. Though systems are already very much a part of what we teach, students need explicit support to add this layer to their thinking. Research has shown that given the appropriate support students can make connections across disciplines and contexts using the systems lens (Goldstone and Wilensky 2012). This can only happen through explicit instruction about the use of the systems lens in coordination with science and engineering practices and disciplinary core ideas.

References

Arias, A. M., and E. A. Davis. 2016. Making and recording observations. *Science and Children* 53 (8): 54–60.

Daane, A. R., L. Wells, and R. E. Scherr. 2014. Energy theater. *The Physics Teacher* 52 (5): 291–294.

Fick, S. J. 2018. What does three-dimensional teaching and learning look like?: Examining the potential for crosscutting concepts to support the development of science knowledge. *Science Education* 102 (1): 5–35.

Fick, S. J., A. M. Arias, and J. Baek. 2017. Unit planning using the crosscutting concepts. *Science Scope* 40 (8): 40–45.

Fick, S. J., and J. Baek. 2017. Supporting students to understand watersheds by using multiple models to explore elevation. *Science Scope* 40 (6): 24–32.

Fick, S. J., A. M. McAlister, J. L. Chiu, and K. M. McElhaney. 2019. Analysis of students' system models in an *NGSS*-aligned curriculum unit about urban water runoff. Poster presented at NARST Annual International Conference, Baltimore, MD.

González, N., L. C. Moll, and C. Amanti, eds. 2005. *Funds of knowledge: Theorizing practices in households, communities, and classrooms*. Mahwah, NJ: Lawrence Erlbuam Associates.

Goldstone, R. L., and U. Wilensky. 2008. Promoting transfer by grounding complex systems principles. *Journal of the Learning Sciences* 17 (4): 465–516.

Hmelo-Silver, C. E., R. Jordan, C. Eberbach, and S. Sinha. 2017. Systems learning with a conceptual representation: A quasi-experimental study. *Instructional Science* 45 (1): 53–72.

Hmelo-Silver, C. E., L. Liu, S. Gray, and R. Jordan. 2015. Using representational tools to learn about complex systems: A tale of two classrooms. *Journal of Research in Science Teaching* 52 (1): 6–35.

National Research Council (NRC). 1996. *National science education standards*. Washington, DC: National Academies Press.

National Research Council (NRC). 2005. *How students learn: History, mathematics, and science in the classroom*. Washington, DC: National Academies Press.

National Research Council (NRC). 2009. *Keck futures initiative: Complex systems: Task group summaries*. Washington, DC: National Academies Press.

National Research Council (NRC). 2012. *A framework for K–12 science education: Practices, crosscutting concepts, and core ideas*. Washington, DC: National Academies Press.

NGSS Lead States. 2013. *Next Generation Science Standards: For states, by states*. Washington, DC: National Academies Press. *www.nextgenscience.org*.

Passmore, C., C. V. Schwarz, and J. Mankowski. 2017. Developing and using models. In *Helping students make sense of the world using next generation science and engineering practices*, ed. C. V. Schwarz, C. Passmore, and B. J. Reiser, 109–134. Arlington, VA: NSTA Press.

Yoon, S., E. Klopfer, E. Anderson, J. Koehler-Yom, J. Sheldon, I. Schoenfeld, I. D. Wendel, H. Scheintaub, M. Oztok, C. Evans, and S. Goh. 2016. Designing computer-supported complex systems curricula for the *Next Generation Science Standards* in high school science classrooms. *Systems* 4 (4): 38.

Chapter 8

Energy and Matter: Flows, Cycles, and Conservation

Charles W. (Andy) Anderson, Jeffrey Nordine, and MaryMargaret Welch

Diamonds are forever. Or so James Bond and your local jeweler may want you to believe. But in fact, diamonds can burn. It's not easy to burn a diamond—you'd need to heat it in air to about 900°C (over 1600°F)—but at that temperature, diamonds will burn, leaving the world with one less diamond. When a diamond burns, where does it go? In the living world, small acorns grow into large oak trees without using up the soil they live in. Where did all that wood come from? The answers to these questions involve quantities that truly are forever: matter and energy.

One of the great achievements of science is the development of the matter and energy conservation laws. These laws state that during physical and chemical changes, certain quantities do not change, no matter what happens. A conserved quantity does not spontaneously appear or disappear, and this fact often provides a powerful starting point for understanding even the most mysterious and complex phenomena. When a diamond burns, the diamond is gone but its matter is not. Similarly, conservation laws require that the energy released during burning was not produced by burning but in fact already existed.

This is not obvious; our intuition tells us that when something burns, the matter goes away and energy is produced. When energy or matter seems to appear or disappear, the conservation laws insist that we ask questions such as, "Where did the matter go?" and "Where did the energy come from?" These questions help us gain deeper insight into how and why phenomena occur.

Chapter 8

Matter and energy are unique in that they appear in *A Framework for K–12 Science Education* (the *Framework*; NRC 2012) and the *Next Generation Science Standards* (*NGSS*; NGSS Lead States 2013) both as disciplinary core ideas (DCIs) and as crosscutting concepts (CCCs). The primary difference, as we see it, is in the ways these important concepts are used. The DCIs focus primarily on *mechanisms* involving matter and energy—explaining change—whereas the CCC focuses primarily on *conservation* of matter and energy, tracing what stays the same. That is, even though different science disciplines focus on different mechanisms for changes in matter and energy (e.g., photosynthesis, boiling, water cycling), they all rely on the idea that matter and energy are conserved. The classical conservation laws, separating energy conservation from matter conservation, apply with great precision to phenomena involving physical and chemical changes from the atomic-molecular to the global scale.[1] The conservation laws enable us to "take a step back." They provide us with strategies for making sense of systems and phenomena even when we don't know all the details. In particular, the conservation laws are powerful for two purposes.

Conservation laws as rules. Scientific models, explanations, hypotheses, and engineering designs must conform to the conservation laws. The conservation laws function as rules that constrain the range of possibilities for how systems behave. These rules provide a basis for evaluating the viability of ideas. For example, there may be rich debate about the best diet for people to eat, but we can all agree that humans cannot produce their own energy or matter (we must get them from the food we eat) and that this matter and energy are not simply "used up" when we go about our daily activities—both must go somewhere.

Because the conservation laws are accepted across disciplines, they can serve as a sort of "scientific Rosetta Stone." Even though an astrophysicist may know very little about the biochemical mechanisms involved in photosynthesis, she would know that something is wrong with any explanation of photosynthesis that implies that matter or energy either go missing or appear out of nowhere.

Tracing matter and energy as heuristics. Heuristics are "rules of thumb" that people can use to get started on difficult problems. Tracing matter and energy is often a valuable heuristic. Matter and energy conservation are frequently a good place to start when little else is known. No matter how vexing the phenomenon or system under consideration, matter and energy conservation can suggest questions that lead to deeper insight about how phenomena occur and systems operate.

For example, several brands of wristwatches are advertised as never needing a battery and never needing to be wound. Conservation of energy prompts the question "How

1. In nuclear processes, the matter and energy conservation laws cannot be separated in the way we discuss them in this chapter. Conservation laws are critical to understanding nuclear changes, but applying mass-energy equivalence is beyond the scope of the *NGSS*.

is energy transferred to the watch from the surroundings?" Answering this question involves investigating what parts of the environment interact with the watch in ways that result in enough energy transfer to power the movements of the watch. Similarly, the fact that a nail gains mass while it rusts is a clue that there must be some interaction with the environment that transfers mass to the nail. By tracing matter and energy through systems as they interact and undergo changes, we can gain important insights into how the world works. Fully understanding the function of a watch or the rusting of a nail also requires specific disciplinary knowledge of things such as torque or the electronegativity of atoms, as well as knowledge of science and engineering practices (SEPs), such as Planning and Carrying Out Investigations and Analyzing and Interpreting Data. Although the CCC of energy and matter may help spur and frame an investigation, all three dimensions are necessary to fully investigate and explain phenomena.

Conservation laws are so powerful and pervasive that they are used by all scientists and by scientifically literate citizens; in this way, they cut across all scientific disciplines. Yet, scientists in different disciplines use different terms and representations as they apply the conservation of matter and energy. In this chapter, we identify common uses of the conservation laws in different disciplines and in people's practical actions. Then we discuss how students can build understanding of these laws across disciplines as they move through elementary, middle, and high school, and we illustrate instructional strategies that can help students use these powerful principles to interpret and explain phenomena.

How Do Scientifically Literate People Use the Conservation Laws?

Like other crosscutting concepts, the true utility of the energy and matter conservation laws emerges when they are blended with SEPs, DCIs, and even other CCCs to predict and explain how phenomena happen in the natural and designed world. People who are successful in using conservation laws as rules and heuristics engage in three strategies: (a) defining systems, (b) identifying matter and energy, and (c) connecting systems at different scales.

"Like other crosscutting concepts, the true utility of the energy and matter conservation laws emerges when they are blended with SEPs, DCIs, and even other CCCs to predict and explain how phenomena happen in the natural and designed world."

Defining Systems

The conservation laws are closely connected with the SEP of Developing and Using Models and the CCC of systems and system models. In fact, the conservation laws themselves are commonly stated in terms of systems. The law of conservation of matter states that the mass of a system can only change by the amount of mass transferred to it from some other system. The same is true for energy, as described by the *Framework*:

> *That there is a single quantity called energy is due to the remarkable fact that a system's total energy is conserved. Regardless of the quantities of energy transferred between subsystems and stored in various ways within the system, the total energy of a system changes only by the amount of energy transferred into and out of the system (NRC 2012, pp. 120–121).*

In order to use the conservation laws, it is important to specify the system(s) that are involved in the phenomenon or device (such as an appliance or a machine) under consideration. In the simplest terms, a system is the part of the universe we are interested in, and we can draw an imaginary boundary between what is in our system and what is outside of it (see Chapter 7 for more about systems and system models).

When we define a system such that no matter crosses this imaginary boundary, we call it a *closed system*. For example, Classroom Snapshot 8.1 (p. 177) focuses on a soda can. If we define our system boundaries as the walls of the can, we can say that a sealed can is a closed system because no matter leaves or enters. In an *open system*, mass can cross the system boundary. If you open the soda can and place it on a scale for a few days, the reading on the scale will decrease as the soda goes flat (i.e., dissolved carbon dioxide [CO_2] bubbles out of the soda) and water in the soda evaporates. After a few days, there is less mass in the soda can than there was before. In the context of the conservation of matter, systems are either closed or open. In a closed system, the total mass of the system remains the same; in an open system, the total mass of the system changes by the amount that enters or leaves the system.

In the context of conservation of energy, the concept of an *isolated* system is useful, even though isolated systems do not exist in the real world. An isolated system is one in which there is no energy transferred across the system boundary, but the systems we encounter are never truly isolated. Consider the sealed soda can once more. No matter crosses the system boundary, but we can easily heat or cool the can of soda, which is evidence that energy crosses the system boundary; therefore, the soda can system is closed but not isolated. Some containers are designed to reduce energy transfer, like ice chests or vacuum-insulated containers that can keep our drinks cold (or hot) for hours, but not forever. Energy transfer across a system boundary is impossible to prevent, since it involves stopping interactions like collisions between molecules (Nordine and Fortus 2016, pp. 62–63). As a result, energy conservation is exceptionally difficult to observe through experimentation. Hot drinks tend to cool down, motion tends to stop, and electric devices cease to operate.

The conservation laws really show their power when they are used as rules for constructing models of how open systems and devices behave. Conservation laws help put limits on what is possible and what is not, and they help recognize whether models have fully accounted for all relevant systems and interactions. If we observe that the energy in a system is decreasing over time, we know there must be some mechanism by which our system interacts with its surroundings such that more energy leaves the system than enters it. Related to this, if a device seems to continue operating indefinitely with no apparent energy input, we know our model must specify some energy input that is equal to the amount of energy transferred to the surroundings.

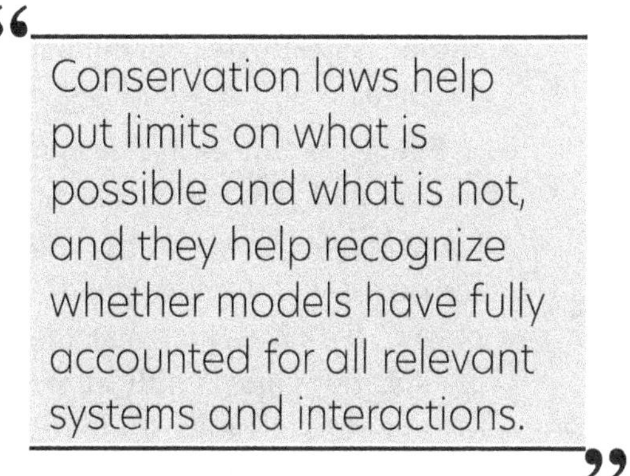

Identifying Matter and Energy

The CCC of energy and matter focuses on *matter and energy as conserved quantities.* Because they are conserved, we can use them for accounting purposes. Just as tracing money can play an important role in understanding our economy, tracing matter and energy can play an important role in understanding phenomena. In order to use matter and energy for accounting proposes, we need to make clear distinctions between matter and energy and other entities that are not matter or energy. There is extensive research that documents students' struggles to identify manifestations of matter and energy (Fortus and Nordine 2017; Jin and Anderson 2012; Mayer and Krajcik 2017; Mohan, Chen, and Anderson 2009; Smith et al. 2006). Many of these struggles are rooted in the multiple ways that *matter* and *energy* are used in everyday language and in science. Here are a few in-depth examples of the challenges:

Matter and energy in everyday language. *Matter* and *energy* are common words in our everyday language. People use these words all the time but in ways that don't correspond with their scientific meanings.

Matter as solid and liquid "stuff." As students learn to apply the matter conservation law and trace matter through systems, a key challenge is often *expanding their notions of matter.* In particular, young students have trouble accepting that gases can be as massive and substantial as solids and liquids. This notion is built into our everyday language. For example, *Thesaurus.com* identifies 20 synonyms for *matter* when used to mean "substance." Most of these synonyms (e.g., *material, thing, body, entity, stuff, substantiality, corporeity*) are associated with solids and liquids, but none are associated with gases.

Matter conservation is a useful rule only if we identify *all* the matter in a system and *all* the matter that crosses system boundaries—that's all the solids, liquids, and gases, including the ones that change state or become new substances through chemical changes. Tracing matter works as a strategy only if we recognize that *materials* can be created or destroyed, but not *matter.* A fire destroys wood (a material), but the matter that it is made of still exists, mostly as gases in the atmosphere.

Energy as causes and resources. In contrast to matter, a key challenge in applying conservation laws and tracing energy through systems is *restricting our notions of energy.* For example, *Thesaurus.com* lists 79 synonyms for *energy* in two broad categories: one associated with technology and the physical sciences (used in the sense of "generated power," with 25 synonyms), the other associated with living systems (used in the sense of "spirit or vigor," with 54 synonyms). Most of these synonyms suggest that energy *causes* events to happen or that energy is a *resource* that enables organisms and devices to do their work. For example, the stimulants in coffee or energy drinks give us "energy" that is consistent with many of *Thesaurus.com*'s synonyms (e.g., *spirit, stamina, vitality, animation*) but in scientific terms is not energy at all.

Many of *Thesaurus.com*'s synonyms that do have scientific meanings (e.g., *conductivity, current, force, friction, gravity, horsepower, kilowatts, pressure, voltage, wattage*) are causes or resources but NOT synonyms for energy in scientific terms. So, in our everyday language, all of us are accustomed to using *energy* in an expansive way that makes it difficult to use the energy conservation law effectively.

Matter and energy in science. Matter and energy can be hard to trace because they *appear* so different in different systems and phenomena. This was a problem, too, for 18th- and 19th-century scientists who developed different labels for the manifestations of matter and energy that they encountered in their disciplines. We live today with the legacy of this historical development; scientists still label manifestations of matter and energy differently in different disciplines. Thus, we all face the challenge of applying the conservation laws in systems where the concepts of matter and energy have a variety of scientific labels and the words *matter* and *energy* have additional colloquial meanings. Meeting this challenge requires connecting models of matter and energy in systems at different scales.

Connecting Systems and System Models at Different Scales

Constructing models that trace matter and energy can be very useful for making sense of systems. These models are typically constructed at different scales (i.e., macroscopic, atomic-molecular, and ecosystem or global scales), and different energy and matter representations are useful at each scale.

Macroscopic Scale

This scale is the "everyday" scale. Conservation of matter and energy is challenging at this scale because both appear in many different manifestations, and because both scientific and everyday language include many different labels for those manifestations (see the previous section Identifying Matter and Energy). Applying the conservation law of matter requires identifying substances and changes in substances and mass; applying the energy conservation law requires identifying the variety of different manifestations of energy.

Furthermore, the rules governing those manifestations can seem arbitrary: Why is air a form of matter but not sound? Why can people get energy from food but not from water or sleep? In order to see how those apparently arbitrary rules are based on underlying principles, students need to connect their macroscopic observations of phenomena with atomic-molecular models.

Atomic-Molecular Scale

Atomic-molecular models enable students to see the hidden continuity in manifestations of matter and energy that seem to appear or disappear when they observe phenomena at the macroscopic scale.

Matter: Identifying Continuity of Atoms and Molecules Through Changes

The solids, liquids, and gases in the world around us are mixtures of substances, or materials consisting of one kind of molecule. The DCI of Matter and Its Interactions focuses on how materials and substances change (Krajcik and Mayer 2017). The CCC of energy and matter, on the other hand, focuses on what *doesn't* change. Atomic-molecular models make it clear what doesn't change:

- Molecules stay the same during physical changes in matter. When water evaporates, for example, all the water molecules are still there, now moving freely as a gas.
- Atoms stay the same during chemical changes in matter; the atoms are rearranged into new molecules.

Chemical equations provide a concise way of keeping track of all the atoms involved in a chemical change. For example, the chemical equation for methane burning ($CH_4 + 2O_2 \rightarrow CO_2 + 2H_2O$) shows how one carbon atom, four hydrogen atoms, and four oxygen atoms are rearranged from one set of molecules (methane and oxygen) to another set of molecules (carbon dioxide and water). This rearrangement happens quintillions of times whenever we light a gas stove (HS-PS1-7).

Many students follow the procedure for balancing chemical equations without realizing that they are using conservation of matter as a rule. Since atoms are not created or destroyed in chemical changes, we know that the molecules of the products in a chemical reaction MUST have the same atoms as the molecules of the reactants. Chemical equation balancing is a way to check to make sure the numbers are correct.

Measuring Matter: Mass as Fundamental

In everyday life, we use a variety of measures for the "amount" of a material we have—mass, weight, volume, and apparent size are a few such measures. In order to use the matter conservation law, we must recognize that one of those measures—mass—is fundamental. Other measures, such as volume, can change without the amount of matter changing (e.g., through thermal expansion or a change of state). However, mass is different: *If the mass of a system changes, then matter MUST have moved into or out of that system.* Thus, we can always use mass changes to detect movements of matter and to know exactly how much matter has moved.

Atomic-molecular models make it clear why mass is the fundamental measure of matter: *The mass of a system is the total mass of all the atoms in the system.*Therefore, changes in a system that rearrange the atoms—like rolling a ball of clay into a sausage shape or the thermal expansion of a balloon—don't change the mass, but if atoms move into or out of the system then the mass must change (MS-PS1-5).

Energy in Fields and Particle Motions

The *Framework* and *NGSS* provide indications of how energy is manifested at the macroscopic and atomic-molecular scales:

- At the macroscopic scale, energy manifests itself in multiple ways, such as in motion, sound, light, and thermal energy (NRC 2012, p. 121).
- These relationships are better understood at the microscopic scale, at which all of the different manifestations of energy can be modeled as either motions of particles or energy stored in fields (which mediate interactions between particles). This last concept includes radiation, a phenomenon in which energy stored in fields moves across space (HS-PS3-2).

At the atomic-molecular scale, the confusing welter of macroscopic forms of energy is radically simplified. Energy is manifest as the motion of atoms and molecules and in the fields that mediate their interactions (Nordine and Fortus 2016); so, when atoms rearrange in a chemical reaction, changing fields can change the speed of the atoms. As these faster or slower atoms interact with their surroundings, energy is transferred either to or from the environment.

Ecosystem and Global Scales

Scientists commonly use *pool-and-flux models* to trace matter and energy through large-scale systems such as ecosystems, weather systems, human agricultural and industrial systems, and global systems. Figure 8.1 is an example of this kind of model; it follows the conventions of using boxes or images to represent matter pools and using arrows to represent fluxes that move matter from one pool to another.

Figure 8.1. Quantitative global carbon cycling diagram

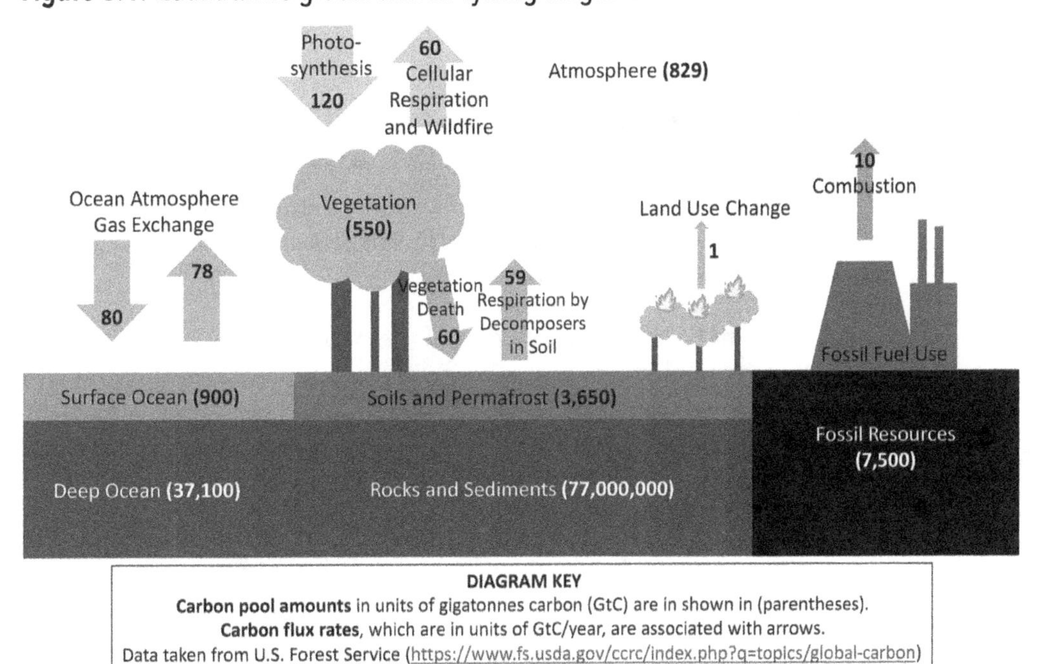

Source: U.S. Forest Service 2019.

Earth's global systems, represented in Figure 8.1, are virtually closed systems—the amount of matter entering or leaving global systems is negligible. In addition to carbon cycling, represented above, the *NGSS* mentions cycling of materials in Earth's interior, water cycling, and global atmospheric circulation. In all of these closed systems, the conservation laws lead to a basic pattern: *matter cycles; energy flows.* Energy from two sources—sunlight and radioactive decay in Earth's interior—drives movements of matter and changes in matter through convection, changes of state, and chemical changes. This energy ultimately leaves Earth systems as infrared radiation going into space. Within Earth systems, the same atoms are recycled over and over as they move and combine into different molecules.

Regional and local Earth systems, such as ecosystems, local weather systems, watersheds, and human agricultural and industrial systems, are open systems. Both matter and energy enter and leave these systems.

Quantitative Reasoning About Large-Scale Systems

The conservation laws provide important rules and heuristics for understanding all of these large-scale systems because they lead to an ironclad relationship between fluxes and pool sizes: *Changes in pool sizes are determined by the balance of fluxes into and out of the pools.* Scientifically literate people can learn to use this relationship to understand quantitative predictions of global environmental issues such as climate change (HS-ESS2-6).

How Students Can Learn to Use the Conservation Laws Over Time

The conservation laws can be stated simply, but as the previous discussion shows, that doesn't mean they can be applied simply. People who successfully use the conservation laws as rules and trace matter and energy as heuristics have mastered these three associated strategies for applying the conservation laws:

1. *Defining systems* and constructing models of the systems that illustrate the conservation of matter and/or energy
2. *Identifying matter and energy*, including the many forms of evidence and descriptive terms that scientists use to describe matter and energy and to trace matter and energy through systems
3. *Connecting scales*, or tracing manifestations of matter and energy through system models at different scales, from atomic-molecular to global

> "Building expertise in applying these strategies is a long and intellectually arduous process, taking all of children's K-12 years and beyond. Yet, K-12 students can make substantial progress down the road toward expertise."

Building expertise in applying these strategies is a long and intellectually arduous process, taking all of children's K–12 years and beyond. Yet, K–12 students can make substantial progress down the road toward expertise. In this section, we describe some of the key milestones in this process during students' elementary, middle, and high school years.

Elementary School: Laying the Foundations for Conservation Rules and Heuristics

The *NGSS* do not advocate explicitly teaching the conservation laws to K–5 students. Although elementary students could learn to apply conservation laws to some phenomena, it is unlikely that this is the best way for them to spend their time. Elementary students can learn to recount, for example, that "energy is never created nor destroyed," but they generally do not have all the experiences they will need in order to use the conservation laws systematically. Elementary students can, however, make substantial progress as they become more proficient in these three strategies: defining systems, identifying matter and energy changes, and connecting scales.

Children's learning during the elementary school years is particularly focused on the second strategy—identifying matter and energy. As we will discuss momentarily, this involves a growing awareness of the manifestations of matter and energy in phenomena (i.e., recognizing more aspects of the systems involved), but formal definitions of systems are more appropriate for the middle and high school years. It's important to note that the vast majority of focus during the elementary years is on building ideas about matter, not energy. Energy ideas really only appear in earnest in fourth grade, and even then in limited scope. Similarly, the *NGSS* recommends that students' introduction to atomic-molecular models (which are more useful for applying conservation laws) wait until middle school.

Matter: Tracing Materials and Measuring Amount

Although it is generally not productive to teach elementary school children a formal definition of matter or formal rules for applying the matter conservation law, they can learn to trace matter through increasingly complex phenomena. In particular, this growing understanding involves developing their capacities for tracing materials and measuring amounts of matter.

Tracing Materials: In some circumstances, when there is observable continuity of matter for a phenomenon, young children learn to trace matter. For example, Piaget and his colleagues (Driver 1985; Piaget 1951) showed how young children come to understand that pouring water from a tall, narrow glass to a short, wide glass does not change the amount. These children are successfully recognizing the continuity of water as a material in the system of water and the two glasses.

However, tracing matter gets more difficult in more complex systems.

- It is nowhere near as clear to children that when water freezes there is still the same "amount" of ice as there was water before. Children in elementary school can study this system and see how water can change to ice and back again, as well as how the mass[2] and volume of water remain the same after these changes. These experiences can help them develop the argument from evidence that ice is another form of water and that freezing and melting do not change the amount of water.
- Tracing materials for changes of state involving gases—evaporation and condensation—is more challenging still, especially since it contradicts the notion in everyday language that matter is solid or liquid "stuff." When a puddle evaporates, why should children believe that the water is still there—in the air, invisible—and that it weighs just as much as the water did before it evaporated? To trace materials in this system, children need to see how water vapor, like ice, is another form of water and how they can trace water as it goes to and from the air. (The *NGSS* advocates waiting until middle school to teach children atomic-molecular explanations for these processes; see performance expectation 5-PS1-1.)

2. Elementary students should not be expected to distinguish between mass and weight, so students could equivalently monitor the "weight" of the water.

- In chemical changes, some materials cease to exist and new materials are created. Tracing matter through these changes is more productive when students learn to use atomic-molecular models in middle school.

Measuring Amounts of Matter: The question of "how much" of a material there is can be complicated, particularly if children are comparing different materials or different states of the same material. A balloon gets larger when it is moved from a cold to a warm place. Does that mean there is more air inside the balloon in the warm place? A liter of water forms more than a liter of ice when it freezes. Does that mean there is more ice than there was water? Children in elementary school can make substantial progress toward scientific answers to these questions.

Piaget's conservation tasks show a predictable progression in how children develop more sophisticated ideas about "amount." For example, Elkind (1961), replicating Piaget's earlier work, asked children of different ages about what happened when a ball of clay was rolled out into a sausage shape. For eight-year-old children:

- 72% said the "amount" of clay stayed the same after it was rolled out. The other children generally relied on perceptions (e.g., that the clay was longer or thinner) to make judgments about how the ball was different.
- 44% said the "weight" of clay stayed the same.
- 4% said the volume of clay stayed the same (agreeing that "They both take up the same amount of space," p. 221).

The important takeaway from results like these is not about specific ages at which students accomplish specific tasks. It is that judging or measuring "how much matter" is in a system is both conceptually and procedurally complicated, so children in elementary school need multiple opportunities to compare and measure amounts of materials.

The Piaget tasks did not involve measurement, but measurement and quantification play an essential role in arguments from evidence that involve conservation of matter. These can involve both qualitative comparisons like the Piaget tasks and quantitative comparisons in which children measure volume and mass/weight (not differentiated at the K–5 level; see NRC 2012, p. 96, p. 108). In addition to learning how to make these measurements accurately, children in grades K–5 can work toward the understanding that *mass is fundamental,* as discussed in the section on macroscopic conservation:

- If the volume of a system changes, then it is possible that materials in the system expanded or contracted.
- If the mass of a system changes, then matter *must* have moved into or out of the system.

Precise and inexpensive digital scales make it possible for teachers and children to measure and reason about even small changes in mass. Classroom Snapshot 8.1 illustrates how Ms. Ramirez uses these measurements to engage students in tracing matter through changes of state (addressed in *NGSS* PE 5-PS1-2).

CLASSROOM SNAPSHOT 8.1
Tracing Changes of State With a Soda Can

Ms. Ramirez's fifth-grade students have been observing different forms of matter and thinking about matter within a system. Today, Ms. Ramirez is engaging her students in sensemaking activities about what happens to matter when water condenses on a cold soda can.

Ms. Ramirez says, "Do you think I can change the weight of this cold soda can without opening the top to pour anything in or pour anything out? I mean, if I just let it sit here in the open, do you think I can get it to change weight?"

Her students turn and talk to their table partners to discuss how to change the mass of the can. Jack suggests they might add clay from the art bin to the can. Ms. Ramirez says, "Yes, you are correct that adding a solid like clay would change the can. But I mean, what if we didn't physically add anything? Do you think the can could change weight if we just let it sit here? Let's remember our discussion about a system. Do you think the clay is a part of the system? What should we consider our system?"

The students engage in a discussion and agree that the system they want to focus on is the can, the contents of the can, and the air surrounding the can. Ms. Ramirez says, "Now that we have defined our system, how do you think we could change the weight of the can in this system?" Ramona says, "Let's do an experiment to find out." Ms. Ramirez asks the student to help her design the experiment.

Students busily write their ideas on their whiteboards, and then the class does a gallery walk to share those ideas. Finally, they come to consensus about what to investigate. Tomorrow, they will take the cold soda can and weigh it immediately after it comes out of the refrigerator during their morning math lesson. Then they will let it sit out until they have science class in the afternoon and weigh it again.

The next day, Ms. Ramirez takes the can from the refrigerator, wipes off the outside of it, and then weighs the can. She records the mass of the full, closed can on the board. Then the children get to work on their lesson. Throughout the morning, though, they write down their observations of the can, including remarks about the moisture drops gathering on the outside of the can.

During science class, Ms. Ramirez calls for the students' attention once again and asks them to write down their observations independently. She asks students what

Continued

Classroom Snapshot 8.1 *(continued)*

they notice. Most students share that they notice water on the outside of the can. Ms. Ramirez says, "Where do you think the moisture on the outside of the can came from?" This starts a conversation among the students:

Samantha: "I think that it came from inside the can. Some of the water inside leaks out when it's cold."

Ahmad: "I think it came from the air. Water in the air got on the outside of the can."

Jenna: "But there isn't any water in the air in this room; if there was, it would feel like it is raining."

Manuel: "I think it's because the can was cold. Water forms on cold things when you leave them out. I remember this summer when it was hot outside, my mom had a glass of ice water and she left it on the table. The same thing happened. I don't know why, but I saw this happen before."

Ahmad: "I saw water on the grass in the morning even though it wasn't raining. Somehow water can get on things even when you don't see water in the air."

Isaac: "There is always water in the air, even when it isn't raining. I think that water in the air sticks to the can because the can is cold."

Ms. Ramirez: "We have several ideas about where the moisture came from. How could we collect evidence about these ideas? Talk with your partners and tell me your thoughts."

Student teams talk and share their ideas with the class. Some teams suggest weighing the can again. Ms. Ramirez weighs the can again and writes this mass on the board. Then she says, "Where did the extra weight come from? Using the drawing, explain what you think happened."

Isaac draws the following illustration and shows it to the class, which starts another conversation:

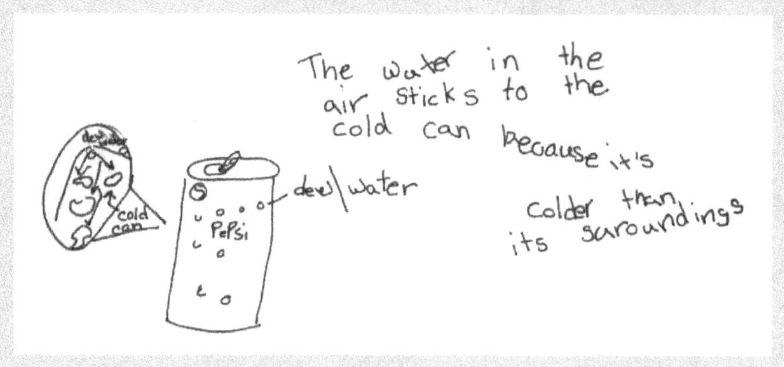

Continued

Classroom Snapshot 8.1 *(continued)*

Samantha: "I think that maybe Isaac is right; it came from somewhere outside the can. What happens if you wipe off the moisture and weigh the can again?"

Ms. Ramirez uses a towel to wipe the can dry and weighs it again. "It still weighs the same as it did when it came out of the refrigerator. What do you think that means?"

Samantha: "I think that means that there is no water seeping out of the can."

Ms. Ramirez: "Let's try one more consideration for our soda can. What if I open the can but do not pour out any of the soda? Just let it sit here. What do you think will happen to the weight? Draw your initial ideas."

Students write down their ideas and share at their tables. Ms. Ramirez pops the top of the can and lets it sit for 15 minutes. Then she says, "No liquid has been poured out of the can. We did not remove the pop top. Let's weigh the can and write down the mass. Will there be a change in mass? Why or why not?" Ms. Ramirez weighs the open can and writes this weight on the board. "Draw what you think is happening to the weight," she says. The students begin a discussion, to be continued later, of how the can could lose weight even if no liquid was poured out.

Recognizing Manifestations of Energy as a Basis for Energy Conservation

Children in elementary school should engage with many different phenomena, and all of those phenomena will involve energy, since all phenomena involve energy changes. As they encounter the phenomena in upper elementary, they have the opportunity to discuss energy and its various manifestations (the DCI). For example, students at the elementary level can begin to explore simple phenomena in which energy is transferred—such as colliding balls or putting a room-temperature rock into warm water—and to ask questions such as "What components are involved?" and "Where does energy come from?" and "Where does energy go?" (see Crissman et al. 2015). In such investigations, students begin to attach the idea of energy to its different manifestations and recognize that energy is transferred between objects as they interact. This is enough. In middle and high school, students will learn to use energy ideas more explicitly to make sense of an increasing variety of systems.

Middle School: Using Conservation Rules and Heuristics to Analyze Phenomena

Middle school students begin to coordinate all three strategies—defining systems, identifying energy and matter changes, and connecting scales—as they explain phenomena using conservation laws as rules and heuristics. While they begin to use matter conservation in a more quantitative way (e.g., measuring the mass of reactants and products in a chemical reaction), they use energy conservation qualitatively—meaning they recognize that energy is neither created nor destroyed, but they do not calculate amounts of energy.

Defining Systems: Defining System Boundaries and Distinguishing Changes in Matter From Changes in Energy

Middle school students study systems in which both energy and matter change, including living systems, Earth systems, and technological systems. Three-dimensional engagement with these systems requires students (a) to distinguish between changes in matter and changes in energy and (b) to define systems and system boundaries carefully.

Distinguishing Between Changes in Matter and Changes in Energy

Middle school students commonly believe that engines "consume" fuel or that humans "burn off" fat when they exercise or that those processes convert matter into energy. The idea that every atom of the fuel or fat is still present as invisible gases is much less intuitive. The example of growing plants shows a similar pattern:

- When asked about sources of matter, most middle and high school students correctly identify soil nutrients and water as sources of matter for growing plants, but they leave out a gas (carbon dioxide) that is a primary source of matter. Students often describe gas exchange separately from plant growth, saying that plants "breathe in" carbon dioxide and "breathe out" oxygen, exchanging one colorless, odorless gas for another.
- When they are asked about energy at the beginning of life science courses, most middle and high school students correctly identify sunlight as a source of energy for growing plants. However, before taking biology courses, 95% of students *incorrectly* identify soil nutrients as a source of energy, and 94% of students *also* identify water as a source of energy. (For students who have completed traditional high school biology courses, 84% identify soil nutrients as a source of energy, and 81% identify water as a source of energy.)[3]

These responses make perfect sense for students with restricted notions of matter and expansive notions of energy like those in *Thesaurus.com*. Students commonly think that soil nutrients and water are solids and liquids that provide materials for plant growth, whereas carbon dioxide is an ephemeral gas. Notions of energy as a cause or resource also make it natural to identify anything that contributes to the "spirit or vigor" of growing plants as an energy source, but these expansive notions make it impossible to trace energy through living systems with scientific accuracy.

So, instruction at the middle school level needs to help students expand their notions of matter and refine their notions of energy. Students who believe that fuels or fat are

---

3. These data come from assessments administered by the *Carbon TIME* project to 4,773 middle and high school students.

converted to energy can learn that substantial amounts of mass can end up in gases. They can begin to ask questions such as, "Where does the carbon in the CO_2 that plants 'breathe in' go?" and "Where does the carbon in the CO_2 that people breathe out come from?" Students can begin to recognize that many of the things that help people feel "energetic" (e.g., caffeine) or that help plants grow vigorously (e.g., soil nutrients) are not actually sources of energy, meaning that energy is more difficult to obtain than students might assume.

Defining Systems and System Boundaries

Helping middle school students trace matter and energy through systems also involves helping them analyze the boundaries of systems and the movements of matter and energy across system boundaries. This involves recognizing "invisible" forms of matter and energy that exist in systems and cross-system boundaries. In the examples above, many students assume that the gases in the air are too ephemeral to have much of an effect on the solids in plant and animal bodies. Instruction can help students understand that even materials that are not dense, like air, can still be massive; in fact, *a lot* of matter is exchanged between the atmosphere and the bodies of plants and animals.

Similarly, students see phenomena every day in which energy seems to be "used up" and disappear: Moving objects come to a stop; hot objects cool down; light is absorbed; engines run out of gas; animals die without food. What's happening to that energy? For students who think of energy as a cause or resource (see the Identifying Matter and Energy section on p. 185), it makes sense to say the energy is gone. It can no longer cause events to happen or serve as a resource for organisms. These students need to recognize "invisible" forms of energy that remain in the system (e.g., thermal energy—the kinetic energy of atoms and molecules) or that leave the system (e.g., infrared radiation from objects that are not hot). As we will discuss, atomic-molecular models can provide powerful tools for understanding these invisible forms of energy.

An important idea undergirding both matter and energy conservation is that both entities must always be *somewhere*. That is, they cannot be transferred *from* one system without being transferred *to* another, and vice versa. When matter is exchanged between two systems, the mass of one system increases and the other decreases by the same amount. The same is true for energy, though the evidence of increase or decrease in the energy of a system can look different, as energy can be manifest in different ways. Likewise, the systems involved in energy transfer can be difficult to identify, and students need support in learning to identify these systems and in recognizing how they change in the process. Classroom Snapshot 8.2 (p. 182) illustrates how middle school students can use a tool for identifying the systems involved in energy transfers even when they are not obvious.

CLASSROOM SNAPSHOT 8.2
Developing System Models to Trace Energy

At the beginning of the period in eighth-grade science, Mrs. Gladwell holds two repelling magnetic carts, one red and one blue, close together on a track and releases them from rest. Both carts start moving away from each other.

From the back of the room, Tom exclaims, "Neat!" with his characteristic dry humor, hardly impressed by a phenomenon he's long been familiar with.

Mrs. Gladwell, who has asked students to think about the phenomenon using an energy lens, asks, "What's our puzzle, class?" Silence. Recognizing a need for some more prompts, she continues. "Let's think about the energy transfers here. Think about the red cart. Is there energy transferred *to* the cart or *from* it?"

"To the cart," answers Sarah.

"How do you know?" Mrs. Gladwell further prompts Sarah.

"Because it speeds up," Sarah says.

"And what about the blue cart? Is energy transferred *to* it or *from* it?" Mrs. Gladwell says.

Kim chimes in, "To it."

Mrs. Gladwell repeats her earlier question, "OK, what's our puzzle?"

For the past several weeks, Mrs. Gladwell's class has been studying energy by learning to construct models that represent energy transfers between systems as they interact during phenomena. They have learned that if an object speeds up, this is evidence of energy transfer to it; meanwhile, slowing down is evidence of energy transfer from the object. In a recent lesson, students constructed models of the energy transfers between colliding coins and billiard balls. By observing these interactions, students recognized that anytime one ball or coin sped up, the other one slowed down. To represent energy transfer in the collisions, they constructed models that they call energy transfer models, which look something like this:

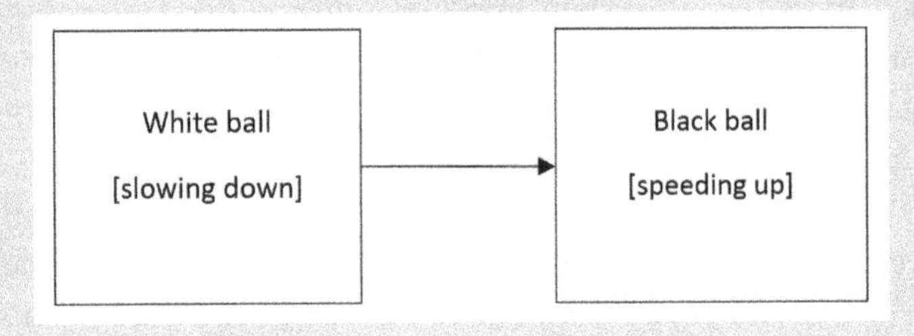

Continued

Classroom Snapshot 8.2 *(continued)*

Over time, and with practice across a range of phenomena, students have come to agree on some common features of their energy transfer models. A box represents a system or an object that is involved in the phenomenon being investigated, the arrow between the boxes represents energy transfer between systems/objects, and the brackets inside of the boxes describe the changes to the system/object that are associated with the energy transfer.

Mrs. Gladwell asks students to work together for a few minutes to think about how they might draw an energy transfer model for the repelling magnetic carts. She then asks students to come to the board and draw their ideas, even if they feel they are stuck. Two groups volunteer to draw their models. Amy and Beth's model looks like this:

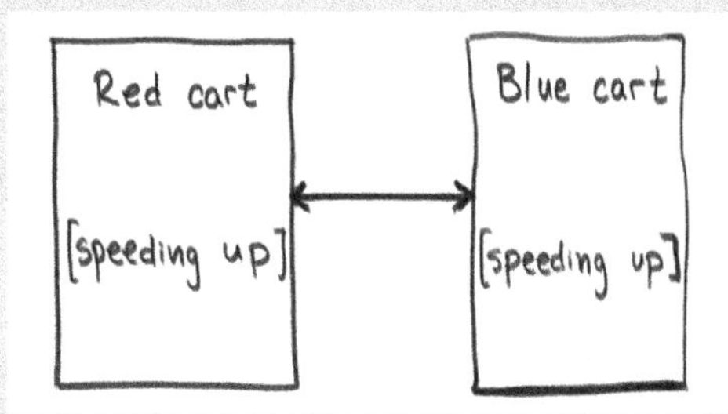

"Will you tell us about your model?" asks Mrs. Gladwell.

"Well, we thought about what was involved, and we decided that the track and your hand weren't really involved since neither really makes the cart go faster. This leaves just the two carts. We know each has to have an arrow pointing toward it, since both speed up, so we made a double-headed arrow to represent this."

"But wait a second," Leo chimes in. "You can't have a double-headed arrow, can you?"

"Hmm ... why not?" asks Mrs. Gladwell.

"Because we never do," Leo responds immediately.

"But why not?" Mrs. Gladwell pushes. "Who has an idea?"

After a few seconds of silence, Leo speaks up once more, "Well, the arrow represents energy transfer, right?"

"Right ... " says Mrs. Gladwell, hoping for more.

Leo continues, "And transfer means that it comes from one system and goes to another. But with a double-headed arrow, it means energy goes into both carts at once without coming from any other system."

Continued

Classroom Snapshot 8.2 *(continued)*

"Yeah," says Beth. "We thought about that and we weren't sure what to do. We knew energy had to be going to the carts but weren't sure where it was coming from."

"That was exactly our problem!" interjects David, who has drawn the second model on the board, which looks like this:

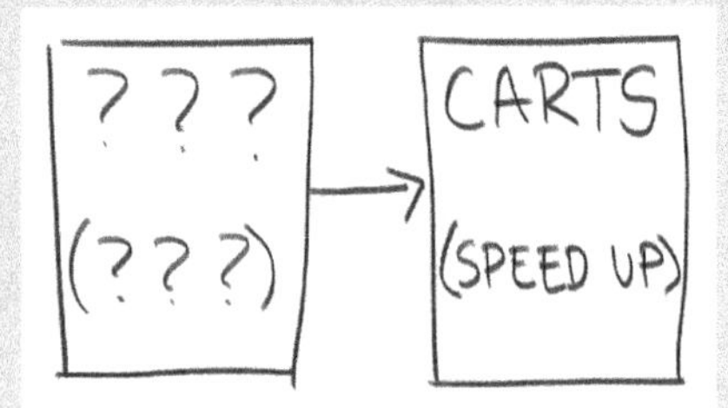

He continues, "We knew that energy was transferred to the carts and that this energy had to come from someplace, but we have no idea where. Hence all the question marks."

"Before we continue, tell us about your box labeled 'CARTS,'" Mrs. Gladwell says. "Amy and Beth had a box for each cart, whereas you had both carts in one box."

"Well, we just thought that would be the easiest way to draw it. We knew both carts did the same thing; and we knew that since they both speed up, they both have energy coming from someplace, and we figured that it must be the same place. So, we thought we could just draw them in the same system. Why, is that wrong?"

"Not wrong at all," responds Mrs. Gladwell. "Both models show energy transfer to both carts, and both identify the same process—speeding up—for the carts. Both models give the same information, even though you represented the cart systems differently. There is no one best way to make your models!"

"But our model is missing a box," says Beth.

"Where would you put it? Will you come show us?"

Beth goes to the board and erases the middle part of the arrow, barely squeezing a third box in between the existing ones.

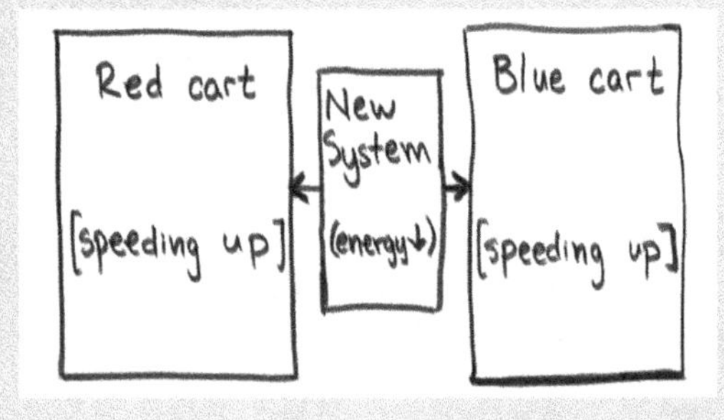

Continued

Classroom Snapshot 8.2 *(continued)*

"Why did you add that?"

"Well, this way, we show that even though we don't know what it is, we know there must be some other system transferring energy to both carts at the same time."

"OK, so now we know our puzzle. Using our models and the idea that energy can only be transferred *from* one system *to* another, we have realized that there is something about this phenomenon we haven't included—something that is critical for fully explaining what is going on here. Today, we will begin exploring what that system is, how it transfers energy to the carts, and how it changes in the process. But before we get started, get into your lab groups and discuss these questions: (1) What do we already know about this new system? (2) What questions do you have about this new system?" The students go to their lab tables and get to work.

In Classroom Snapshot 8.2, students grapple with the idea that energy is transferred from one system to another without any loss of energy through a series of activities, which includes coming to consensus on how to represent energy transfer. Note the importance of marrying energy conservation with the CCC of systems and system models to give students boundaries to frame their ideas.

Identifying Matter and Energy: Working Toward Accurate and Principled Identification

Students' difficulties with tracing matter and energy separately through systems are closely connected with challenges in identifying changes in matter and energy. Middle school students should begin to identify and describe manifestations of matter and energy with more specificity. This means, in particular, recognizing that all solids, liquids, and gases are forms of matter and have mass. Energy transfers do not affect the mass of objects or everyday systems.

Foods and fuels are especially important examples of this distinction. Foods and fuels are important because almost all living systems, as well as human technologies that use fossil fuels or biofuels, rely on a single energy source—the oxidation of organic materials (i.e., materials with reduced carbon, indicated by C-C and C-H bonds). These organic materials all originated in photosynthesis. Chapter 12 covers this process in detail. Here, we will focus on the complementary oxidation processes.

Almost all middle school students can identify foods and fuels fairly accurately (with the exception of soil nutrients—see p. 180). They also recognize that living systems get energy from food and engines get energy from fuel. So far, so good. However, almost all middle school students rely on "force-dynamic" explanations of what organisms do with food and what engines do with fuel: They explain that the food and fuel are "used up" or converted into energy (Jin and Anderson 2012).

Instruction in middle school can help students see the value of the conservation laws as rules and heuristics. Matter conservation tells us that the matter in foods and fuels *must* still be matter—solids, liquids, or gases—after the foods and fuels are used and can guide students to learn more about gaseous products. Energy conservation tells us that the energy manifest as motion or heat existed before the food/fuel was oxidized and that it continues to exist afterward. It was not "created" in the process of using fuel and does not "run out" or "fade away" after the fuel is used. This energy can—and usually does—leave the system via heat transferred to the surrounding environment, but not as material waste such as carbon dioxide (see *NGSS* MS-LS1-7). High school students will learn to trace matter and energy through living systems in more precise ways.

Connecting Scales: Tracing Matter and Energy Using Atomic-Molecular and Large-Scale Models

Middle school students can use the conservation laws as rules and heuristics by tracing manifestations of matter and energy at the macroscopic scale. Atomic-molecular models can help them understand how and why the conservation rules make sense. Large-scale models can help them trace matter and energy through ecosystems and global systems.

Using Atomic-Molecular Models to Trace Matter and Energy Through Chemical and Physical Changes

When changes of matter involve invisible gases, it is difficult for students to observe that the matter is "still there." Atomic-molecular models provide some very simple rules to explain how the matter continues to exist:

- During physical changes in matter, molecules stay intact.
- During chemical changes in matter, atoms stay intact.

Instruction at the middle school level can help students master these rules and apply them consistently. For example, students learn to explain how the changes in mass when water condenses and evaporates on a soda can (see Classroom Snapshot 8.1 on p. 177) are caused by changes in the motion and arrangement of water molecules, and how those changes are associated with changes in the kinetic energy of the water molecules (MS-PS1-4). For the foods and fuels example, students learn to explain how the atoms in foods and fuels are rearranged into new molecules (but all the atoms are still there) and how these changes release energy to be used by engines and organisms (MS-PS1-4, MS-LS1-7).[4]

4. Sometimes, curriculum materials suggest that energy is released when chemical bonds in high-energy reactants are broken (e.g., glucose, gasoline, ATP). This is not true. Breaking the bonds of reactants ALWAYS requires energy, and forming the bonds of products ALWAYS releases energy. Therefore, the oxidation of foods and fuels releases energy when the bonds of product molecules are formed, not when the bonds of reactant molecules are broken (HS-PS1-4, HS-LS1-7).

Using Large-Scale Models to Trace Matter and Energy Through Ecosystems and Earth Systems

Instruction can help middle school students see how they can understand patterns in large-scale systems by tracing matter and energy. Middle school students can understand how water condenses and evaporates not only on soda cans but also on regional and global scales, with changes of state and cycling of water driven by energy from the Sun (MS-ESS2-4). Elementary students see the plants and animals in ecosystems as "actors" that depend on other organisms for the materials they need to survive; middle school students can study plants and animals as systems that transform matter and energy as the matter and energy move through food webs (MS-LS2-3; Mohan, Chen, and Anderson 2009).

High School: Principled Use of Conservation Rules and Strategies

High school students continue their progression to principled use of matter and energy conservation as rules and heuristics.[5] The outcomes they achieve are described in the section How Do Scientifically Literate People Use Conservation Laws? (p. 167). This section builds on the previous sections by describing how instruction can help students achieve those outcomes.

Defining Systems: Learning to Be Strategic

Students can learn to be strategic in defining closed and open systems at all scales, defining system boundaries that make it easier to trace how matter and energy move into and out of systems and change inside the systems. Classroom Snapshot 8.2 (p. 182) illustrates this learning process, which continues through high school.

Strategic choices of system boundaries can support both conceptual clarity and quantitative reasoning. For example, in the mealworm investigation referenced in Classroom Snapshot 8.3 (p. 189), the mealworms gain mass as they grow while their food source (a slice of potato) loses mass (see Figure 8.2). No surprises there—just what students expect, and many high school students are happy to stop there, satisfied that they understand what is happening (see Dauer et al. 2014).

However, using the conservation laws as a heuristic can lead to other questions: What about

Figure 8.2. Observations during a mealworm investigation

Animals Observations and Patterns

Investigation: Mealworms eating and breathing

Key observations and patterns

- Mealworms gain mass
- The potato loses mass
- The potato loses more mass than the mealworms gain
- Mealworms breathe CO_2 out into the air

5. The previous section on learning in middle school is, unfortunately, relevant to most high school students. Students of any age who experience a traditional "learning about" curriculum are unlikely to master middle school performance expectations (e.g., Jin and Anderson 2012).

a larger system, including the potato, the mealworms, and their waste? That system loses mass, so the conservation laws tell us there *must* be something going on here besides the mealworms eating and growing. Since no solids or liquids left the system, there *must* be gases leaving the system. By tracing the movement of matter across carefully defined system boundaries, high school students can construct arguments from evidence that support a deeper understanding of cellular respiration. (See Classroom Snapshot 8.3 for a continuation of this story.)

Identifying Matter and Energy: Learning to Be Principled

Students become more rigorous in recognizing manifestations of matter and energy at all scales and in distinguishing between scientific and colloquial language. In particular, they recognize how the many different forms of energy can be recognized as manifestations of fields and motions of particles and materials (see HS-PS3-2).

For example, instruction can help students connect changes in the speed or temperature of objects to changes in kinetic energy associated with the speed of particles. Similarly, students can recognize that phenomena such as the stretching of a spring, falling objects, or attracting magnets all involve energy manifest in fields between interacting particles, which is often referred to with the umbrella term *potential energy*. Students also recognize that waves across the electromagnetic spectrum all transfer energy through interactions between electrical and magnetic fields, which is often referred to as radiation (HS-PS4-4).

Connecting Scales: Learning to Use Quantitative Models

Students can learn to use quantitative models at all scales to make predictions, interpret and analyze data, and construct arguments from evidence. They recognize and use connections among conserving atoms in chemical changes in atomic-molecular systems, conserving mass in macroscopic systems, and analyzing how fluxes change pool sizes in large-scale systems (HS-PS1-7, HS-ESS2-6).

Matter conservation at the macroscopic and atomic-molecular scales is connected by a precise quantitative rule: *The mass of any system is the mass of all the atoms in that system.* Through instruction, students can appreciate the power of this rule and use it successfully. At the atomic-molecular scale, students can master algorithms for chemical equation balancing, but it is critical that they understand why those algorithms work. Chemical equations express matter conservation in mathematical terms; the algorithms assure that every atom in the reactants is accounted for in the products, and vice versa. Physical modeling, where students manipulate atomic-molecular models and follow what happens to each atom, can play an important role in helping students see the connections between chemical equations and tracing individual atoms through chemical changes (HS-PS1-7). Similarly, students can learn to relate the energy in chemical bonds (as field energy between particles) to the energy absorbed or released during chemical changes (HS-PS1-4). Classroom Snapshot 8.3 illustrates a teacher working with her students on an important chemical change: cellular respiration.

CLASSROOM SNAPSHOT 8.3
Explaining Cellular Respiration[6]

In her ninth-grade biology class, Ms. Callahan is working with her students to develop explanations of how matter moves and changes and how energy changes during cellular respiration in a cow's cells (connecting macroscopic observations with atomic-molecular models and using principles of conservation of matter and energy).

Establishing the Problem

Ms. Callahan begins with reminders about what the class has been working on (mealworm investigation data, molecular modeling kit) and asks if everyone is feeling confident. She says that at the end of the unit she wants students to be able to "say not only for school, but for life, [that] this is exactly what happens when the organism moves." She adds, "We're going to actually figure out what's going on in this cow's muscle cells. Get ready to explain."

Next, the class reviews the results of an earlier investigation: mealworms eating and breathing. Ms. Callahan highlights how the product of CO_2 might have something to do with the missing mass students found in their mealworm investigation evidence.

Private Writing

The students work on an Explanations Tool, which combines a graphic organizer for tracing matter and energy with a paragraph that students write giving an overall explanation of the process. Ms. Callahan assigns students to start their personal writing by saying, "All right, so now it's your turn to figure out some explanations for this. I want you to be specific. Use your evidence. Use your thoughts. Start putting all these things together." Students work for 10 minutes.

Partner Work to Share Ideas

The students work in pairs. Ms. Callahan instructs her students: "Don't just throw your paper at your partner and have them look at it. Talk to them. Communicate and work your way through it. Get out a different-colored pen or pencil. I want you

6. See HS-LS1-7; this Classroom Snapshot is based on Covitt et al. 2019.

Continued

Classroom Snapshot 8.3 *(continued)*

to circle any areas you have in common, and then if you want to add items in that's fine. You have a lot more in common but maybe still some differences, which could be interesting. So talk to each other. Use your words. Let's go. Four minutes."

Consensus-Seeking Discussion

The class comes together for a discussion. Sometimes, the discussion is very specific to the Explanations Tool; other times, it's related to the tool, delving into additional content students are curious about (e.g., tracing water through urine and milk in the body; reviewing functions of organs, including kidneys, gall bladder, and pancreas; discussing why urine is yellow; discussing ATP). Ms. Callahan uses talk moves (Michaels and O'Connor 2012) to scaffold students in figuring things out. Sometimes, she solicits short responses, but often she asks for extended explanations.

Ms. Callahan (after a student says that a cow's food is "grass"): "Grass and then where is it going? Someone raise your hand and tell me the next step. Emma."

Emma: "It's, like, chewed up. It gets started breaking down stuff."

Ms. Callahan: "OK. So, it gets started breaking down stuff. Add on to what Emma is saying. What does it mean to get started breaking down stuff? Logan."

Logan: "The saliva in the cow's mouth begins to break the grass down as it's chewing."

Ms. Callahan: "OK. Is it breaking down the glucose?"

Multiple students: "No. Not until you get into the digestive system."

Ms. Callahan: "So, Riley, tell me what happens next. We've got grass in the mouth. There's some saliva going on. It's breaking down the grass. What happens next?"

Reviewing Students' Written Explanations

The close of discussion scaffolds students in checking whether they have written good explanations. Ms. Callahan queries students about confidence and consensus.

Ms. Callahan: "Do you have an arrow showing oxygen or O_2 going into the cow's cells?"

Students: "Yes."

Ms. Callahan: "Is that pretty universally confident? You're good with that?"

Students: "Yes."

Ms. Callahan: "Excellent. All right. Coming out, do you have CO_2? Did you make it very clear that you're separating the ideas of matter and energy?"

Continued

Classroom Snapshot 8.3 *(continued)*

Students: "Yes."

Ms. Callahan: "At no point did you say glucose was converted into energy?"

Students: "No."

Ms. Callahan: "Good! I'm pretty excited about some of the common ground we have and that you're all in agreement on things that are going in and out after using your molecular modeling kits.[7] Which is great. So, we're good!"

7. Students used molecular model kits to illustrate how molecules are rearranged in a prior lesson.

Large-Scale Modeling

At the large scale, high school students build on their experiences with developing models for cycling materials such as water (MS-ESS2-4) to develop more complex models driven by chemical changes. In particular, they focus on the cycling of carbon through ecosystems and global systems (HS-LS2-5, HS ESS2-6) and on how carbon cycling affects flows of energy through Earth's atmosphere—climate change (HS-ESS2-6). This is a fitting final application of the energy and matter conservation laws, as students use them to study one of the most important socioscientific issues of their lives.

Summary

Energy and matter are unique in that they appear in the *Framework* and the *NGSS* both as DCIs and as a CCC. While the DCIs focus on mechanisms of change in matter and energy, the CCC focuses on *energy and matter as conserved entities.* For all phenomena involving physical or chemical changes, the amount of energy and the amount of matter must stay the same. The conservation laws are especially powerful for two purposes:

- As rules, our models and explanations of phenomena must always follow the conservation laws.
- As heuristics, tracing matter and energy generates good questions to ask about phenomena.

To be useful for making sense of phenomena, conservation rules and heuristics must be applied in conjunction with SEPs (e.g., modeling) and DCIs (e.g., chemical bonding). In order to successfully use conservation laws for three-dimensional sensemaking, students need to master three related strategies: (1) define boundaries and fluxes in closed and open systems that enable tracing matter and energy; (2) identify manifestations of matter and energy in phenomena; and (3) connect models of matter and energy at atomic-molecular, macroscopic, and global scales.

Matter and energy are discussed and represented differently in different disciplines, but conservation laws are applicable for a broad range of phenomena across disciplines. This is the power of this crosscutting concept.

Conservation rules and heuristics are not obvious and must be built over many years. Students' ideas about matter and energy are initially shaped by their everyday language and experience. Students in elementary school often think of matter as solid and liquid "stuff" (not including gases) and of energy as causes of phenomena or resources for living things. These are useful ideas, but matter and energy defined in these ways do not seem to be conserved. As they master more rigorous models and practices, students can initially trace matter and energy through macroscopic systems in local contexts, then build abilities to use matter and energy as conserved entities within and across systems at multiple scales.

Acknowledgments

This material is based in part on work supported by the National Science Foundation (Grant Nos. DRL-1440988 and DUE-1431725).

References

Covitt, B. A., C. M. Thomas, Q. Lin, E. X. de los Santos, and C. W. Anderson. 2019. Relationships among patterns in classroom discourse and student learning performances. Report presented at the Annual International Conference of NARST, Baltimore, MD.

Crissman, S., S. Lacy, J. C. Nordine, and R. Tobin. 2015. Looking through the energy lens. *Science and Children* 52 (6): 26–31.

Dauer, J. M., J. H. Doherty, A. L. Freed, and C. W. Anderson. 2014. Connections between student explanations and arguments from evidence about plant growth, ed. E. A. Holt. *CBE—Life Sciences Education* 13 (3): 397–409.

Driver, R. 1985. *Children's ideas in science*. London: McGraw-Hill Education (UK).

Elkind, D. 1961. Children's discovery of the conservation of mass, weight, and volume: Piaget replication study II. *The Journal of Genetic Psychology* 98 (2): 219–227.

Fortus, D., and J. C. Nordine. 2017. Motion and stability: Forces and interactions. In *Disciplinary core ideas: Reshaping teaching and learning*, eds. R. G. Duncan, J. S. Krajcik, and A. E. Rivet, 33–53. Arlington, VA: NSTA Press.

Jin, H., and C. W. Anderson. 2012. A learning progression for energy in socio-ecological systems. *Journal of Research in Science Teaching* 49 (9): 1149–1180.

Mayer, K., and J. S. Krajcik. 2017. Matter and its interactions. In *Disciplinary core ideas: Reshaping teaching and learning*, eds. R. G. Duncan, J. S. Krajcik, and A. E. Rivet, 13–32. Arlington, VA: NSTA Press.

Michaels, S., and C. O'Connor. 2012. Talk science primer. TERC. Cambridge, MA: TERC. *https://inquiryproject.terc.edu/shared/pd/TalkScience_Primer.pdf*.

Mohan, L., J. Chen, and C. W. Anderson. 2009. Developing a multi-year learning progression for carbon cycling in socio-ecological systems. *Journal of Research in Science Teaching* 46 (6): 675–698.

National Research Council (NRC). 2012. *A framework for K–12 science education: Practices, crosscutting concepts, and core ideas.* Washington, DC: The National Academies Press.

NGSS Lead States. 2013. *Next Generation Science Standards: For states, by states.* Washington, DC: National Academies Press. *www.nextgenscience.org.*

Nordine, J. C., and D. Fortus. 2016. Energy. In *Disciplinary core ideas: Reshaping teaching and learning,* eds. R. G. Duncan, J. S. Krajcik, and A. E. Rivet, 55–74. Arlington, VA: NSTA Press.

Piaget, J. 1951. *The child's conception of the world.* London: Routledge.

Smith, C. L., M. Wiser, C. W. Anderson, and J. S. Krajcik. 2006. Implications of research on children's learning for standards and assessment: A proposed learning progression for matter and the atomic-molecular theory. *Measurement: Interdisciplinary Research and Perspectives* 4 (1–2): 1–98.

Chapter 9

Structure and Function

Bernadine Okoro, Jomae Sica, and Cary Sneider

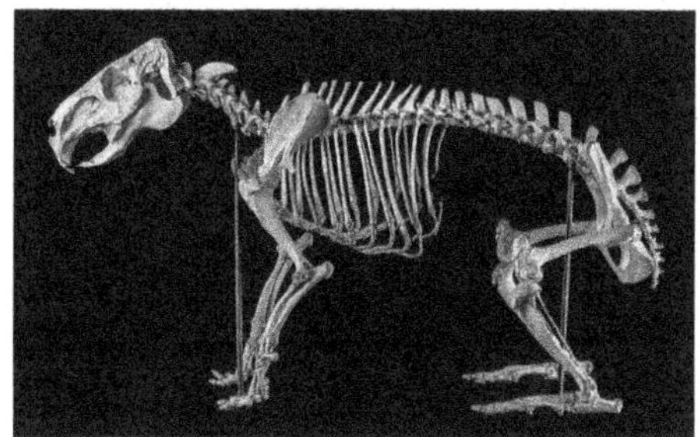

The structure of an animal's spine and ribs bears a striking similarity to the framework of a steel truss bridge. That is not a random coincidence, as they have a similar function. Engineers often borrow ideas from nature to meet human needs, whereas scientists often use engineering methods to understand the functioning of natural systems.

What Is the Crosscutting Concept of Structure and Function and Why Is It Important?

You are probably sitting in a chair right now. Perhaps it's a straight chair in a coffee shop, an adjustable desk chair, a recliner in front of the television; or maybe you're propped up by several pillows in bed. In each case, the structure that is supporting you has been designed to meet your need for comfort so you can carry out one of the activities of your daily life. By "structure," we mean not only the shape of the object you are sitting on, but also the material that it is made from, whether the material "gives" under your weight or is firm so your body conforms to its shape. Structure also refers to how the object is put together. Take a moment to count the number of pieces that make up your chair and notice how those pieces have been joined for maximum strength.

Now look around at the other structures in the room. There is perhaps a lamp or an overhead light allowing you to see; better yet, maybe there is a window enabling you to read this book with natural light, reducing your impact on the environment by using less energy. There are probably tables, rugs, or other things in the room, each of which has a specific function. Then there is the entire building in which you are located, which has a larger function—to meet the needs of several or perhaps hundreds of people. Unless, of course, you are sitting under a tree, which has evolved a number of structures over millions of years to support the functions of growth, reproduction, and survival from environmental threats.

The concept of structure and function cuts across every field of science and engineering. It is a generative concept that provides a means for engineers to analyze a system to see why it may be malfunctioning, as well as a means to generate ideas for solving a problem. In science, understanding the structure and function of a natural system is a common goal. Scientists seek to understand such things as the structure of the nucleus of an atom and its function in holding an atom together, or the puzzling structure of a black hole and its function in shaping a galaxy. For a medical doctor, a large part of professional training involves learning the structures and functions of the human body at every level of scale, from the molecular to the cellular to the level of tissues and organs. The concept of structure and function provides the kinds of insights doctors need to understand how their patients' bodies work and how to repair them when they don't.

Given the importance of structure and function in science and engineering, it is not surprising that it is also one of just seven crosscutting concepts (CCCs) in the *Next Generation Science Standards* (*NGSS*; NGSS Lead States 2013a), which defines the concept in Appendix G as follows:

> *Structure and Function are complementary properties. The shape and stability of structures of natural and designed objects are related to their function(s). The functioning of natural and built systems alike depends on the shapes and relationships of certain key parts as well as on the properties of the materials from which they are made" (NGSS Lead States 2013b, pp. 96–97).*

In this chapter, we share our experiences with helping our students learn to use the CCC of structure and function as a productive thinking tool that helps them become more flexible and effective thinkers, whether they are investigating the natural world through science or solving a problem using an engineering design process. Since structure and function is such an important concept in the world of engineering, our examples will emphasize engineering activities. Nevertheless, we do so with an awareness that engineering is impossible without also taking the relevant science into account.

How Can Structure and Function Serve as a Productive Thinking Tool?

In educational standards of prior generations, an idea like structure and function might have been listed as one of many ideas for students to learn *about*. In the *NGSS*, structure and function and the other crosscutting concepts play a very different role—as ways of thinking that can help students overcome common difficulties when encountering new ideas and perspectives and when learning to apply new skills.

This section of the chapter uses three in-depth Instructional Applications—two at the high school level and one at the middle school level—to illustrate different ways students can learn to use the CCC of structure and function as a productive thinking tool. Subsequent portions of the chapter are intended to illustrate how to develop students' capabilities to use this CCC over the K–12 spectrum and in conjunction with a wide range of science and engineering disciplines. All of the featured Instructional Applications are drawn from our personal experiences.

The first, Instructional Application 9.1, is an example of how structure and function can help students "think outside the box" by framing new questions about problems in engineering and phenomena in science and applying their learning to new situations. The second, Instructional Application 9.2 (p. 202), concerns the difficulty that students have in connecting the structure of matter at the microscopic level with the structure of matter at the macroscopic level—a core concept that undergirds the entire field of chemistry. The third, Instructional Application 9.3 (p. 208), addresses the common misconception that there is no difference between science and engineering by providing an example of the ways in which they are different and by demonstrating how bridging between science and engineering can help people become more effective problem solvers.

INSTRUCTIONAL APPLICATION 9.1

Structure and Function in Buildings by Bernadine Okoro[1]

The students in my third-period spring semester engineering design class consisted of the same cohort of students who took robotics the previous semester. Ninety-nine percent of the school population receives free or reduced-price lunch, and a large

1. Bernadine Okoro, an experienced chemical engineer with master's degrees in teaching and in producing film and video, taught science for 11 years at various schools in Washington, D.C. She also worked at the National Science Foundation as an Albert Einstein Distinguished Educator Fellow and served on the writing team that developed the *Next Generation Science Standards*. This Instructional Application from her high school teaching experience illustrates how the concept of structure and function can help students see a familiar structure in an entirely new light.

Continued

Instructional Application 9.1 *(continued)*

percentage are English language learners. It was a small class, consisting of 3 girls and 10 boys, which gave me time to ask and respond to questions, give my students immediate feedback, and change groups frequently.

Being a newcomer to teaching engineering, I had spent months piecing together foundations for a curriculum. Although I had confidence to teach the course, given my chemical engineering background, I still had to find enough instructional materials to keep the students engaged. One of my most successful units that year came from a workshop I had attended at the National Building Museum in Washington, D.C., called Designing for Disaster. The goals of the unit were to introduce students to the people, processes, and choices involved in designing and constructing disaster-resilient buildings and communities and to increase their awareness of how people affect their surroundings and how their surroundings affect them. Students would also be introduced to forces such as gravity, compression, tension, and shear, and they would illustrate how these forces affect structures.

The lesson featured a large image of the Supreme Court of the United States (Figure 9.1), which had been in the news recently, and which the social studies teacher told me she had recently discussed with my students. I assigned the students to work in groups and discuss three questions: (1) What is this building? (2) What activities take place inside? and (3) How is the building structured to meet those needs?

Figure 9.1. The United States Supreme Court

After seven minutes, the students reported what they had learned from their social studies lessons. They stated that it was the Supreme Court building, that it was part of the judicial branch of the federal government, and that very few court cases are heard at the Supreme Court level. However, they stumbled on the third question. Students' answers focused entirely on what people did in the building rather than how the building itself was structured to serve various functions. I was not surprised since I know that many students struggle to think about a familiar subject in new and different ways. So, I changed the subject to our own school building with questions such as, "How does the shape and material of our school building help it serve its function and meet the needs of the people who use it?" As students warmed up, I shifted the topic back to the Supreme Court building, and they had a lot more to say about it.

Continued

Instructional Application 9.1 *(continued)*

This time, the students noticed the massive columns, sculptures, and overall shape that made it look like a building from ancient Greece. When they began to ask questions, I shifted gears again and described a few interesting facts to enrich the discussion about how choices of materials and structures can be used for different purposes. For example, Vermont marble was used on the outside of the building to give a grand impression, as were two great bronze doors, each weighing 13,000 pounds, with sculptures depicting the history of the legal system.

I ended the lesson by explaining that when architects design a new building, they start by learning about the function the building must serve before deciding on size, shape, and materials. To learn about the building's function, they ask questions, such as: How many people will the building need to house? What will they need to do in the building? How long will it have to last? How much money is available for construction? For homework, I assigned the students to reflect on their own homes with the following questions: What structures do you see? How well are the structures designed to meet the functions they have to fulfill?

The next day, the students opened the discussion with descriptions of structures from their neighborhoods. I transitioned to the next lesson by asking the students to think about what might happen if there were a severe storm or earthquake. They'd all seen images on television of the aftermath of earthquakes and tornadoes. I asked them to think about what made the difference between the buildings that survived those events and those that collapsed. Students then viewed a TED Talk by Peter Haas titled "When Bad Engineering Makes a Natural Disaster Even Worse" (Haas 2010). The students concluded from the video that buildings that suffered the most damage from the earthquakes in Haiti and Chile were not properly structured to resist the forces from those earthquakes.

I explained that good engineering meant that a building could withstand two types of forces that architects call "loads." Dead loads include permanent parts of the structure itself, such as columns, beams, nuts, bolts, windows, and doors. Live loads include the weight of temporary elements, such as people, equipment, and furniture. I explained that it's important for architects and engineers to use the same words across different structures so they don't make any mistakes, as a mistake in a building can be fatal.

Next, I asked the students to work together in teams of two to walk around the room identifying live and dead loads of our school building. I took care to pair students with higher and lower oral proficiencies so they could help each other, but I also encouraged them to use whatever language was easiest for them. I enjoyed hearing the Spanish speakers categorize the walls, floor, ceiling, windows, and doors

Continued

Instructional Application 9.1 *(continued)*

as *muerto* and the furniture and each other as *vivendo* and argue about what's "live" and what's "dead."

To introduce what can happen to buildings if they are not engineered to survive natural disasters, I showed them several images of damaged buildings. They examined the images individually for a few minutes and then discussed them in small groups. Some students thought the damage might have been caused by tornadoes. Others noted the damage to foundations, which was a clue that the damage was caused by earthquakes. (Although earthquakes are not common in Washington, D.C., these students had experienced a significant earthquake that caused damage to many government buildings in 2011.)

To illustrate the different types of forces that a building needs to withstand during an earthquake, I had the entire class use their hands to experience the forces of tension, compression, and shear. I demonstrated as I gave the following instructions: "Now place your palms together, with your elbows bent, and press your palms against each other. You should be experiencing a pushing force. This force that you are experiencing is called compression. Next, lace your fingers together and try to pull them apart as hard as you can, but don't let go. The force you are feeling is called tension." To introduce the shearing force, I handed out sheets of paper from the recycle box. I told students to tear a sheet in half and explain what they did in order to tear it. Although it was challenging for them to explain an action they took for granted, they could see with some help that they had to pull the sheet of paper in two opposite directions at once. I explained that the force they witnessed is called a shearing force. I illustrated what a shearing force could do during an earthquake by showing the image of a large apartment building damaged in the 1989 earthquake in the San Francisco Bay area (Figure 9.2). By looking at the image, the students could see that when the ground moved sideways, the building resisted the sideways force and the walls on the ground floor gave way.

Figure 9.2. San Francisco apartment building damaged during 1989 earthquake

I asked the students how this building could have been better engineered to prevent the damage. They noted that the upper part of it was undamaged, and they could see a car parked inside the building on the ground floor, indicating that it was

Continued

Instructional Application 9.1 *(continued)*

a garage. They thought that the parking garage must have made the walls on the ground floor weak, and placing a garage on the ground level was not such a good idea in a city that has a lot of earthquakes.

I explained that the wall between the concrete foundation and the rest of the building is called a shear wall. During an earthquake, the ground moves sideways, while a house tends to remain stationary, resisting the sideways motion. A strong shear wall between the house and the foundation is needed to allow the house to move with the ground. Their job, during the rest of the period, would be to design and build a section of a shear wall called a bracing panel. I assigned work groups of three and covered the tables with newspapers. Students took a moment to reflect on the materials they could use—six craft sticks, one 4" × 4" piece of cardboard, and a small container of glue. I also reminded them about the engineering design process they learned to use during the previous semester, which involved listing their criteria and constraints and thinking of a number of different solutions and how they could test their designs to see if they solved the problem.

We spent the rest of the 45-minute period building walls. I encouraged the students to look at one another's solutions, explaining that engineers are always interested in how other engineers solve the same problem. The next day, we displayed the bracing walls together. The students discussed which ones they thought would be the strongest and then shared their ideas about how to test them. They decided to build three more identical panels to make a box, place heavy books on top, and give the books a sideways push, just like in a real earthquake.

At the end of the unit, I asked the principal to visit us so the students could show what they had learned. The students started with what they learned about the Supreme Court building. The words *structure* and *function* were clearly part of their vocabulary by then. They explained that the structure of the Supreme Court building, with all its marble and huge doors, was intended not just to allow for daily functions of the court but also to communicate the power and importance of the Supreme Court. They explained how architects needed to consider all of the functions of a building before deciding on its structure. Then they showed off their surviving shear walls, explaining how the structure of diagonal braces functioned to keep a building together as it moved back and forth during an earthquake. When the principal asked the students if they had any ideas about how their own school might be strengthened to withstand a stronger earthquake, they had several ideas and questions about what might be possible. That last part of the discussion showed me how important it was to improve students' understanding by emphasizing structure and function in many different ways throughout the unit, as my students could now use this CCC as a productive thinking tool to generate ideas and questions when encountering an entirely new problem.

Continued

Instructional Application 9.1 *(continued)*

Since I taught that unit in 2014, I've spent many hours reflecting on ways I could have done a better job. Recalling that my students had no way to understand what I was asking them to do when I showed them an image of the Supreme Court, I realize that I should have communicated the basic concept first. If I were to teach this again, I would start with a simple object such as a pencil, asking students to think about how the shape and material of each part of the pencil helps the system as a whole serve its function. We would then expand the system to include a sheet of paper and consider the larger function of the paper-and-pencil system, and so on, before progressing to something as complex and overwhelming as the U.S. Supreme Court building. What I hope that would accomplish is to help my students see their everyday world as a marvelous array of structures that people have designed and built to make our lives as safe, comfortable, and meaningful as possible.

INSTRUCTIONAL APPLICATION 9.2

Navigating Micro and Macro Structures of Candy by Jomae Sica[2]

The 2009 *Oregon Science Standards* included engineering design for the first time. When these standards were adopted, I started my journey in developing engineering projects that could be used in mainstream chemistry and biology courses. In the beginning, this was a tremendous undertaking as I did not have an engineering background. Moreover, there were very few teachers developing resources at the time, and there were no engineering resources available through the major textbook publishers. Many of the projects I used in those early days were the "guess and check" type of engineering, which left students feeling frustrated, as they had no idea how to gauge what would produce a successful product. Furthermore, I felt that I was not modeling how real engineers go about doing their work. However, as I learned more about engineering and implemented data-informed engineering

2. Jomae Sica is an experienced biology and chemistry teacher in Beaverton, Oregon. She is also a creative curriculum developer, having worked with colleagues to develop a number of integrated STEM programs and as a teacher leader, who served as president of the Oregon Science Teachers Association. As a chemistry teacher, Jomae has long been aware that high school students struggle with bridging between the microscale of atoms and molecules and the macroscale of chemical properties and reactions. In this Instructional Application, she describes a unit she developed that combines a strongly motivating engineering design challenge with controlled experimentation and the CCC of structure and function to help students bridge between the worlds of the micro and macro.

Continued

Instructional Application 9.2 *(continued)*

(as described in the *NGSS*, which Oregon adopted in 2014), I saw an exponential increase in student engagement. Students who were normally disengaged in class were asking if they could come in during lunch or after school or take their projects home to continue working on them. It was then that I realized how important engineering was to the mainstream curriculum and that it needed to play a prominent role in my classroom.

Many students that I teach dream of eventually working at one of our top local companies, such as Nike or Intel. I often ask them if they know how these companies decide what materials to use to make athletic apparel and processing chips. The field of materials science, in which scientists and engineers continually try to improve our products by studying, developing, and testing new materials, embodies the CCC of structure and function. It is a vast field that encompasses everything from pharmaceuticals and food to technology and apparel. Nike continually strives to develop and test materials for clothing and shoes that provide athletes with the ability to reach their peak performance by staying cool and absorbing the shock of impact while simultaneously being sustainable and fashionable. Intel employees in this same field strive to increase processor speed while continually shrinking its size to produce devices that are smaller, and thus make it easier to integrate technology into our daily lives. These are two completely different businesses and products, but both products' functionality is determined by the structure of particles that make them up.

The concept of structure and function is prominent in the bonding unit I developed with fellow chemistry teachers. The purpose of the unit is to help students connect materials science with something familiar—the structure and function of candy. The unit begins by showing students pictures of three different types of candies as the phenomena. Then I ask them to tell me what is similar or different about the properties of the candies, focusing on taste, texture, and hardness. Candy is a highly engaging topic for teenagers because most of them consume a lot of this product. Therefore, they have funds of knowledge and background to build on, yet few have made candy themselves or have read the ingredients in their favorite candy. I have them consider the major ingredients in each candy and ask them to write one question about how a substance in the candy might affect its properties. They share their questions among their table groups and then select the question they are most curious about to share with the class. As we progress through the lessons in our unit, we slowly build up the background knowledge required to answer their questions. For instance, students might ask, "Why are Dum-Dums hard while Skittles are soft and chewy?" or "Why are Smarties and Skittles tart?" Both questions are grounded

Continued

Instructional Application 9.2 *(continued)*

in understanding intramolecular and intermolecular forces, which is the focus of the unit. The unit culminates in an engineering project where students design a candy with specific properties.

During the unit, students engage in two investigations where they test bulk properties as a way to infer the level of intramolecular or intermolecular forces within the substance. In the inquiry activity about intramolecular forces, students are asked to predict the properties of a sugar substitute. They quickly realize they will need to investigate the solubility, conductivity, and melting point of several other substances and find a pattern in the types of elements and their attractions in order to make a data-informed prediction. At the conclusion of this investigation, the students find that all substances with ionic bonds dissolve in water, have high melting points, and are good electrolytes, whereas substances with covalent bonds generally do not dissolve in water (with some exceptions like sucrose), have low melting points, and are poor electrolytes.

Continuing with their investigation of intermolecular forces, students research bulk properties that indicate the level of attraction between molecules and choose one property with which to design their investigation. All groups test the same five liquids for their properties so their data can be analyzed for patterns based on the polarity of the molecule. When comparing data sets, it becomes clear that substances that are polar exhibit high levels of surface tension, capillary rise, heat capacity, and density, whereas nonpolar substances have low levels of these properties (although groups may have tested these properties in different ways). In both investigations, the structure of matter at the submicroscopic level determines how properties function at the macroscopic level.

The unit culminates with students designing a candy with specific properties, and then explaining the rationale for their design in terms of the relationship between the structure of molecules and the properties of the candy. First, students read a Request for Proposals (RFP); identify the problem, criteria, and constraints; and do an initial investigation to design a candy that would be similar to a Starburst. They are assigned to research groups that seek to alter one ingredient or preparation method from a base recipe that will affect the final taste, texture, or hardness of the candy (Figure 9.3). For example, the group that will test emulsifiers makes a batch with butter, soy lecithin, and mustard seed to see which will do the best job of evenly distributing the fat and sugar throughout the candy. Since there are normally

Continued

Instructional Application 9.2 *(continued)*

8–10 groups in my classroom, each of the factors being tested—fat content, emulsifier type, pH, and cooking temperature—is duplicated to ensure high-quality results that can be utilized by all of the teams in the design phase.

Figure 9.3. Chemistry student making her first batch of candy

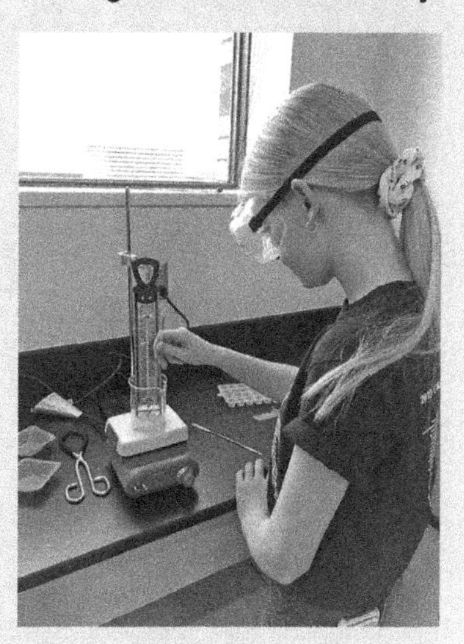

Students return the next class period to test their candy for taste, texture, and hardness. This is by far their favorite aspect of the project. To ensure safety, we utilize glassware and other supplies that have not been used for other chemistry labs and are designated as "food safe." I always enjoy seeing the looks of delight or disappointment as my students try the samples of the three variations they created. For example, the groups testing tartness add 1, 2, and 3 grams of citric acid in each batch to see the effect on pH; by the bitter grimaces, it is always obvious which batch contains the maximum amount. When all members of each group have tasted all three batches, the group records their qualitative tasting notes and decides which variation(s) they will recommend to other classmates during the subsequent board meeting.

After taste-testing, it's time for students to dig into research and utilize their chemistry knowledge gained in the unit to explain how the structure of the molecules involved affect the taste, texture, and hardness of their initial prototypes. For the group that altered fat content, they work through a series of questions and resources to learn how the fat content affects the chewiness of the candy. Using what was learned earlier in the unit about the solubility of nonpolar molecules, students explain their results. They reason that higher fat content in their candy makes the candy less likely to stick to their teeth since fats are nonpolar molecules and the enamel in our teeth is composed of the polar minerals calcium and phosphorous. The emulsifier group learns how an emulsifier has a hydrophobic and hydrophilic end, which allows the fat and the sugar to be equally distributed. This prevents separation of the molecules, making the candy have a uniform taste and appearance. Both groups are reminded to explain how molecular structure explains the macroscopic properties of the candies.

Continued

Instructional Application 9.2 *(continued)*

Figure 9.4. Communicating results of candy investigations

Source: Jomae Sica

This figure shows a poster explaining how fat affects the chewiness of candy (left) and students gathering information from peers during the gallery walk (right).

In the next class period, students prepare whiteboards to communicate the results of their investigations in a Claim, Evidence, Reasoning format using the data and research from the previous class. Students do a gallery walk through the boards to collect the recommendations from different research groups and then return to their original group to generate a candy recipe that fulfills the criteria from the RFP and is based on the recommendations from their peers. (See Figure 9.4.) Their candy formulation must include a budget and a food label to aid in determining pros and cons of their solution as they compare it to the cost and nutritional information in a Starburst.

Much to their delight, nearly all of the students are pleased with their final candy formulation. In one class I taught, a group proclaimed that they were quite proud of themselves, as their texture perfectly matched that of a Starburst. Most students go with the recommended 2 grams of citric acid, which gives a pH of 3.5, resulting in what students tend to feel perfectly complements the lemon flavor without overpowering the sweetness or causing them to cringe when it touches their tongue. Most also choose a cooking temperature between 210°F (99°C) and 220°F (104°C), which leaves a sufficient amount of moisture in the candy so it can be bitten into easily. Creatively, some groups decide to mix the shortening and butter because, although many groups identify butter as the best emulsifier, they feel using *all* butter imparts a strange taste to what is supposed to be a citrus candy. Most of all, they realize how

Continued

Instructional Application 9.2 *(continued)*

important it is for food engineers to understand the structure and function of atoms and molecules in order to design the foods we consume every day.

For the final part of the project, students pick a culture and research a candy that is representative of or unique to that culture. My students come from diverse backgrounds, and this activity is a fun way to relate what they learn in the project to something that is culturally relevant to them. Not only must they provide evidence that the candy is culturally significant, but they also need to find out the ingredients and the preparation methods of the candy so they can use this information to explain how altering the molecular structure during cooking or other processes affects the candy's bulk properties. Through this activity, I learn about my students' travels and family traditions as they explain the origin and properties of various sweets. By reading their projects, I become *their* student, learning from their worldly knowledge and demonstrating that in a student-centered classroom, information flows in both directions.

During the second iteration of this unit, I added research questions that explicitly related the structure of the ingredients to the properties of the candy. That helped me take a big leap forward in making the CCC of structure and function explicit for students. Oftentimes, it is easy for the CCC to become the subliminal message that goes over the heads of students because it is not as concrete as the science and engineering practices (SEPs) or disciplinary core ideas (DCIs). That is why I find it is essential to continue making these connections more apparent. The explanations that students have been able to provide for their results have demonstrated to me that they are capable of synthesizing multiple ideas to explain broad topics like structure and function as long as they are provided with appropriate scaffolding, such as word banks and sentence frames. An example of these scaffolds for the part of the project where students model and explain how an emulsifier works in their candy mixture is shown below:

Word Bank

polar, nonpolar, hydrogen bonds, London Dispersion, hydrophobic, hydrophilic, emulsifier, stable, separate

Choose one of these sentence frames to get you started, or create one of your own.

- _____ contains _____ and tends to …
- _____ is characterized by several distinct features such as …
- _____ is _____ and functions as ...

Continued

Instructional Application 9.2 *(continued)*

This Instructional Application brings out a second important aspect of structure and function—that it operates at different scales. Specifically, the structure of a material at a microscopic scale affects the function of the material at the macroscopic scale. In this case, the CCC of structure and function provides students with insight into the core ideas of the discipline and complements their use of SEPs and DCIs. The engineering challenge of designing a new kind of candy makes the experience come alive for the students. Each time the CCC is applied in a new situation, students' understanding of it deepens and becomes more flexible and applicable to a wider variety of phenomena.

INSTRUCTIONAL APPLICATION 9.3

Bridging the Science-Engineering Divide by Cary Sneider[3]

This five-day lesson on electromagnets began by showing the students an image of a large electromagnetic crane picking up scrap iron and steel and explaining that the crane operator threw a switch that caused the metal to stick to the crane, then turned off the switch to allow the metal to fall in a different location, such as into the back of a truck. (Today, I would look for a short video online to illustrate that idea.) I then gave each team of three students a large steel nail, a length of insulated wire, paper clips, and a battery to make an electromagnet. The purpose of this initial activity was for the students to become familiar with the materials, to see how to make an electromagnet, to explore what an electromagnet does, and to raise some initial questions. (See Figure 9.5 for images of different electromagnets.) Although all of the students were familiar with magnets, some were surprised that they could make a device that can attract metal objects without using permanent magnets.

The next day, each team received a plastic bag with a large and small nail, three lengths of wire, a battery, and a slip of paper with instructions. Unbeknownst to the students, half of the class had a *science* assignment to investigate the factors that

3. Cary Sneider's experience as a teacher stretched from Maine to California, from Central America to Micronesia, and from fifth-grade classrooms to PhD committees. Much of his five decades as a science educator involved curriculum development and teacher education. As a lead for engineering on the *Next Generation Science Standards*, he has devoted considerable time to communicating the value of engineering as a way of thinking that is complementary to science. In this instructional approach, he uses a middle school activity to illustrate how the CCC of structure and function applies equally well to science and engineering and how these two disciplines are on the one hand distinct, and on the other inseparable.

Continued

Instructional Application 9.3 *(continued)*

determine the strength of an electromagnet's magnetic field and the other half of the class had an *engineering* assignment to design an electromagnet that is as strong as possible.

Given the nature of the materials, the CCC of structure and function naturally provided a "lens" for examining the arrangement of the materials and carrying out the assigned task (Fick 2018). What is most interesting is that the students with the different assignments used the lens in different ways. Because the class had learned about controlled experiments in previous classes, the students with the science assignment took data by wrapping each of the three lengths of wire around one of the nails and measuring the function of the electromagnet by noting how many paper clips it picked up, using the same procedure for each different length of wire. They did the same experiment with both nails. The students with the engineering assignment were also systematic, but they tried many more arrangements of the nail and wire, sometimes using the sharp end to pick up the paper clips, sometimes the head of the nail, and sometimes both. Two of the engineering groups joined forces and used both of their batteries and all of their wire to make one electromagnet, picking up all of their paper clips.

Figure 9.5. Different electromagnets

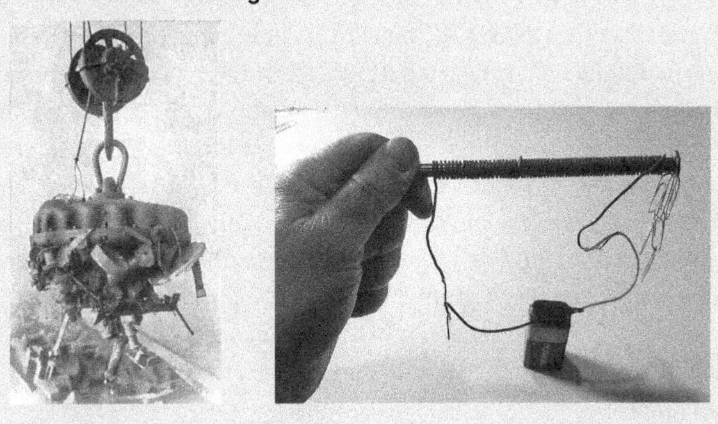

The electromagnetic crane at left is structured to pick up as much ferromagnetic metal as possible. The electromagnet at right is structured to determine the effect of each variable (electrical current, turns of wire, and size of steel core) on the strength of the magnetic field.

As groups reported out the next day, some of the students had puzzled looks. Finally, one student challenged another group: "You guys changed too many different things. How could you determine which was most important thing that made the magnet stronger?"

Continued

Instructional Application 9.3 *(continued)*

"We didn't need to," a member of the group responded. "We started out that way, but we kept thinking of new ways to use the materials to make the magnet as strong as possible. Our magnet picked up a lot more paper clips than yours did."

Eventually, the students figured out that they had different assignments. I confessed that I had given out different assignments so they could see how scientists and engineers used the same crosscutting concept—structure and function—for different purposes. I used a Socratic approach to engage the students in discussion, leading with questions about the purpose of their team, what structures they explored, how they measured the outcomes of their explorations, and the nature of their conclusions.

Although I hadn't planned on it, the students who had the science assignment insisted that the engineers took the wrong approach. They argued that by skipping the step of controlled experiments, the engineers "didn't know what they were doing." What they should have done is *first* find out which variables made the biggest difference, and *then* apply their knowledge to try out different designs to create the strongest electromagnet. The students with the engineering assignments pushed back, saying they thought that would be a waste of time.

I was pleased that the students were growing in their awareness that structure and function can be used as both a lens and a bridge between science and engineering and that they had increased their understanding of the DCI regarding the strength of the magnetic field generated by an electric current. By comparing the different assignments, the students also learned more about some of the key differences between the practices of scientific investigation and engineering design and how they can complement each other. However, I was troubled with how the students tackled the engineering design activity. They were highly motivated, with enthusiastic "high fives" when they put the batteries together and their electromagnet picked up all the paper clips. Yet, the students were so fixated on the goal of picking up as many paper clips as possible that they were not concerned about understanding the system. On the other hand, I suppose that too much enthusiasm is not the worst problem I can imagine.

I concluded the day's lesson by sharing my perception that both groups of students had some good points. The group that had the science assignment was learning a lot about how the structure of the electromagnet affected its function. In their investigations, some teams spread out the turns of wire along the entire nail, whereas others bunched up the coil on one side. Some teams compared the effectiveness of the large and small nail, whereas one team used both nails. That's a great way to gather scientific information, which can later be applied to designing a strong electromagnet.

Continued

Instructional Application 9.3 *(continued)*

In other words, I supported the idea that "a scientific approach" to engineering is an excellent way to develop the best possible solution. I also complimented the teams that had the engineering assignment on their collaboration with the other teams and their creative approaches, such as using both ends of the nail at the same time to pick up the paper clips.

I explained that scientists and engineers often work together, using the idea of structure and function as a bridge between the two fields. In the case of building an electromagnet, they might start working together to define the engineering design problem, then "cross the bridge" to science in order to scientifically investigate the variables involved. Once they have a good handle on the science, they can cross the bridge back again and apply what they learned to design the strongest possible electromagnet by combining the engineers' creative ideas with the scientists' discoveries. Structure and function could also be used as a lens for coming up with different questions and ideas.

I asked the students to think about the day's lesson and to come in the next day with a plan that would help them learn more about how the structure of the electromagnet affects its function *and* how to make a stronger electromagnet. The next day, I reorganized the teams so those who had the science assignment were now working with students who had received the engineering assignment. I showed the image of the electromagnetic crane again (see Figure 9.5) and challenged the teams to apply what they learned to envisioning how such a device could be made as strong as possible. I asked them to sketch what they thought was inside the huge magnet and then test their idea of a strong magnet by giving them a battery and a variety of steel shapes and wire. We ended the unit by having the teams share their experiences and new insights. After each team's presentation, I asked them to share how they used the concept of structure and function as a lens to think about their task in different ways or as a bridge to move back and forth between science and engineering in order to accomplish their goal.

The teams were very engaged in the final activity. Some ideas were common to all teams, such as using more electric current; however, there were large differences in their drawings and in the shapes of their final electromagnets. I was especially pleased that, with some prodding, each team was able to provide a scientific explanation, using their own terms and current thinking, for why they thought their design would work as expected.

In all three instructional approaches described in this chapter, students used the CCC of structure and function as a lens to look at the subject matter in a new way. In Instructional Application 9.1 (p. 197), students considered the various ways buildings are structured to serve different functions and keep people safe. In Instructional Application 9.2 (p. 202), students were guided to examine how the molecular structure of food determined its macroscopic properties, including taste and texture. In Instructional Application 9.3 (p. 208), students used structure and function as a lens to examine how different structures of an electrical circuit generate a magnetic field.

The students also used the CCC as a bridge between scientific investigation and engineering design. This was evident in the building unit, in which students applied their understanding of forces to design and build a shear wall. It was also clear in the chemistry activity, in which students' scientific knowledge of molecular bonding helped them design candies with specific macroscopic properties. The idea of a bridge between science and engineering was extended in the electromagnet activity in order for students to become explicitly aware of how scientific investigations can help support and inform an engineering design process.

How Can We Build Understanding of Structure and Function Over Time?

Learning requires many different experiences over an extended time frame. No single lesson is a silver bullet; however, experiences do build on one another, and more engaging and memorable experiences make a deeper impression. As we've illustrated, the CCC of structure and function plays out in architecture, chemistry, and designing electrical circuits. Each of these examples is in the physical sciences, as most of our teaching experiences have been concentrated in those fields. However, we don't want to leave the impression that structure and function should be limited to physical science. Therefore, the purpose of this next section is to point out ways that structure and function can provide insights into every discipline and subdiscipline of science and engineering, starting with the life sciences.

In biology, structure and function at the molecular level helps explain essential processes, such as photosynthesis and respiration, as well as the properties of tissues and functions of organs in plants and animals. At the macro level, structure and function can be applied to the functioning of whole organisms and whole ecosystems. In agriculture, structure and function is essential in thinking about the best ways to prepare a field for planting, weeding, and harvesting while avoiding erosion during heavy rains. It's also important in processing of foods, such as separating the kernels of wheat from the husk and milling it into flour to make bread. In that case, structure and function concerns both the plant structures that need to be taken into account and the structures of the machines that are needed to accomplish different functions along the journey from sheaves of golden grain to freshly baked bread. Medicine is another area where structure and function is essential, from the development of pharmaceuticals to the design of surgical equipment and procedures. Biomedical engineering requires an understanding of the structure and function of both the human body and technologies to meet medical needs. For example, consider the negative pressure ventilator (once known as an iron lung) that helps people who have been paralyzed breathe by taking on the role played by the diaphragm to expand the chest cavity and allow the lungs to draw in air through the trachea.

A series of curriculum units called Biology in a Box (Riechert and Post 2010) introduces students to the ways that engineers have used natural structures to inspire technological solutions (a process known as *biomimetics*) and invites students to do similar design activities themselves. For example, From Skeletons to Bridges—which inspired the pictures at the beginning of our chapter (p. 195)—is based on the metaphor that "bridges are simply well-designed skeletons." This unit would be an excellent complement to the unit on building structures described in Instructional Application 9.1 (p. 197), as it can help students develop a deeper understanding of compression, tension, and bending as they construct bridge structures using strands of dry spaghetti, attempting to achieve a maximum span while maintaining structural integrity. Biomimetics can also help biology teachers make the leap from traditional science teaching to a blended science-engineering approach because activities engage students in learning biology simultaneously with developing skill in applying an engineering design process to solving human problems. For example, in the From Skeletons to Bridges unit, students learn how the evolution of skeletal structures over millions of years provides productive ideas for engineers charged with designing bridges that are both strong and light.

In geology, the science of plate tectonics is all about the structure and function of huge crustal plates and the dynamic movements of materials under the surface that explain such diverse phenomena as earthquakes, volcanoes, mountain formation, and the journeys of entire continents across the planet over hundreds of millions of years. The structure and function of natural materials also helps explain why different natural resources are found in different locations. Petroleum, for example, was formed when organic material settled to the bottom of a sea or lakebed. Now large deposits are found both under the ocean and on land that was covered by oceans in the distant past. Petroleum engineers apply their understanding of the formation of petroleum to start their search for new oil deposits, using a number of instruments to probe underground structures to see where petroleum is more likely.

The primary message of this part of our chapter is that the CCC of structure and function applies in all areas of science and engineering, and it should be made explicit so students develop a deep and more flexible understanding that they will naturally use when they encounter a new area to study or a new problem to solve. A secondary message is that structure and function is nearly always connected to other CCCs, SEPs, and DCIs. For example, the need to apply structure and function is typically initiated by an SEP, such as defining a problem, developing and using a model, explaining a phenomenon, or developing a solution to a problem. The CCC of structure and function must also be applied to a DCI, as the concept of structure and function only derives meaning within a given context. Regarding its connection to other CCCs, structure and function is commonly paired with cause and effect, as people naturally think about how a given structure enables a specific function. For example, the relationship between the molecular structures in the candy unit and the properties of different candy formulations is explained by the strength of different kinds of bonds caused by different atomic and molecular structures. Other pairings are common, as well. For instance, when a petroleum engineer investigates the structure of a deposit of oil, the next step is to envision the surrounding strata as a system that contains the oil and to develop a model in order to estimate the amount of oil available.

Structure and Function Across the Grade Bands

Given our collective experience, we have focused on examples of structure and function at the middle and high school levels. However, we don't wish to leave the impression that curriculum developers and teachers should wait until middle school to introduce this CCC.

Guidance for how students are expected to use structure and function at increasingly sophisticated levels as they mature is offered in Appendix G of the *Next Generation Science Standards*. In order to communicate what these activities would look like in the classroom, here are brief quotes from the table related to structure and function in Appendix G (NGSS Lead States 2013b, p. 87), along with a few examples of instructional materials that enable students to use the CCC of structure and function at an appropriate grade level.

Grades K–2: Excerpt

"In grades K–2, students observe [that] the shape and stability of structures of natural and designed objects are related to their function(s)." Naturally, the youngest students are not expected to use the phrase *structure and function*, but they should have a chance to recognize that different shapes and materials have different effects in nature and can be used by people for different purposes.

Grades K–2: Instructional Materials

The Best of Bugs: Designing Hand Pollinators (Higgins 2011) involves students in designing an object that a farmer or gardener can use to pollinate flowers in cases where natural pollinators may not be available. When engineering, the students need to think about the biological structures of the flower ("as nature did") and to carefully examine flowers so they can design a hand pollinator that will reach where the pollen is located for that flower. The structure of the flower affects the structure of the hand pollinator if it's going to function to actually reach the pollen and then transfer it.

Grades 3–5: Excerpt

"In grades 3–5, students learn that different materials have different substructures, which can sometimes be observed; and substructures have shapes and parts that serve functions." As illustrated in the following unit, substructures generally refer to what students can see, such as the parts of fish ladders that help them swim upriver.

Grades 3–5: Instructional Materials

In Fish Friendly Engineering, students learn about the challenges that salmon face when they encounter a dam that bars their way to their spawning grounds. They brainstorm possible ways to modify the dam and play an animated PowerPoint game to learn about various ways that engineers have developed "fish ladders" to help the salmon survive. The activities help students understand the importance of various structures and substructures of a human-modified river system. Fish Friendly Engineering is a free activity available from *www.teachengineering.com*.

Grades 6–8: Excerpt

"In grades 6–8, students model complex and microscopic structures and systems and visualize how their function depends on the shapes, composition, and relationships among its parts. They analyze many complex natural and designed structures and systems to determine how they function. They design structures to serve particular functions by taking into account properties of different materials, and how materials can be shaped and used."

Grades 6–8: Instructional Materials

Vertical Farms: Fresh Food for Cities (Hacker et al. 2017), from the International Technology and Engineering Education Association (ITEEA), begins by explaining the environmental challenges caused by massive migration from farms to cities that is still ongoing. Growing food miles from where it will be consumed affects the environment in numerous ways, from straining freshwater supplies to atmospheric pollution from transportation, in some cases over thousands of miles. To counter these problems, students learn about a new movement to grow food close to where it will be consumed through hydroponics—engineered systems for growing food without soil. Middle school students learn about the science and engineering of hydroponic systems, then use the knowledge and experience gained from these hands-on activities to design vertical farms—structures that can be attached to the walls of an apartment building, serving as a source of fresh food for residents and neighbors. By using fundamentally different structures to serve the same functions as traditional flat fields, vertical farms save energy while allowing for the preservation of natural habitats that would otherwise be turned into farmland. This curriculum is part of the Engineering for All project, which is available free of charge to educators in states that are members of the STEM ITEEA Consortium. See the following website for more details: *www.iteea.org/STEMCenter/Research/136896/EfA136957.aspx*.

Grades 9–12: Excerpt

"In grades 9–12, students investigate systems by examining the properties of different materials, the structures of different components, and their interconnections to reveal the system's function and/or solve a problem. They infer the functions and properties of natural and designed objects and systems from their overall structure, the way their components are shaped and used, and the molecular substructures of their various materials."

Grades 9–12: Instructional Materials

Hands-on activities in which students have opportunities to construct and test solutions to problems are just as important at the high school level as at the middle and elementary levels. Designing structures for buildings, formulating new recipes for candy with predictable properties, and constructing and testing electrical circuits, described previously in this chapter, are just three of the many kinds of hands-on activities that allow students to become familiar with structure and function in a variety of disciplines. At the high school level, unlike in the lower grades, cause and effect plays a more important role, as students are expected to apply theory and other aspects of mathematical, computational, and scientific thinking to how different structures can be expected to behave in different circumstances.

Summary

The introduction to this chapter invites readers to look around at their immediate environment and see that structure and function is everywhere. We can hardly understand either the natural world or the designed world that we interact with every day without encountering thousands of structures whose functions enable us to survive and thrive.

As teachers, we wanted the three Instructional Applications in this chapter to communicate how we use structure and function in the classroom to provide our students with productive thinking tools. Each Instructional Application had a different purpose.

We chose buildings as the first Instructional Application because buildings offer a very clear example of how different shapes and materials serve different purposes. In that Instructional Application, students used structure and function to analyze current buildings and to design a building element to improve safety. As a lens, the concept of structure and function helped students think "outside the box" by generating questions they might not have otherwise considered, such as "What might happen to endanger the people who will live in the building I'm designing?" The second Instructional Application illustrated how structure and function can help students envision how changes at the molecular scale can affect the bulk properties of substances—a core concept that undergirds the entire field of chemistry. The third Instructional Application illustrated the differences between the practices of science and engineering; it also showed how structure and function can serve not only as a lens but also as a bridge between the two fields, allowing people to investigate and solve meaningful problems. We also selected these examples of classroom practice because the engineering activities were highly motivational, which made students open to applying structure and function as a means of helping them accomplish their goals.

The subsequent portion of this chapter emphasized that in order to build students' understanding of structure and function over time, it is important for science teachers to recognize that the CCC of structure and function serves a wide variety of purposes in *every* discipline. It can serve as a lens to probe deeply into just about any subject or as a bridge to connect not only science and engineering but also different fields of science and the humanities. In some cases, students may investigate a structure with the goal of determining its function; in others, students may observe a function and ask what structure might be responsible.

The last part of the chapter concerned teaching structure and function across the grades. In this section, our intention was to emphasize that students need to start learning about structure and function at the earliest grade levels and that teachers need to gradually increase the challenge as students mature so that at any given stage the level of difficulty is slightly beyond their comfort zone, allowing their productive thinking tools to stretch and grow.

References

Fick, S. J. 2018. What does three-dimensional teaching and learning look like?: Examining the potential for crosscutting concept to support the development of science knowledge. *Science Education* 10 (2): 5–35.

Haas, P. 2010. When bad engineering makes a natural disaster even worse. Lecture presented at TEDGlobal 2010, Oxford, England. *www.ted.com/talks/peter_haas_when_bad_engineering_makes_a_natural_disaster_even_worse.*

Hacker, M., D. Crismond, D. Hecht, and M. Lomask. 2017. Engineering for all: A middle school program to introduce students to engineering as a potential social good. *Technology and Engineering Teacher* 77 (3): 8–14.

Higgins, M. 2011. *The best of bugs: Designing hand pollinators.* Engineering Is Elementary. Boston: Museum of Science.

NGSS Lead States. 2013a. *Next Generation Science Standards: For states, by states.* Washington, DC: National Academies Press. *www.nextgenscience.org.*

NGSS Lead States. 2013b. *NGSS Appendix G: Crosscutting concepts.* Washington, DC: National Academies Press. *www.nextgenscience.org.*

Riechert, S. E., and B. K. Post. 2010. From skeletons to bridges and other STEM enrichment exercises for high school biology. *American Biology Teacher* 72 (1): 20–23.

Chapter 10

Stability and Change

Brett Moulding, Kenneth Huff, and Kevin McElhaney

Stability and change is an important science and engineering concept used to make sense of the universe, the world, and our own bodies. It is central to making sense of science phenomena and engineering problems by focusing observations on specific aspects of a system that cause it to change or remain stable over time. Stability and change can comprise the fundamental lens that defines the very nature of phenomena (as something that shifts from a stable state to a state of change, or vice versa) and problem solutions (as intentional efforts to achieve either stability or change in a system). In this chapter, we (1) describe how stability and change is important for helping students recognize and make sense of phenomena, (2) provide a brief history and context of stability and change as a crosscutting concept (CCC), (3) explain how stability and change helps students make sense of systems, (4) discuss the complementary nature of stability and change relative to practices and other crosscutting concepts, and (5) illustrate how stability and change may function differently across disciplines.

Teachers can focus student attention on how stability and change relate to time, highlight key distinctions between different states of a system, and support the investigation of causal relationships within systems (Moulding et al. 2020). The following example provides insights into how the concept of stability and change helps students investigate a locally relevant phenomenon. Niagara Falls appears to be a stable geological feature for most students and is relevant for those in western New York. Middle school students in Williamsville, near Niagara Falls, were presented with information describing changes that caused the location of Niagara Falls to move by more than a meter per year for the last 560 years. After students gathered and evaluated information for evidence to support an explanation for the causes of changes to Niagara Falls, the teacher provided each student group with rocks found at the site of the falls—limestone, shale, and dolostone. Students explored by rubbing and hitting the rocks together, and they discovered that the rocks differed in hardness. The students began to develop questions about why the location of the falls changes and how differences in the rocks are part of the mechanism for the changes. These questions became the basis of an investigation to construct an explanation for the cause of changes in Niagara Falls. The investigation continued with students building models to explain changes in the falls. Students used models to communicate the changes in the location of the falls over time. In this investigation, students used the concept of stability and change to explain why an apparently stable waterfall is changing position annually and discovered that what seems to be stable on one time scale constitutes change on another time scale. Using the lens of stability and change helped provide a clearer focus for students to investigate the rate of change of the location of the falls.

Why Is Stability and Change a Crosscutting Concept?

The broad ideas that underlie stability and change as a crosscutting concept in *A Framework for K–12 Science Education* (the *Framework*; NRC 2012) and the *Next Generation Science Standards* (*NGSS*; NGSS Lead States 2013) are not new to U.S. science education. The *Benchmarks for Science Literacy* from 1993 included a set of four common themes across all of the sciences: (1) systems, (2) models, (3) constancy and change, and (4) scale (AAAS 1993). From these four, constancy and change are consistent with the *Framework*'s CCC of stability and change. In addition, the *National Science Education Standards* (NRC 1996) identified five unifying concepts and processes: (1) systems, order, and organization; (2) evidence, models, and explanation; (3) change, constancy, and measurement; (4) evolution and equilibrium; and (5) form and function (NRC 1996). From these five, change, constancy, and equilibrium are represented as part of the *Framework*'s CCC of stability and change.

The *Framework* describes the CCC of stability and change as follows: "For natural and built systems alike, conditions of stability and determinants of rates of change or evolution of a system are critical elements of study" (NRC 2012, p. 84). The *Framework* also describes how stability and change meet a set of three criteria for inclusion as one of the seven CCCs. In the following sections, we describe how the CCC of stability and change meets each of these three criteria.

Criterion 1: It Has "Explanatory Value Throughout Much of Science and Engineering."

Most phenomena can be characterized in terms of the stability of a system or how changes in systems cause a phenomenon to occur. Phenomena are often characterized as a change in the components of a system or as changes in the relationships among interacting systems. When systems are not changing, investigations focus on the conditions necessary for the system to maintain stability. Stability and change can be used to explain phenomena across many disciplines and contexts. Here are a few examples:

- The longer a golf ball falls, the faster it moves. The **change** in the speed of a falling golf ball is a result of Earth's gravitational force on the ball. When the ball is at rest on the ground, it is in a **stable** state due to balanced forces.
- The water in the Chesapeake Bay is clearer when oyster populations are higher. When oyster populations are **stable,** so is water clarity. **Changes** in the oyster population cause a **change** in the amount of particulates filtered out of the water, affecting its clarity.
- More sugar will dissolve in hot tea than in iced tea. At a constant temperature, the concentration of dissolved sugar achieves **stable equilibrium**. Adding thermal energy to the water changes the number of molecular collisions that occur, resulting in more sugar crystals becoming individual sugar molecules.

Criterion 2: "These Concepts Help Provide Students With an Organizational Framework for Connecting Knowledge From the Various Disciplines."

Often, phenomena from different science disciplines share underlying conceptual structures; CCCs can bring these fundamental commonalities to the forefront of student thinking. The CCC of stability and change, in particular, highlights aspects of phenomena that are characterized by distinctions between conditions of stability and conditions of change. For example, a stone statue may appear stable for decades; but when the pH of rainwater changes, the statue begins to weather in only a few years. This phenomenon is analogous to how changing the percent of carbon dioxide in the atmosphere changes the pH of water in the oceans, altering the rate of coral growth and disrupting the stability of a reef system. Viewing both of these phenomena through the lens of stability and change helps students focus on the conceptual understanding of the nature of stability and change across systems, improving students' explanations of these phenomena.

Criterion 3: The Concepts Are "Selected for Their Value Across the Sciences and in Engineering."

Stability and change is as important to engineering as it is to science. Engineered solutions can have both intended and unintended consequences on the natural world. Stability and change provides a lens through which to examine these changes, informing the refinement of solutions that minimize the scope of these changes. For example, when a road is built through an ecosystem, scientists monitor the ecosystem for unintended potential disruptions to stability. Conversely, engineering solutions can be aimed at minimizing the impacts of changes that occur as part of natural processes, such as the earthquake-proofing of human-made structures.

Using Stability and Change to Describe Systems

Stability and change has particular utility for describing how a system functions and behaves. When science instruction describes phenomena from a systems perspective, students learn to investigate changes to systems and how to determine ways to maintain systems we want to remain stable.

Students use stability and change to investigate how changes in a system cause a phenomenon to occur. For example, leaves change color at a predictable time in the fall. Investigating this phenomenon leads students to an understanding that reaching a seasonal change in the number of hours of daylight triggers changes in the previously stable amount of chlorophyll in a leaf, affecting its color. On a broader time scale, the stability of the forest ecosystem depends on the annual cycle of leaves changing color and dropping off the trees in the fall and regrowing in the spring. This process—leaves falling to the forest floor, decaying, and enriching the soil and the organisms living in the soil—has occurred for millions of years and is part of how stable forest ecosystems operate. The cycling of matter from the trees is essential for the stability of the forest ecosystem.

As such, stability and change is a lens students can use to investigate how leaves change color and to focus attention on how changes to the proportions of components in the leaf system result in changes in the color of leaves. It is also a lens students can use to investigate the annual cycles that affect the organisms in ecosystems.

Other Aspects of Stability and Change: Equilibrium, Rate, and Scale

Stability and change is useful for describing systems in equilibrium. The concept of equilibrium is based on dynamic stability; that is, the rate of change in one direction is offset by an equal rate of change in the opposite direction. The idea of equilibrium in systems is an important concept for students to use to understand phenomena and ideas about chemical reactions, ecosystems, constant motion of objects, plate tectonics, and other ideas where change is offset by opposing changes. For example, a chemical reaction at equilibrium is one in which products are being formed from reactants at the same rate as reactants are being formed from products. Chemical systems adjust in response to changes in the concentration of reactants or products. Systems at equilibrium continue to react even though the net change to the quantity of reactants and products approaches zero.

Stability and change is an important concept for describing homeostasis in living systems. Homeostasis refers to the self-regulating processes that are used by biological systems to maintain stability by adjusting to changing environmental conditions. Homeostasis is regulated by feedback loops that adjust specific components of a system in response to changes to the environment. For example, stomata regulate the movement of gases into and out of leaves. When the amount of water moving past the guard cells changes, the cells inflate or deflate resulting in the stomata opening and closing. The feedback mechanism that regulates breathing in humans is complex and regulated by multiple triggers. For instance, higher levels of carbon dioxide change our breathing patterns to be faster and deeper. Both equilibrium and homeostasis are examples of ways systems can appear stable while their components continue to change.

The rate of change of a system can itself be a stable aspect of a system. A system that changes at a constant rate can be described as being in a state of stability. The system experiences a disruption of this stability when the rate of change is changed. Many phenomena can best be understood based on the rate of change of the components of the system in which a phenomenon is occurring. For example, the phenomenon of climate change is really about the rate of climate change. The climate has changed over geologic time but typically at a slow rate. The phenomenon, then, is human beings changing the composition of the atmosphere that results in the acceleration of climate change. Patterns in the rate of climate change can be analyzed using graphs, charts, and simulations in order to provide evidence of the causes of phenomena.

Changes in a system are often characterized in terms of time, but these changes can also be characterized in terms of distance, quantity, proportion, or size. The scale at which change occurs is an important attribute of systems and should be considered when investigating phenomena. For example, when a substance changes state from a solid to a liquid due to the addition of heat energy, such a change is usually characterized by a change in the arrangement and relative motion of particles in the substance. Furthermore, objects that appear stable at one scale may not be stable at other scales of time, distance, or size. The Mississippi Delta (like Niagara Falls) appears to be a stable landform, but on a scale of thousands of years, the delta has increased in size by hundreds of square miles. The population of Canada geese seems to be stable when taken as a whole, but local populations can change quickly when agricultural practices change.

Stability and Change Is a Useful Tool for Thinking Across Disciplines

Stability and change is a particularly useful concept that functions in similar ways across science disciplines. However, there are some important differences in how it is used within different disciplines, given the different space and time scales at which phenomena occur. For example, geological and astronomical changes occur across millions of years and vast expanses, whereas many physical science phenomena in K–12 are studied at smaller space and time scales.

Stability and Change as a Lens to Make Sense of Phenomena

An important theme in this book is that CCCs are useful lenses to make sense of phenomena (Miller and Krajcik 2015). The concept of stability and change helps students focus on what is changing and/or remaining stable in a system, distinguish between these states of stability and change, and explain the reasons for states of change and stability.

For example, a simple phenomenon like water droplets forming on the outside of a glass of ice water may seem complex unless students focus on how the stable state of gaseous water in the air is disrupted and changed into liquid water on the glass. In this case, students can use stability and change to focus simple observations on complex investigations in various ways, including: (1) measuring changes of energy between the systems of the air and glass, (2) observing changes in the number of droplets due to changes in humidity, (3) comparing the rate ice is changing from a solid to a liquid within the glass to the rate water droplets are forming on the glass's surface, (4) graphing changes in the rate water is condensing on the glass if salt is added to the ice water, or (5) characterizing the stability of the system when the air temperature is close to the temperature of the water in the glass. Without the benefit of seeing the phenomenon with the lens of stability and change, students may struggle to understand the causes of the phenomenon.

Science and Engineering as Investigation and Design

Students use investigation and design to make sense of phenomena. The water droplet example in the previous section illustrates how phenomena constitute central aspects of authentic science and engineering. The report *Science and Engineering for Grades 6–12: Investigation and Design at the Center* by the National Academies of Sciences, Engineering, and Medicine (NASEM) states: "Engaging students in learning about natural phenomena and engineering challenges via science investigation and engineering design increases their understanding of how the world works. Investigation and design are more effective for supporting learning than traditional teaching methods" (NASEM 2019, p. 4). The NASEM report conceives of "Investigation" (with a capital *I*) as more than the practice of planning and carrying out investigations, and it conceives of "Design" (with a capital *D*) as more than the practice of designing solutions. Consistent with the view of Investigation and Design described in the NASEM report, stability and change serve the broader context of science Investigation and engineering Design.

Student Investigations and Designs are at the center of science learning and should regularly engage students in three-dimensional performances (what students are expected to do to make sense of phenomena). Stability and change is a particularly important lens to focus students' sensemaking about systems. Students should engage in Investigation and Design to determine (1) what the causes of changes in a system are, (2) how the structure of a system functions to maintain stability, (3) how long a system will remain stable, and (4) how changes to one component of a system affect other components of the system. These stability and change aspects of systems apply across all disciplines and dimensions of science.

Stability and Change Is Used Across Disciplines and Core Ideas

Stability and change functions in similar ways across all science disciplines; however, there are some differences. Changes studied in one discipline may not be regarded as meaningful changes in another discipline. For example, when a rock falls from a steep mountainside, the change can be investigated in physics in terms of acceleration of the rock, changes in energy as the rock falls, and/or the force of impact. Geologists study the mountain on a different time and size scale and may not consider a single rock falling as a significant change. Regardless of the scale of change across the science disciplines, scientists quantify changes, analyze the causes of the changes, and use evidence from investigations to explain these changes to systems. The next sections illustrate the use of stability and change across science and engineering.

Physical Science

The physical science core ideas (PS) emphasize developing models that support explanations of the changes across matter, energy, and forces. PS1 focuses on the stability and change of substances from a physical perspective (e.g., changes of state, stability of the composition of matter undergoing physical changes) and a chemical perspective (e.g., chemical reactions and equilibrium, changes at the molecular level that affect properties of substances at the bulk scale, using evidence from observing chemical change to identify unknown substances). PS2 focuses on stability and change in the motion of objects (e.g., the stability of motion under balanced forces; changes in motion from unbalanced forces; whether forces are gravitational, electrical, or magnetic). PS3 focuses on how energy is changed from one form to another while still being conserved in a system (e.g., energy from sunlight changes to heat energy; motion energy of the wind changes to electrical energy in a turbine). PS4 focuses on how the stability and change of waves is used to make sense of the transfer of information from one place to another (e.g., standing waves produce sound, while changing wave patterns can encode complex information).

Physical science Investigation and Design focus on the conditions necessary for systems to be stable (e.g., balanced forces result in stable motion of an object, whereas unbalanced forces accelerate an object at a constant rate) and the causes of changes in systems. Often in physical science, changes are occurring while the system maintains stability in some other way (e.g., burning wood changes cellulose and oxygen into carbon dioxide and water while the identity and mass of the atoms remain stable; ice changes into liquid water when thermal energy is added, while the water molecules maintain a stable structure). Stability and change that occur at the bulk scale are caused by unseen changes at the molecular scale. Changes in the motion of objects are caused by forces that cannot be seen; stable motion is due to balanced forces acting on the objects.

Life Science

The life science core ideas (LS) emphasize changes related to the structure of organisms, how energy changes matter in the process of the carbon cycle in both organisms and ecosystems, how stable ecosystems are affected by changes in the environment, and how organisms change through adaptation resulting from selective pressure. LS1 focuses on how stability and change of the structure of cells, tissues, organs, and organisms affect components of living organism functions to meet the needs of an organism (e.g., the processes of photosynthesis and cellular respiration change carbon dioxide and water into sugars to store energy in a stable state to eventually be used as the building blocks of the organism). LS2 focuses on how changes to the physical or biological components of a stable ecosystem affect populations of organisms living in the ecosystem (e.g., the sudden introduction of an invasive species can disrupt the stability of native species in an ecosystem). LS3 focuses on the relationship between genes and observable traits in organisms (e.g., changes to these genes [mutations] disrupt the stability of organismic processes and may result in harmful, beneficial, or neutral changes to an organism's structure). LS4 focuses on how sudden or gradual changes in environmental conditions may result in increased numbers of some species, the emergence of new species over time, and the extinction of other species (e.g., the environment selects organisms with specific traits that aid survival and the ability to reproduce).

Life science Investigation and Design focus on the conditions that cause change (e.g., environmental disasters affect an ecosystem, resulting in changes to organisms) or that maintain stability in systems (e.g., homeostasis). In life science, changes are occurring at the molecular, cellular, organ, organism, and ecosystem levels, yet homeostasis is a mechanism by which these levels retain or re-establish stability. Change that occurs at the ecosystem level may affect stability at the cellular level.

Earth and Space Science

The Earth and space science core ideas (ESS) address changes that occur both quickly and very slowly within and among the geosphere, atmosphere, biosphere, and hydrosphere, extending to the solar system and, ultimately, the universe. The concept of stability in Earth and space science helps students investigate changes happening so slowly that the system appears to be stable; however, change is inevitable given enough time. ESS1 focuses on the apparent stability of entities in the universe and their patterns of motion, as well as gradual changes in Earth's surface from tectonic motion, weathering, and erosion. The core ideas include changes that account for the origin of the universe, time, and evolution of the universe on scales of billions of years. ESS2 focuses on how stability is disrupted when changes in air masses cause weather, the cyclic stability of regional climates, and how changes to the composition of the atmosphere cause rapid climate changes to once stable ecosystems. ESS3 focuses on how human activity contributes to rapid changes in Earth's systems (e.g., climate change, landslides, erosion) and how natural changes to Earth systems (e.g., volcanic activity, tsunamis, earthquakes) have an influence on human activity.

Earth and space science Investigation and Design focus on the observation of changes on very large scales. Many changes in the apparent position of objects in the sky are the result of Earth's rotation and revolution. Seasonal changes and daily changes are due to the motion of Earth and the relative position of the Sun and Moon. Cyclic changes cause seasonal changes that affect life on Earth. These changes are observed as cycles and result in the patterns of change that are recognized as a type of stability.

Engineering, Technology, and Applications of Science (ETS)

Stability and change constitute an important lens for framing engineering design problems and understanding their impact on human society and the natural world. Engineering solutions often address human needs by achieving or preventing stability or change. Sometimes these solutions have unintended natural or social consequences, which also constitute disruptions to previously stable system states. ETS1 focuses on the process of achieving successful solutions to problems through defining and delimiting problems and systematically generating, comparing, testing, and refining solutions. These solutions aim to change the way the system operates to either restore stability to the system (e.g., restore the vegetation to stabilize a hill to prevent or slow erosion) or effect a particular change in the behavior of that system (e.g., design a medical treatment for a person with a genetic condition). ETS2 focuses on the relationships among science, engineering, and technology and the impacts (intended or unintended) on Earth's natural or social systems. Science, engineering, and technology are interconnected in an ongoing cycle of change where scientific discoveries (e.g., refraction of light) inform engineering solutions (e.g., lenses) that yield new technologies (e.g., microscopes), which, in turn, enable new scientific discoveries (e.g., cellular structures). New technologies can cause fundamental and unintended changes in natural systems, profoundly affecting how people live (e.g., urban development disrupting a once stable water system by inhibiting the natural flow and filtration of rainwater).

> “Engineering solutions often address human needs by achieving or preventing stability or change.”

Interestingly, most engineering design performance expectations, from elementary to high school, do not explicitly include CCCs. Instead, many draw from core ideas in both ETS1 and ETS2 to highlight the influence of science, engineering, and technology on society and the natural world. These influences include people's needs and demands for new technologies, the burden technology development places on Earth's natural and economic resources, and unanticipated consequences of technology on society. While these factors are continuously changing, specific technologies can rapidly accelerate these changes and have especially profound impacts on human activities (e.g., gene editing methods, the internet).

Stability and Change Works Well With Other Crosscutting Concepts

> "When instruction couples stability and change with other CCCs, the couplet becomes a "compound lens," constituting a more precise and powerful tool for students' sensemaking."

In many cases, more than one CCC is needed for students to successfully make sense of a phenomenon. When instruction couples stability and change with other CCCs, the couplet becomes a "compound lens," constituting a more precise and powerful tool for students' sensemaking.

Crosscutting concepts are powerful tools to focus students on specific aspects of phenomena. Any combination of crosscutting concepts could potentially be used to help students make sense of a phenomenon or design problem. Teachers and/or curriculum developers can intentionally select couplets of crosscutting concepts to focus on specific aspects of the system being investigated. The decision that teachers, curriculum developers, and students make in selecting crosscutting concepts helps direct the investigation. For example, an investigation of a compost pile can focus on changes in the energy or changes in the matter. The crosscutting concepts can be used to focus instruction on specific changes in a system in ways that support student investigations.

Most phenomena can be fundamentally conceived as occurring either because of a change in a system or as a stable process. In a similar way, most engineering solutions can be viewed broadly as a way to effect a particular change or to achieve a state of stability. Because of this, stability and change is an especially important concept for coupling with other CCCs. Table 10.1 includes examples of how stability and change couples well with each of the other CCCs.

Table 10.1. Coupling of stability and change with each of the other CCCs

Crosscutting Concept	Examples of How Stability and Change Couples With Other CCCs
1. **Patterns** of change are used to predict how systems will change in the future.	Observing patterns of **change** in the rate ice cubes melt when placed on different surfaces (e.g., countertop, towel, wood, metal) provides evidence to support the selection of materials to use in the design of a device to defrost frozen food.
2. Science explanations seek to identify **cause-and-effect** relationships in systems.	Engineering solutions are often designed to **change** the nature of a cause-and-effect relationship. Road surface materials are designed to minimize how seasonal temperature changes affect the **stability** of road surfaces.
3a. The **scale** of change determines if a system is stable or if the change is significant.	The position of the North Star relative to Earth does not observably **change** during a human lifetime; but at longer time scales of thousands of years, the star Polaris would no longer be observed to be directly north of Earth's axis, due to planetary precession.
3b. Changes in the **proportion** of components in a system affect how the system operates.	**Changing** the proportion of ions in the blood can disrupt an animal's cellular processes.
3c. Changes in the **quantity** of the components of a system are used to describe the stability of a system.	**Changing** the number of people on Earth changes the amount of fossil fuels that are burned and the levels of carbon dioxide in the atmosphere, thereby changing climate.
4. **Systems** are stable or changing. Changes to a system affect how the system operates.	**Changes** in ocean temperatures affect the behavior of weather systems because warm ocean water has more energy than colder water. The rate of airflow increases over warm water and decreases over cold water.
5a. **Energy** is conserved so the quantity of energy is stable, but it moves from one system to another. Energy transfers are changes from one system to another system.	A cup of hot tea **changes** temperature as heat energy is transferred from the cup of tea to the surrounding systems. All of the energy from the cup of tea can be accounted for because heat energy is conserved in a system.
5b. **Matter** changes when energy is added to or removed from a system.	Clouds appear when warm moist air moves up in the atmosphere. Gaseous water (which does not reflect light) **changes** to liquid water droplets or ice crystals (which reflect light) to make the clouds visible.
6. The relationship between the **structure and function** of a device or system can determine its propensity to change under different conditions.	Certain materials will be structurally **stable** when exposed to specific environmental conditions such as high temperature, physical stress, or chemical exposure. These materials will function in desirable ways to achieve specific engineering solutions.

Stability and Change Across Science and Engineering Practices

Science and engineering practices describe what students do to engage in the broader endeavors of Investigation and Design. Students use core ideas to support their reasoning while using the CCCs to focus on specific aspects of phenomena.

The practice of asking questions is more productive when specifically focused on how changes to a system result in an observable phenomenon or on what conditions result in a system being stable. Gathering data and information can be helpfully constrained by focusing on factors that promote either stability or change. When students plan investigations, stability and change can focus students' cause-and-effect reasoning on variables responsible for systems shifting between a state of stability and a state of change. When students analyze data, stability and change can inform the ways they look for patterns in data that reveal system stability or change or that distinguish different rates of change.

Students construct explanations describing the causes of changes in systems and the stability of systems, and they develop arguments for how the evidence supports or refutes explanations. Students design solutions that help make systems more stable or cause systems to change at a desired rate. In both science and engineering, students seek the causes of changes and implement approaches for controlling those changes.

Models are especially helpful for understanding how stability and change affect systems. Students use models to construct explanations of how changes occur in systems. Students also use models to predict when systems that are changing will reach a stable state and when stable systems will change. Models can represent changes that are occurring among matter, energy, and/or forces at scales that are difficult to comprehend. Developing models is an effective way for students to communicate changes occurring in systems, rates of change, and how the systems are designed to be stable.

INSTRUCTIONAL APPLICATION 10.1
Three-Dimensional Science Performances: A Design Vignette

Deliberate instructional design is essential to effectively engage students in using stability and change to investigate phenomena and engineer solutions across disciplines. Consider a curriculum unit addressing the phenomenon of urban water runoff. The following vignette illustrates how stability and change is an essential lens to productively engage with the phenomenon from the Earth science and engineering perspectives. We draw from an existing curriculum unit for upper elementary students called the Water Runoff Challenge (Chiu et al. 2019).

Continued

Instructional Application 10.1 *(continued)*

Urban water runoff occurs when urban development replaces absorbent natural ground surfaces, such as dirt or grass, with impermeable surfaces, such as asphalt, concrete, and rooftop materials. Instead of flowing to groundwater sources, water runs downhill along ground surfaces. The runoff water causes undesirable accumulation of water in urban areas. Additionally, rainwater collects surface pollutants, bringing them to natural water sources. These pollutants can negatively affect ecosystems.

The investigation of urban water runoff is driven by a community problem: Students have observed an increase in standing water levels following the construction of a new school parking lot. Stability and change can frame students' interpretation of the relationship between the parking lot construction and water flow. Students hypothesize that the parking lot has had an unintended consequence—somehow disrupting what was previously a stable system of water flow.

Students then proceed to gather information available on the internet about the impact of urban development on water flow. In synthesizing this information, students must articulate more precisely how the construction of the parking lot has changed the water system and the level of standing water. They develop models that focus their explanations on how changes in surface materials affect the stability of water flow.

Having generated an explanation for the mechanism that disrupted the stability of the system, students are ready to engage in the engineering challenge: to generate solutions to help restore the previous stable state. Students generate a range of solutions using different surface materials and use models (either physical or computer-simulated) to test and compare them. For instance, different surface materials have different costs, water permeability, and wheelchair accessibility. Students explore ways to combine different solutions in a way that mitigates the impact of the parking lot on the movement of water. These solutions can be compared by examining trade-offs among the main design variables (cost, permeability, and wheelchair accessibility). Stability and change provides a lens students use to interpret the test data and determine whether a solution achieves the goal of restoring a previous system state. Finally, students communicate their solutions to peers using evidence from tests to argue for their design solutions.

This unit integrates Earth science and engineering. It uses stability and change as a lens to guide investigation during all the unit's phases, including defining a problem, obtaining information, developing and using a model, and generating, testing, optimizing, and communicating solutions. Throughout the investigation, the teacher uses the concept of stability and change to focus students' thinking

Continued

Instructional Application 10.1 *(continued)*

on the explanations for the phenomenon and solutions to the engineering design. Moreover, stability and change frame possible extensions of this unit to life science contexts. For instance, students could subsequently investigate causes of sudden changes in the pollution levels of a nearby stream, affecting populations of species within that ecosystem.

Building Understanding Across Grade Bands

Very young children begin to explore stability when playing with blocks or climbing on rocks, and they use concepts of change as they notice changes in the day-to-day appearance of the Moon. In the early grades, stability and change are useful in helping students connect prior experiences to science ideas. Once students have developed these ideas, they can begin to add more precise science language to support their thinking. In the early grades, students add more examples of how stability and change are used to describe their everyday observations. They begin to develop questions about stability and changes in systems, for example: "How does pumping my legs help the swing go higher?" or "How can I best build a block tower that stays balanced?" or "Why does grass change with the seasons?" These questions help focus children's investigations of everyday phenomena. In the early grades, change is an important concept students use to tell others about their observations of phenomena, find patterns of changes, and explore the causes of the changes. The systems used at the early grades are simple and generally about how changing one component affects the stability of the system. As students build experience over time, they are expected to reason about systems with greater numbers of components and interactions.

In the upper elementary grades, stability and change supports discussions about how to measure change over time, with students observing that change may occur at different rates. In these grades, students use graphical displays to find patterns of change. Students begin to develop an understanding of the nature of change, realizing that some changes can be reversed and other changes cannot. Examining questions about how stability and change can be used to describe systems across many phenomena broadens students' understanding of science. For example, students can investigate seasonal changes of a frog population in a pond ecosystem, shadow length related to the Earth-Sun system, and surface water temperature in a lake system, as well as how these systems exhibit stability from year to year.

In middle school, stability and change helps students focus on the relationship between changes in matter and changes in the input of energy to systems. Students focus on the stability of matter cycling in systems and why changes at the molecular scale can result in observable changes (e.g., water molecules changing from a gas to a liquid form clouds, molecules of oxygen in the air reacting with iron change the color of the iron to orange). Students apply the law of matter conservation to investigate chemical and physical changes in matter and learn that the identity and mass of the atoms involved in the change are stable regardless of the type of change. Students describe the changes in systems using models that explain the stability of the system and the causes of changes to the system. For instance, changing the concentration of salt in water disrupts the stability of cells in a tissue because changing the environment changes the direction water moves across the cell wall. An observable phenomenon for this is when a carrot stick is placed in freshwater. Here, it becomes crisp. However, when it's placed in saltwater, it becomes limp.

Middle school students are expected to think more abstractly about phenomena; therefore, they address stability and change with respect to a broader range of time and distance scales than they do in elementary school (e.g., seasonal changes in sunlight and precipitation determine the stability of ecosystems, balanced forces in the solar system maintain the stable orbit of planets around the Sun and the Moon around Earth). The theory of plate tectonics can help students explain how local changes to Earth's surface at varying time and spatial scales are caused by changes in the movement of tectonic plates.

High school students use even more complex models of systems and are able to investigate more subtle issues of stability in terms of mechanisms that facilitate changes, such as evolution, Le Chatelier's principle, and plate tectonics that may result in sudden changes or gradual changes. These students recognize that science aims to construct explanations of how things have evolved over time to have the structures they currently exhibit. Thinking this way requires students to use models of the rate of change and conditions under which the systems are stable or change gradually. For example, students' ability to understand rates of change enables them to distinguish between sudden changes in geoscience processes (e.g., volcanoes, earthquakes, floods) and evolutionary processes (e.g., invasive species, climate changes).

In high school, students are expected to apply the ideas of dynamic equilibrium in chemistry and changes due to biological and geological evolution over very long time scales. Students understand that feedback mechanisms are essential to maintaining the homeostasis of systems. The concept of random change being a measure of the relative probability of an event becomes a useful way to describe stability. High school students develop an understanding of second-order change (such as acceleration) and are able to describe the stability of isotopes of elements in terms of half-life, which follows exponential laws. The scale of these changes ranges from nanoseconds to millions of years.

The progression of the concept of stability and change across grade bands is often subtle, but it leads students to a deeper understanding of why stability depends on the

> "The progression of the concept of stability and change across grade bands is often subtle, but it leads students to a deeper understanding of why stability depends on the time scale at which change is observed."

time scale at which change is observed. Students' increasing sophistication with the CCC of stability and change depends a great deal on their experiences with phenomena at different scales and the extent to which scale and stability and change are interconnected. High school students understand that things that are stable at a short time scale may not remain stable over a longer time scale. The scale at which change occurs becomes an essential component of understanding temporal stability and change over time. Older students are expected to use a wider range of representational tools to describe stability and change in a system. More specifically, elementary students use simple timelines and direct causal relationships to illustrate the change, whereas middle school and high school students use graphs, charts, and correlational relationships to determine the rate of change and the causes of changes.

The progress across grade levels for stability and change can be exemplified by the different ways students apply this crosscutting concept to make sense of a phenomenon. For example, students use the idea of chemical change to investigate the phenomenon that toast tastes different from untoasted bread. In second grade, students toast bread and observe that the color and flavor are changed and that they cannot change the toast back into regular bread. In fifth grade, students weigh bread, toast it, and then weigh it again. They taste the bread and determine that a chemical and/or physical change has occurred. Students use the law of conservation of matter to argue that some of the matter in the bread has moved to the air. In middle school, students investigate the properties of bread before and after toasting. They toast the bread until it turns black. They compare the properties of the new substance to the properties of the bread before toasting. In high school, students investigate the taste of bread before and after toasting it. They find the taste has changed and gather information about why this change has occurred. Students find that the change is due to the reaction of oxygen with the bread, which caused a chemical change (known as a Maillard reaction). The new substance produced interacts differently with their olfactory cells and taste receptors.

Challenges

There can be challenges to students' learning about stability and change. Although we cannot present a comprehensive review of research on these challenges, we highlight three salient examples that have arisen throughout this chapter: rate of change, conservation, and scale.

First, as discussed earlier, the rate of change is central to characterizing the nature of stability and change in phenomena. Research in both mathematics and science education illustrates the extent to which students can struggle to distinguish a quantity from the rate of change of that quantity (e.g., Leinhardt, Zaslavsky, and Stein 1990). For example, in physics education, research (McDermott, Rosenquist, and van Zee 1987) documents difficulties high school students have in distinguishing the concepts of position, velocity, and acceleration (an example of second-order change). Challenges with rate have also been documented in research on students' understanding of related concepts such as chemical equilibrium (e.g., Banerjee 1991). The challenges with concepts such as rate and equilibrium have the potential to interfere with understanding and interpreting stability and change for many phenomena.

Second, students often struggle with using the concept of conservation when describing changes in both matter and energy in systems. Conservation is closely related to stability and change because it describes an aspect of a system that remains stable even while individual components of a system may change. For example, it is often difficult, especially for younger students, to understand that matter can be changing forms during the process of evaporation, but the number of particles remains stable (Stavy 1991). In many contexts, students believe if matter is changing, it must be disappearing and reappearing, instead of understanding that matter is stable. In a similar way, students struggle to understand that when metal rusts, it gains mass from the air. This struggle occurs because students have difficulty interpreting chemical changes that involve gases, which are often invisible (Driver 1985). Similarly, students struggle with understanding energy transfer, conservation, and transformation due to difficulties in the abstract nature of energy. For example, even students who recognize the principle of energy conservation struggle to apply it to a bouncing ball; instead, they intuitively reason that energy is lost rather than transformed (Black and Solomon 1983). The ability to apply these principles of conservation of both matter and energy is critical to explaining stability and change in many systems.

Third, as previously discussed, what constitutes either stability or change is closely tied to the scale at which the phenomenon occurs. Stability and change frequently occur on space and time scales that are imperceptible. Research studies document challenges students at all levels have with understanding phenomena at these imperceptible distances (e.g., Tretter et al. 2006) and time scales (e.g., Cheek 2012). These challenges present obstacles to students explaining stability and change at such scales for geologic, evolutionary, or astronomical phenomena.

Instructional Strategies

Many instructional strategies are effective at supporting students' use of the CCC of stability and change. We discuss three important strategies that support classroom instruction to help students use stability and change to make sense of phenomena: (1) using stability and change to illustrate the relevance of phenomena, (2) using stability and change to prompt three-dimensional student performances, and (3) using stability and change to focus formative assessment.

Using Stability and Change to Help Students See the Relevance of Phenomena

An important instructional strategy is to engage students with science phenomena and engineering challenges that are locally and/or culturally relevant to them. Stability and change can highlight the specific relevance of phenomena to students. Relevant phenomena for students in upstate New York (e.g., changes in weather due to a lake effect snow) are different from relevant phenomena for students in Hawaii (e.g., changes to the path of lava flows). Culturally relevant phenomena are an important way to ensure that science is personally meaningful to students. In the elementary grades, these phenomena can be related to social studies instruction (e.g., how changes in the buffalo population affected Native American cultures) and fine arts instruction (e.g., how musical instruments make sound from the stability of standing waves and how these sounds change as the size of the instrument changes). Engaging students in relevant engineering design scenarios involves highlighting things students see daily in the community where they live (e.g., how skyscrapers in Chicago, freeways in Los Angeles, or tunnels under the rivers near Boston may cause changes in formerly stable ecosystems). Engaging students in solving local problems may mean raising issues related to the manufacturing and production process for local industries and agriculture (e.g., impacts of water pollution on commercial fishing in New England or on ranching in the West).

Using Stability and Change to Prompt Three-Dimensional Student Performances

The structure of instruction is important for helping students use stability and change in making sense of phenomena. Stability and change can inform how students engage with phenomena, gather data and information, reason about explanations, communicate their reasoning, and apply learning beyond the classroom. In the following section, we revisit the phenomenon of Niagara Falls to illustrate how prompts can focus the sequence of three-dimensional performances that lead to three-dimensional student learning. The instructional sequence described here generally takes two or three class periods and includes multiple class discussions in which students use the evidence they have gathered to support explanations for the changes in the position of Niagara Falls.

INSTRUCTIONAL APPLICATION 10.2
Investigating Changes to Niagara Falls

Engaging With the Phenomenon

Students begin an investigation by observing a series of time-lapse photographs of Niagara Falls to initiate developing questions about the changes in the position of Niagara Falls over time.

The teacher prompts the students to develop questions. Students respond by developing questions such as

- Do harder rocks weather differently from softer rocks?
- How does the amount of water in the river change the falls?
- How fast do the changes to Niagara Falls happen?

This approach is aimed at engaging learners by piquing their curiosity and motivating learning as children wonder about possible causes for the changes.

Gathering Data and Information

In small groups, students engage in obtaining information for the causes of change. This may include watching a short video on the geology of Niagara Falls; reading a brochure from New York State Parks and Recreation; examining different types of sedimentary rocks found at Niagara Falls, including shale, limestone, and dolostone; and analyzing data tables from the United States Geological Survey (*https://pubs.usgs.gov/bul/0306/report.pdf*) on the rate of recession of Niagara Falls. Using stability and change explicitly in prompts helps focus students' attention on specific evidence of stability and change in data and information. For example:

- Is the rate of change in the position of the falls constant or changing?
- How does changing the amount of water diverted for hydroelectric generation influence the stability of the falls?

Patterns in data bring ideas about the rate of change to the forefront and provide students with a greater understanding of this phenomenon and time scale.

Reasoning About Models and Explanations

Students in small groups use their gathered data and information to develop evidence for the causes of the phenomenon. They evaluate changes in the position of

Continued

Instructional Application 10.2 *(continued)*

the falls and grapple with how the system is stable over short time scales but changes over long time scales. Precise prompts using stability and change can direct student thinking about changes in the shape and position of the falls over time. For example:

- Develop a model explaining changes in Niagara Falls.
- Develop solutions to solve the problems that affect the stability of the viewing areas around Niagara Falls for tourists and park workers.

In constructing models, students specify the components of the system that change over time and reflect on the rate of change. (See Figure 10.1.)

Figure 10.1. Model showing how weathering causes geological changes to Niagara Falls

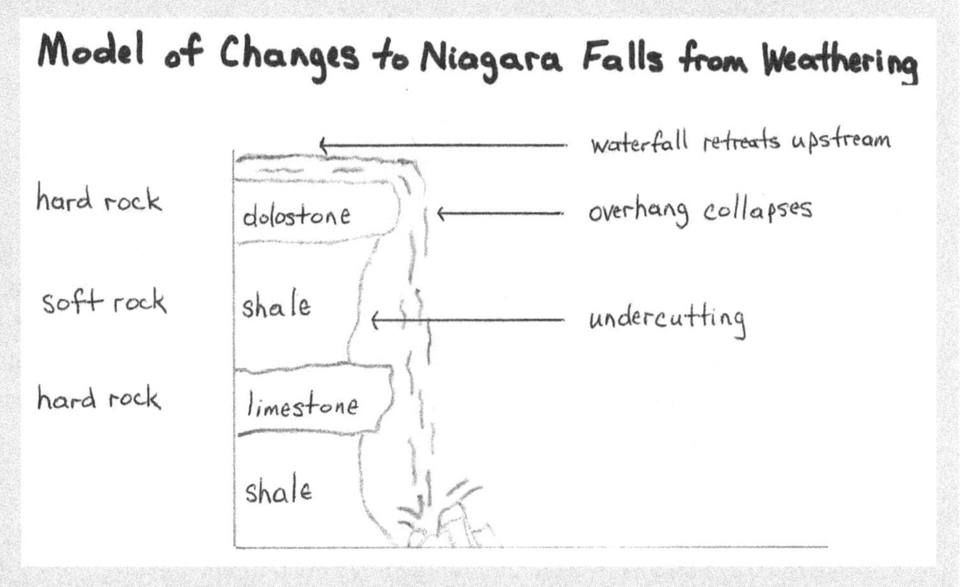

Whole-class discussion is a means to engage with others' ideas with the benefit of generating additional evidence to support explanations. The teacher prompts students with increasingly sophisticated questions to guide meaningful scientific discussions. As a result, students refine their understanding of those phenomena. For example, teacher questions to lead the class discussion can focus on change in the system of Niagara Falls:

- How do changes in climate affect the falls?
- How do changes in the Great Lakes system cause changes in the rate of movement of the falls?

Continued

Instructional Application 10.2 *(continued)*

Student models include differences in rock hardness and incorporate ideas learned via class discussion (e.g., how soft rock at the falls is undercut, causing the falls to retreat up the Niagara River). (See Figure 10.1.) Using the lens of "rate of change" to focus student thinking is one way to support more sophisticated explanations about the Niagara Falls phenomenon.

Use Evidence to Communicate Reasoning to Self and Others

Following the class discussion, students develop an argument for how the evidence they have gathered supports the group and/or class explanation for the causes of the changing position of Niagara Falls. Well-designed prompts help students use a stability and change lens to examine how each line of evidence (e.g., observations of photographs of the falls, data from published timelines of changes in the falls' position, models showing the mechanism of the collapse of the upper layers, differences in the hardness of the rock layers supporting the falls) supports or refutes their explanation of changes in the falls. The written arguments, along with the group models and explanations, are artifacts the student and teacher use to monitor learning.

Students use their individual arguments to support their explanation as they participate in the class discussion. The discussion provides opportunities for students to combine the evidence gathered by different student groups, bringing the entire class to a common understanding of how weathering and erosion of the rock causes Niagara Falls to retreat upstream each year. The teacher introduces investigations of related phenomena to help students use the lens of stability and change in other contexts. For instance, weathering mechanisms explain analogous phenomena, such as changes in the hoodoos found in Bryce Canyon National Park or the changes that resulted in the formation of the Grand Canyon. Students applying what they have learned about stability and change from studying one phenomenon to make sense of a similar phenomenon is an important outcome of science learning.

Apply Science Learning Beyond the Classroom

Students can apply their learning about Niagara Falls to investigate weathering and erosion of analogous phenomena in their everyday lives. One strategy teachers may use to encourage students to bring novel phenomena back to the classroom involves having students maintain a journal of phenomena they observe. Teachers can take a few minutes at the beginning of class to have students share phenomena related to the prior class discussions. Students can use stability and change as a lens for describing the phenomena they have discovered beyond the classroom by highlighting how stability and change characterize the phenomenon. Teachers can also

Continued

Instructional Application 10.2 *(continued)*

maintain poster boards (physical or virtual) that students can use to organize these phenomena according to how changes constitute disruptions to stable systems. This instructional strategy creates a platform for students to share photos or videos of phenomena (e.g., pictures of weathering of bricks on buildings, photos of sidewalks buckling, pictures of potholes in roads), which can be used in various ways. These artifacts are tools that help students broaden their conceptual understanding of stability and change by applying it to everyday life. This promotes student autonomy in using CCCs to make sense of phenomena beyond the classroom.

Using Stability and Change to Focus Formative Assessment

As part of classroom instruction, teachers should continuously monitor their students' improvement in applying the three dimensions of science to understanding phenomena. This vision of proficiency described in the *Framework* includes the ability to use CCCs to help make sense of phenomena. Off-the-shelf assessment tasks will not necessarily be effective at eliciting students' ideas about particular CCCs. Assessment tasks must be carefully crafted in order to prompt students to use a specific CCC. As such, teachers will need to introduce specific language into the design of assessment tasks in order to help their students focus on specific aspects of phenomena. By engaging students in formative assessment with these experiences, teachers can help students improve their ability to use CCCs as lenses for sensemaking over time. For additional discussion on using crosscutting concepts in assessment, see Chapter 15.

For the CCC of stability and change, part of contextualizing phenomena includes choosing a phenomenon for which the distinction between the stable system state and a change in the system behavior is central to understanding it. Moreover, assessment tasks should present consistent use of the language of stability and change in a system. For example, the teacher may prompt students to do the following:

- Construct an explanation for the changes in the Lake Erie ecosystem after 1992.
- Develop a model to show the changes in the North Carolina shoreline over time.
- Develop an argument for how a design solution maintains stability or produces a change in the system.

In addition to promoting student reasoning, these prompts model how to use stability and change as a lens to make sense of a phenomenon or justify engineering design decisions.

Developing models that represent the changes in systems and the causes of changes is one way to assess students' ability to use the concept of stability and change. Models enable students to demonstrate their understanding of system behavior and how system components interact. Causes of both system changes and system stability (e.g., flow of energy, the cycling of matter, balanced or unbalanced forces) can best be described by using models. For example, students can use food webs to explain why organism populations in ecosystems either remain stable over time or change following the occurrence of catastrophic events. Models are also explanatory in nature; models combined with written explanations of phenomena are an effective way to assess students' ability to apply the CCC of stability and change.

Analyzing and interpreting data is another important way to engage stability and change for formative assessment. Prompting students to explain how patterns in data either indicate the stability of a phenomenon or characterize the nature of change can help foreground stability and change.

Summary

The CCC of stability and change is a powerful lens students use to investigate phenomena. Students use stability and change to describe interactions within and among systems and to support evidence-based explanations of phenomena. The concept of stability and change helps students understand complex systems in which the distinction between system stability and system change occurs at scales that give the impression of stability. Stability and change can define the very nature of phenomena (as a shift from system stability to a state of change) and engineering solutions (as designed ways to achieve system stability or change). The concept highlights situations in which the distinction between system stability and change in the system is ambiguous, such as when change occurs on scales that give the impression of stability, when the rate of change is stable, in cyclical phenomena, or in dynamic equilibrium or homeostasis.

Stability and change is a powerful lens for focusing students' sensemaking around all three dimensions of the *Framework*. Moreover, stability and change supports learning the core ideas across the disciplines. For instance, in physical sciences, it highlights matter and energy transformation and interactions among forces. In life sciences, it helps characterize adaptation and processes in organisms and ecosystems. Earth and space science describes interactions within and among systems—the geosphere, atmosphere, biosphere, hydrosphere, solar system, and universe. In engineering, it encompasses the human need to achieve or prevent stability or change. The CCC also focuses on students' engagement in practices (such as analyzing data to provide evidence for the disruption of system stability or developing models to predict these disruptions). Stability and change is especially powerful when coupled with other CCCs. A CCC couplet becomes a compound lens that can more clearly focus students' sensemaking on specific aspects of the phenomenon or engineering solution under study.

Students experience the concept of stability and change from very young ages, such as when balancing blocks or observing changes associated with the seasons. As students advance in age, they are expected to reason about increasingly more complex systems and use an increasingly wider range of tools to make sense of stability and change. Research studies document challenges students specifically have with second-order change, changes that are difficult to perceive visually (such as with energy or gaseous matter), and changes that appear to be stable, such as movement of tectonic plates, dynamic equilibrium, and mechanisms like homeostasis that re-establish stability.

Instructional strategies should specifically target the concept of stability and change. System stability and/or change should be a central feature of the phenomenon or engineering problem to make it relevant and meaningful to students. The concept can help focus instructional prompts by including specific language that highlights the difference between stable states and the nature of change. Instructional prompts can focus students on the specific nature or causes of change as they ask questions; gather evidence; develop models, arguments, and explanations; and engage in science and engineering beyond the classroom. The language of stability and change can help focus formative assessments. Including such language will make assessments more effective than existing, readily available assessments, which are not likely to elicit student reasoning specifically about stability and change. Appropriate assessment tasks can foreground the conceptual boundary between a stable system state and a change in the system behavior so teachers can gauge how well students are able to use the concept for sensemaking.

Throughout this chapter, we have used the Niagara Falls scenario to illustrate the power of stability and change as a lens for sensemaking. Stability and change highlight the distinction between the nature of system stability and changes on geologic time scales. The concept promotes both scientific explanations of changes in the falls, as well as engineering design around how to achieve stability around the falls for tourists and workers. Instructional sequences involving the falls focus on local relevance, gathering information on changes in the falls, identifying mechanisms for change, and identifying patterns in data as evidence for these mechanisms.

References

American Association for the Advancement of Science (AAAS). 1993. *Benchmarks for science literacy*. New York: Oxford University Press.

Banerjee, A. C. 1991. Misconceptions of students and teachers in chemical equilibrium. *International Journal of Science Education* 13 (4): 487–494.

Black, P., and J. Solomon. 1983. Life world and science world: Pupils' ideas about energy. In *Entropy in the school: Proceeding of the 6th Danube seminar on physics education, vol. 1,* ed G. Marx, 43–55. Budapest: Roland Eotvos Physical Society.

Cheek, K. A. 2012. Students' understanding of large numbers as a key factor in their understanding of geologic time. *International Journal of Science and Mathematics Education* 10 (5): 1047–1069.

Chiu, J. C., K. W. McElhaney, N. Zhang, G. Biswas, R. Fried, S. Basu, and N. Alozie. 2019. A principled approach to *NGSS*-aligned curriculum development integrating science, engineering, and computation: A pilot study. Paper presented at the NARST Annual International Conference, Baltimore, MD.

Driver, R. 1985. Beyond appearances: The conversation of matter under physical and chemical transformations. In *Children's ideas in science,* eds. R. Driver, E. Guesne, and A. Tiberghien. Philadelphia: Open University Press.

Leinhardt, G., O. Zaslavsky, and M. K. Stein. 1990. Functions, graphs, and graphing: Tasks, learning, and teaching. *Review of Educational Research* 60 (1): 1–64.

McDermott, L. C., M. L. Rosenquist, and E. H. van Zee. 1987. Student difficulties in connecting graphs and physics: Examples from kinematics. *American Journal of Physics* 55 (6): 503–513.

Miller, E. and J. Krajcik. 2015. Reflecting on instruction to promote alignment to the *NGSS* and equity. In *Next Generation Science Standards: All standards, all students,* eds. O. Lee, E. Miller, and R. Januszyk, 179–192. Arlington, VA: NSTA Press.

Moulding, B. D., K. L. Huff, and W. Van der Veen. 2020. *Engaging students in science investigation using GRC: science instruction consistent with the* Framework *and* NGSS. Ogden, UT: Elm Tree Press.

National Academies of Sciences, Engineering, and Medicine (NASEM). 2019. *Science and engineering for grades 6–12: Investigation and design at the center.* Washington, DC: National Academies Press.

National Research Council (NRC). 1996. *National science education standards.* Washington, DC: National Academies Press.

National Research Council (NRC). 2012. *A framework for K–12 science education: Practices, crosscutting concepts, and core ideas.* Washington, DC: National Academies Press.

NGSS Lead States. 2013. *Next Generation Science Standards: For states, by states.* Washington, DC: National Academies Press. *www.nextgenscience.org.*

Stavy, R. 1991. Using analogy to overcome misconceptions about conservation of matter. *Journal of Research in Science Teaching* 28 (4): 305–313.

Tretter, T. R., M. G. Jones, T. Andre, A. Negishi, and J. Minogue. 2006. Conceptual boundaries and distances: Students' and experts' concepts of the scale of scientific phenomena. *Journal of Research in Science Teaching* 43 (3): 282–319.

PART III
Using CCCs to Teach Key Science Topics

Chapter 11

Using Crosscutting Concepts to Develop the Structure of Matter

Joi Merritt and Kristin Mayer

Understanding that matter is composed of tiny particles that interact is fundamental for describing and explaining concepts such as the states of matter, why phase changes occur, properties of substances, conservation of matter, and chemical reactions (NRC 2012). Thus, developing an understanding of the structure of matter, as well as the interactions of these particles, is essential for explaining phenomena (Krajcik, Duncan, and Rivet 2016). Yet, numerous studies document the difficulties elementary, middle school, high school, and college students have in understanding the particle nature of matter (Hadenfeldt, Liu, and Neumann 2014; Harrison and Treagust 2002).

In this chapter, the term *particles* is used in reference to atoms and molecules, especially at the late elementary to middle school range. This is to reflect the progression in understanding matter at an atomic level as students move through the grade bands reflected in *A Framework for K–12 Science Education* (the *Framework*; NRC 2012) and the *Next Generation Science Standards* (*NGSS*; NGSS Lead States 2013a).

Matter and Its Interactions

The particle nature of matter is a fundamental concept for learning and understanding many physical and chemical processes (Tsarparlis and Sevian 2013). However, studies demonstrate that students have difficulty in connecting observable phenomena to what is happening on a microscopic (particle) level (Treagust et al. 2013). Because it is important for students to understand microscopic-level interactions to make sense of more complex chemistry concepts, as well as concepts in other science disciplines, it is important that students begin to develop a particle view of matter (Treagust et al. 2013).

Studies show that learners have nonscientific conceptions of matter even as they are developing a particle nature of matter view (Ben-Zvi, Eylon, and Silberstein 1986; Driver, Guesne, and Tiberghein 1985; Driver et al. 1994; Lee et al. 1993; Nakhleh 1992; Novick and Nussbaum 1978). For example, a student could attribute observable phenomena to what they think is happening on the particle level. In particular, a student could describe silver as a shiny gray metal. This same student could then also state that atoms of silver are gray in color because a piece of silver metal is gray. Studies also indicate there is a divergence between students' views of matter and the language they use to describe matter and its interactions. For example, students may be able to recite that all matter is made of atoms yet struggle to use the interactions between atoms to explain observable phenomena (Ben-Zvi, Eylon, and Silberstein 1986; Driver et al. 1994; deVos and Verdonk 1996; Johnson 1998; Renstrom, Andersson, and Marton 1990; Taber 2003).

Students' understanding of matter involves not only comprehending that matter is made up of particles, but also understanding how the interactions of these particles on an atomic level help make sense of phenomena on a macroscopic level. Claesgens et al. (2008) found that students could have hybrid reasoning in which they apply observations of phenomena on a macroscopic level to explanations on the particle level. In other words, students can express a particle view of matter, but they describe the phenomena on a macroscopic level. In Figure 11.1, in response to an assessment item, a student drew a representation to explain what happens when bromine undergoes a phase change from liquid to gas. In the space provided to show what is happening on a particle level, the student represented liquid bromine with particles and bromine gas as particles that are even further spread out. However, in the explanation,

Figure 11.1. Student model explaining what happens when bromine changes phases from liquid to gas

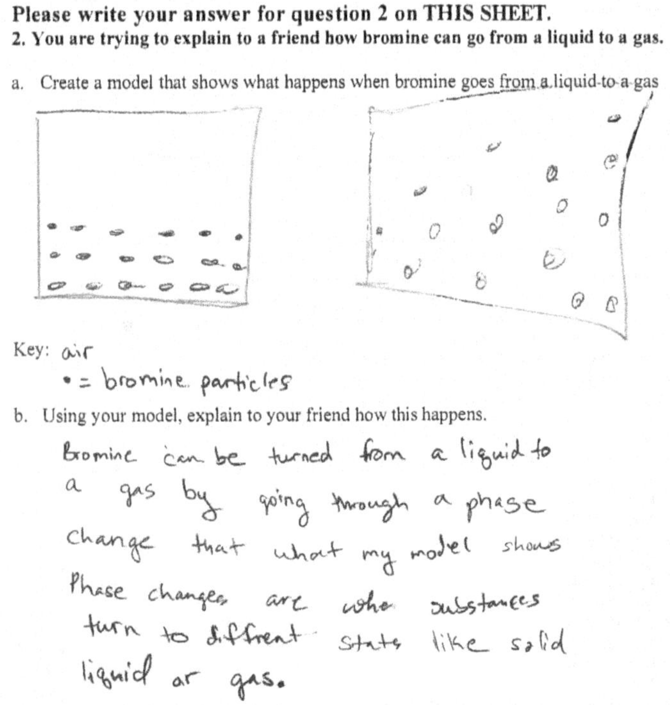

Please write your answer for question 2 on THIS SHEET.
2. You are trying to explain to a friend how bromine can go from a liquid to a gas.

a. Create a model that shows what happens when bromine goes from a liquid to a gas

Key: air
• = bromine particles

b. Using your model, explain to your friend how this happens.

Bromine can be turned from a liquid to a gas by going through a phase change that what my model shows Phase changes are whe substances turn to diffrent states like solid liquid or gas.

the student did not include any ideas represented in the model, instead explaining that a phase change occurs during which bromine goes from a liquid to a gas; the student never described or explained what is happening to the bromine particles. This indicates that students tend to respond to assessment questions without the depth of explanations that would align with scientific explanations. (See Chapter 15 for more about how crosscutting concepts can be used to support student sensemaking in assessments.)

Through three-dimensional learning, we can support students in building integrated understanding of matter by blending this disciplinary core idea (DCI) with science and engineering practices (SEPs) and crosscutting concepts (CCCs) according to the vision of the *Framework* (NRC 2012) and the *NGSS* (NGSS Lead States 2013a). Previous chapters have examined each of the different CCCs and their importance in helping students make sense of phenomena. In this chapter, we examine the power of CCCs in helping students make sense of the structure of matter at the different grade bands. Several CCCs have clear overlap with the DCI of Matter and Its Interactions. For example, the CCC of scale, proportion, and quantity is clearly connected with developing disciplinary core ideas that relate to microscopic particle scale interactions for explaining observable phenomena. Similarly, the CCC of matter and energy overlaps with the DCI of Structure of Matter, both emphasizing the aspects of conservation of matter and energy and tracing changes to matter and energy. In this chapter, we focus on how a range of CCCs are helpful for supporting students' development of the DCI of Structure of Matter.

How Crosscutting Concepts Support Student Understanding of Matter and Its Interactions

CCCs are a powerful dimension in three-dimensional learning. As discussed in Chapter 3, explicitly discussing the CCCs helps students understand the CCCs and use the CCCs as resources. It also supports student learning of the DCIs and SEPs as they encounter and figure out new phenomena. The CCCs help students develop an understanding of DCIs and SEPs with which they are engaging to make sense of phenomena.

Students can use the CCCs to build their understanding of matter and interactions across the K–12 grade bands. Starting in early grades, students focus on making observations of matter. This fits well with developing understanding of the CCC of patterns, and using patterns supports students' ability to identify and describe the observations they make of materials. Students can find patterns in their observations of matter, noting things that are shiny and not shiny or things that pour and don't pour. Furthermore, the CCC of patterns can be blended with the SEP of Planning and Carrying Out Investigations to develop the DCI of Matter and Its Interactions. These three dimensions support one another as students explore materials.

In later grades, students continue to identify patterns as a method for exploring new phenomena and new observations of matter. As students analyze new phenomena, they continue to develop their understanding of the DCI of Matter and Its Interactions while developing and incorporating additional CCCs, such as scale, proportion, and quantity; cause and effect; and structure and function, to build their model of matter and explanations of observations. A focus on CCCs, like cause-and-effect relationships, has the power of addressing some of the established studies that show students struggle to connect the observations with atomic-level causal interactions.

Developing the DCI of Matter and Its Interactions and the DCI of Energy (understanding of bonding at the high school level) through engagement in SEPs and a focus on CCCs like identifying patterns and developing causal explanations supports students in linking atomic-level explanations with macroscopic observations of phenomena. In order to support students in building these abilities across grade levels, teachers must explicitly discuss and support development of the crosscutting concepts and help students use the concepts as tools for analyzing new phenomena and questions they encounter.

“In order to support students in building these abilities across grade levels, teachers must explicitly discuss and support development of the crosscutting concepts and help students use the concepts as tools for analyzing new phenomena and questions they encounter.”

CCCs can support students in engaging in SEPs and understanding DCIs as they make sense of phenomena. The Classroom Snapshots in this chapter provide examples of how CCCs can be used to support student understanding of matter at different grade bands. Across different grade bands, the Classroom Snapshots illustrate the use of different CCCs, as well as different SEPs.

Learning About the Structure of Matter in Elementary School: Grades K-2

At the lower elementary level, students observe and describe matter in different states (Mayer and Krajcik 2017). In the K–2 grade band, students are first introduced to CCCs, where an understanding of the CCC of patterns is needed to make sense of other CCCs. For example, the learning progression across grade bands for the CCC of cause and effect states that in the K–2 grade band, "students learn events have causes that generate observable patterns" (NGSS Lead States 2013b).

Classroom Snapshot 11.1 illustrates how a teacher helped students make sense of the concept of properties using the CCC of patterns. This two-day lesson was part of a larger matter unit called "How Many Uses Can an Object Have?" in which students were learning about properties and the effects of heating and cooling on matter. In addition, the CCC was used to help students understand planning and carrying out investigations, specifically the importance of observations. Though not a focus of the lesson, there was also a discussion of why they sorted by one variable (property) at a time.

CLASSROOM SNAPSHOT 11.1
Exploring the Properties of Objects

Students were provided with two different objects and asked to write different words they would use to describe their objects (e.g., words about size, color, hardness). The teacher pointed out to the class the Observation Word Wall. Students were then asked to share out the words they used to describe their objects. She had different students share one of their words for each object and listed the words on the word wall.

The teacher then put the students in predetermined groups to sort their objects. The teacher purposefully provided objects that could obviously be grouped into different categories (i.e., by color, texture, size, hardness). She reminded the students that they were observing the different objects and could use their observation words from the word wall to describe their objects. The teacher then asked the student groups to put the objects into groups based on how similar they were. Students seemed confused about how to begin, so she initiated the following discussion[1] to help them get started.

Teacher: "Who can give me an example of one way the objects are alike?"

Student 1: "Some are small; some are big."

Student 2 (*from the same group*): "Some are rough and some are smooth."

Teacher: "So, you both gave me two different ways the objects are the same. Class, in each of your groups, I want you to pick one way you can group your objects." (*The teacher addresses the group that shared.*) "You can group your objects as small or big or as rough or smooth, but pick only one of those."

The teacher then led a whole-class discussion to identify how different groups sorted their objects. The teacher selected a student from each group to explain how their objects were grouped. As the students shared their groupings, the teacher listed

1. Dialogue is based on classroom observation notes and teacher reflection.

Continued

Classroom Snapshot 11.1 *(continued)*

them on the whiteboard. Then the teacher held an explicit discussion on patterns with students. This discussion on patterns centered on why it is important to focus on one observation at a time (e.g., focusing on color versus focusing on color and size).

Teacher (*pointing to groupings listed on board*): "We have a lot of different groups here. How did you pick the way you would group your objects?"

Student 3: "We did red and white because we [saw] they were red and white."

Teacher: "Another group chose colors, too—orange and blue. So we can group things by color." (*The teacher writes* color *on the board.*) "Who had a different way they grouped their objects?"

Student 4: "We did bumpy and [smooth]."

Teacher: "How were you able to tell if the object was bumpy or smooth?"

Student 4: "I touched it, and that's how it [felt]."

Teacher: "So you could touch it and you decided if it was bumpy or smooth. Scientists call how something feels *texture.*" (*The teacher writes* texture *on the board next to* bumpy/smooth).

The teacher continued the discussion to identify size and hardness as patterns. She then wrote the words *size* and *hardness* on the board, as well. Finally, she summed up the lesson.

Teacher: "So, you looked for ways that a lot of your objects were alike?"

Class: "Yes."

Teacher: "Scientists look for how things are similar, as well. Like you, they use their senses to tell one object apart from another. They call this 'looking for patterns.' We have talked about different patterns today (*points to* color, texture, size *and* hardness *written on the board*). Tomorrow we will look at even more patterns.

On day two of the lesson, the teacher reviewed what they did the previous day to identify patterns. She then gave individual students different sets of objects and asked them to identify the patterns in their objects. The students identified their different groupings again. This time, the teacher pointed out the patterns they had grouped their objects by—color, hardness, texture, and size. The teacher then explained that scientists call these patterns "properties."

Teacher: Scientists call these patterns *properties*. Properties are observations of an object that help to tell one object from another. Who can give me an example of a property?

Continued

Classroom Snapshot 11.1 *(continued)*

Student 5: "Color."

In addition to learning about properties, students were able to further understand the practice of planning and carrying out investigations by focusing on observations. Discussions about the different groupings and words they chose to describe objects linked to the importance of their observations. It was important that they described their objects so they could sort the objects into different groups. In addition, it set the foundation for understanding variables (without using the term). In a discussion that arose about the reasons they sorted their objects the way they did, students talked about why they could only sort the objects by one property at a time. As the lesson progressed, students were given the opportunity to demonstrate their knowledge via their "own investigation" in which they were to individually sort a different set of objects using three different properties. They wrote down the different ways in which they sorted the objects. One student sorted objects by (1) color, (2) size, and (3) hardness and softness, whereas a different student sorted their objects by (1) color, (2) shape, and (3) size.

As a part of this unit, the CCC of patterns was explicitly discussed to help students understand what properties are. Students initially struggled with identifying the specific property they were using. After explicitly discussing the patterns they were employing, they were able to describe in everyday language what properties are. Once they were able to understand and identify different properties (e.g., hardness, size, color), they could define what a property is, using what they learned from their observations and discussions. Thus, students were able to develop their understanding of properties and deepen their understanding of investigations (an SEP) through the CCC of patterns.

Using Crosscutting Concepts to Teach the Structure of Matter in Elementary School: Grades 3–5

At the upper elementary level, students begin to develop a particle model of matter. This means students are beginning to understand that matter is made up of particles too small to be seen. Students begin to understand that multiple properties can be used to distinguish one material from another. Students are also expected to understand that matter is conserved even when it appears to disappear (e.g., mixtures, phase changes).

Classroom Snapshot 11.2 (p. 254) is from a unit for fifth grade titled "How Can We Prevent Our Water From Making Us Sick?" The unit was developed in response to the issue of *E. coli* found in the wells in part of the county. The unit integrated science and literacy, with students gaining background knowledge via reading and writing about content, including farm pollution, the water cycle, and matter, during their literacy block. The science block engaged students in investigations aimed at supporting them in developing a particle model of matter that helped them understand states of matter, properties, and mixtures. At the end of the unit, students chose a method (e.g., news report, public service announcement, letter to a local farmer) to educate others about the problem and to explain how to prevent water contamination.

CLASSROOM SNAPSHOT 11.2
Developing a Particle Model of Matter

The unit identified the CCC of scale, proportion, and quantity to help students develop a particle model of matter. Students tend to have difficulty understanding how what is happening on a macroscale can be explained using a particle model of matter (atomic scale). The unit engaged students in the SEP of Developing and Using Models to help them explain what is happening on a macroscopic (visible) level to their particle models. This lesson was near the end of the second week of the unit. At this point, students had read about the water cycle and conducted investigations that helped them reach the conclusion that there must be empty space between the particles. In addition, they had investigated the properties of solids, liquids, and gases related to shape (e.g., gases take the shape of their container), volume, and compressibility. In this lesson, students viewed simulations of solids, liquids, and gases to explain the movement of particles in the different states. Students had created particle models of solids, liquids, and gases in previous lessons.

To begin the lesson, the teacher showed students a time-lapse video of ice melting in a glass. The teacher had students write or draw their observations of what happened as the ice melted and explain why they thought it was happening. The teacher let the time-lapse video run several times so he could allow students to write or draw their observations. The teacher then had students share their observations in their table groups.

Teacher: "I am going to play the video one more time and then pause halfway through. I want you to notice how much ice is left." (*The teacher plays video to halfway.*) "How much ice is left?"

Student 1: "None."

Teacher: "What are we left with?"

Student 1: "Water."

Teacher: "We started with a glass full of ice. Now the glass is half full of water. Who can explain why that might be?"

Continued

Classroom Snapshot 11.2 *(continued)*

Student 2: "The ice stacked together."

Teacher: "What state was the ice?"

Class: "Solid."

Teacher: "So, what do you think happened at a particle level?"

Student 3: "The particles moved."

Teacher: "What is between the particles?"

Class: "Empty space."

Throughout the discussion, the teacher asked students how they could think about what was happening to the ice on a particle level as it melted to explain what they saw on a macro level. The teacher then informed the students that they would be looking at simulations of solids, liquids, and gases on a particle level. The teacher let students know that the particles were called atoms in the simulation. The teacher instructed students to think about the particles as what happens to water as it gets heated or cooled to change states. Students were also provided with a table for taking notes about what they saw in the simulation:

Gas	Liquid	Solid

The following is the discussion the teacher held to support students in making sense of the gas simulation. Before starting the simulation, the teacher began with a question.

Teacher: "Remember, each particle is a particle of water in the gas phase. How do you think the particles will move?"

Student 4: "They are going to bounce and hit each other."

Teacher: "Does anyone think anything different?"

Class: "No."

Continued

Classroom Snapshot 11.2 *(continued)*

The teacher showed the simulation, then marked two particles for students to trace.

Teacher: "Now you can watch how the particles move. Do you think they will move in a pattern or randomly? Show of hands. Pattern?" (*A few students raise their hands.*) "Random?" (*The rest of the class raises their hands.*)

The teacher showed the simulation again. The class then came to a consensus that the particles in a gas move randomly.

Teacher: "Substances in the gas phase—do they take up the space of their container?"

Class: "Yes."

Teacher: "What do we call that?"

Class: "Volume."

Teacher: "What would happen to the gas if we changed the container to a balloon? What shape would the gas be? What if I wanted the gas to be a different shape? Talk in your table groups."

Students decided that the gas took the shape of its container and you would have to have a different-shaped container for the gas to take a different shape. The teacher then had students write down what they now knew about gases based on the simulation and previous evidence from other investigations.

Based on the discussion, students typically noted the following ideas about gas:

- Takes the shape of its container
- Takes the volume of its container
- Lots of empty/free space between particles; is compressible
- Particles move everywhere

The teacher then showed the simulation of liquids and solids, supporting students with questions linking to scale. For example, the teacher asked: "What happened when the ice melted? Did the liquid fill the container? What do you think this looks like on a particle level?" He also asked students to think about how the movement and spacing of particles related to the ability to pour liquids like coffee. Students again came up with notes for liquids and solids.

The typical student notes for liquid included the following ideas:

- Not a lot of free space between particles; not easily compressible
- Takes the shape of container it is in, just like the gas does

Continued

Classroom Snapshot 11.2 *(continued)*

- Definite volume
- Particles slide past one another

Typical student notes for solids included these ideas:

- Has its own shape
- Has a set volume
- Not a lot of free space between particles; not easily compressible
- Particles vibrate; can't move past one another

Once students were able to develop an understanding of particle movement in the different states, they were then able to apply these concepts to explaining how waterways get polluted with contaminants (including *E. coli*). By the end of the unit, students were able to create a model to explain how contaminants enter local waterways from farm animal waste pollution.

By observing the simulation (the SEP of Developing and Using Models) and linking it to the time-lapse video and other investigations they had engaged in, students were able to summarize characteristics of solids, liquids, and gases. The teacher used the CCC of scale, proportion, and quantity to help students relate what they were seeing in the simulation to their macrolevel observations. He linked the random movement of particles in a gas to the fact that gases take the shape of their container.

Using Crosscutting Concepts to Teach the Structure of Matter in Middle School: Grades 6-8

At the 6–8 grade band, students continue to build their understanding, moving from a particle model to an atomic-molecular model. This means, instead of just representing matter as made up of particles, they are now identifying the particles as atoms and molecules. By the end of middle school, students should be able to use this model to explain changes in matter (Mayer and Krajcik 2017). Students at this point are not learning what bonds are, but they are understanding that different substances are composed of atoms of different elements in different arrangements, which results in these substances having different properties. In addition, they learn that different substances can react together to form new substances with different properties due to the rearrangement of the atoms of the original substances into new molecules (Mayer and Krajcik 2017).

Classroom Snapshot 11.3 is from a unit in which students were using evidence to develop an atomic-molecular model. The unit focused around students trying to solve a mystery: Who stole the school mascot? The students read about the different suspects and needed to identify the mysterious substance that was left behind by the thief. This section of the unit focused on students learning about properties that are measurable. The teacher identified scale, proportion, and quantity as the CCC to support student understanding because she noticed students had struggled to understand mathematical concepts in previous years. Because they would be using a model, it would allow students to look at the proportion of mass to volume to help them visualize and understand the mathematical calculation of density.

CLASSROOM SNAPSHOT 11.3
Exploring the Density of Matter

On day one of the lesson, students were learning about density. In a prior lesson, students had learned about other properties (e.g., hardness, color, flexibility). They had learned that properties are "identifying characteristics of elements and substances. All substances have their own properties unlike any other substance." Students were provided with six cubes that were the same volume but made of different materials: steel, aluminum, oak, pine, PVC, and nylon. Each group received a different cube to start with. The teacher reviewed with students how to measure the volume of a cube (length × width × height). Afterward, each group measured the volume of the cube they'd been given. The class then shared their results. (All were the exact same size with rounding, which was discussed.)

The students rotated stations to measure the mass of each of the cubes. This led to a discussion of why the mass of the cubes was so different. Students were asked what they thought it meant that cubes taking up the same amount of space (volume) could have very different masses. This led to an explicit discussion about the crosscutting concept of scale, proportion, and quantity, specifically proportion. The teacher prompted students with questions such as, "How can proportion help us understand the relationship between these mass and volume differences?" and "What did you learn about what a proportion is in math?" This led to a discussion about the relationship of one variable to another. Then students were instructed on how to calculate density (mass divided by volume). Instruction was scaffolded based on how comfortable students were with their math skills and algebraic formulas. The class then reviewed their calculations. This was followed by a discussion of what density shows when we think of it as a proportion. Students were able to come to density as the relationship between mass and volume for a substance. Day one of the lesson ended with students wondering if the density remained the same for a substance, no matter its size.

Continued

Classroom Snapshot 11.3 *(continued)*

On day two, students were provided with cubes that were different sizes but made of the same materials. Each group got a different cube. They measured the mass and volume for the cube and then calculated the density. Within groups, students found minor differences among their individual calculations, but they found that their answers were almost the same. Then students compared their results for the different-size cubes to the calculations they had made on day one. Students were surprised to find that the density of the different-size cubes was basically identical. They concluded that no matter the size of the cube, the density was the same. This led to a discussion about why this made density a property.

Using the SEP of Developing and Using Models, students were asked to think about how their particle (atomic-molecular) models would show the difference in density between any two cubes with different densities. That is, they had to consider what would be different about the particles. The teacher supported students in thinking about the mass (m) to volume (v) ratio, asking them to reflect on the mathematical equation that defines density ($d = m/v$). Students were encouraged to use their data as evidence.

Teacher: "When is density a bigger number?"

Student 1: "When the mass is a larger number."

Teacher: "When is density a smaller number?"

Student 2: "When the volume is a bigger number."

Teacher: "What about when density is close to one?"

Student 3: "The mass and volume are almost the exact same number."

The teacher provided students with a representation of a cube and asked them to think about how they could illustrate the density of one of the materials.

Teacher: "How could we make sure to show that one material has more mass than the other? Should all the particles be the same size?"

Student 4: "They should be different sizes."

Teacher: "Why different sizes? How does your data help you to make that decision?"

Student 4: "[The materials] have different masses, so the model should show different-size particles."

Teacher: "Could this also mean more or less particles? Why?"

Student 4: "Yes. [The amount] depends on the size [of the particles]."

Continued

Classroom Snapshot 11.3 *(continued)*

After the conversation, the teacher provided time for students to model the different densities of the different materials. Students were then asked to share their models, and they came to consensus that the higher the density, the more mass per unit volume.

Students may develop different ways to represent more mass. Commonly, students show more particles in a given area or larger particles in a given area. Although more particles or larger particles could both increase density, students' models are a simplification and do not need to include all these relationships. A powerful question that students could explore in later grades is whether density increases by increasing the size of particles or number of particles in a given volume.

Instead of modeling, the teacher could have engaged students in the SEP of Analyzing and Interpreting Data. At grade 7, CCSS.7.RPA.2A and CCSS.7.RPA.2B (from the *Common Core State Standards for Mathematics*) focus on students representing and identifying proportional relationships using (linear) graphs on a coordinate plane. The teacher could have had students graph the data from their different-size cubes, helping them to understand that the graph could be used to determine how much mass is in any given volume of a substance. The students would have needed to graph mass on the y-axis and volume on the x-axis. The teacher could have then helped students comprehend that the slope of the straight line, $m = (y_2 - y_1)/(x_2 - x_1)$, is the density of the substance. This graph would have allowed students to understand density is a proportional relationship of mass and volume (illustrated in the graph of the straight line), which means density remains the same no matter how much of a material there is.

In Classroom Snapshot 11.3, students engaged with questions related to the CCC of scale, proportion, and quantity to understand density. By the end of the unit, students applied what they learned about the different properties to identify the mystery substance and, ultimately, who stole the mascot. They engaged in activities to understand that density is a proportional relationship of the ratio of mass to volume for a material. The SEP of modeling was used to help students visualize what density means (mass per unit volume). Moreover, students could engage with other SEPs, such as Analyzing and Interpreting Data, to understand the same CCC of scale, proportion, and quantity.

Using Crosscutting Concepts to Teach the Structure of Matter in High School: Grades 9–12

In high school, students build mechanistic explanations for the observations they make in middle school. In middle school, they note that atoms can connect and form new substances. In high school, students develop models of atomic structure that can be used to explain why atoms form these connections (or bonds), how energy changes are related to these bonds, and the mechanisms that govern the interactions between atoms and molecules.

Classroom Snapshot 11.4 provides an example of a high school chemistry unit that focused on using the CCC of cause and effect to support student development of causal explanations of observations that came from comparing the boiling points of different compounds. This chemistry class was a "lower level" class for students who struggled with math or science in the past and were not planning on taking the Advanced Placement chemistry course during high school. The class was taught by a teacher who had done significant work on the *NGSS*, including curriculum development, and who had chosen to return to school to apply these ideas to a classroom. The school district had not updated its curriculum and was using a textbook from before the creation of the *NGSS*. Though the curriculum was not an *NGSS* curriculum, the instructor did incorporate curricular resources that were developed for the *NGSS*, in particular the Interactions curriculum. The Interactions curriculum was written as an introductory science curriculum for ninth grade. It is a freely available resource for all teachers. (Visit *https://learn.concord.org/interactions* for more details.)

CLASSROOM SNAPSHOT 11.4
Comparing Boiling Points of Different Compounds

In the beginning of the unit, students made observations about several compounds— some that contained metals and others that had only nonmetals. Students had worked on finding patterns in several previous units, so this was a familiar way to start exploring a new topic. Students found many different patterns in the properties of the compounds they tested and formed questions from their observations.

Although the class worked with a variety of crosscutting concepts in this unit, the teacher was explicit about developing cause-and-effect relationships that can explain observations. Students had developed particle models of matter and models of atomic structure based on evidence in previous units. As part of developing their models of atomic structure, students also noted patterns in how electrons fill in various energy levels and how those patterns are reflected in the periodic table. Because students were familiar with finding patterns, they could identify patterns in the properties of atoms on the periodic table. The class started with the pattern in atomic radius. Moving from left to right across a row on the periodic table, the atomic radius, surprisingly, gets smaller as more protons and electrons are added to the atom; going down a column, the atomic radius increases. The teacher asked

Continued

Classroom Snapshot 11.4 *(continued)*

students to develop a cause-and-effect explanation that could explain the pattern they noticed. This was a struggle for students, so they shared their explanations and built on one another's ideas. To explain the pattern of decreasing atomic radius from left to right across a row, students developed the idea that the additional protons in the nucleus cause a stronger attraction, with the effect of pulling the electrons in closer and reducing the atomic radius. As the class worked together to develop this explanation, the teacher recorded consensus ideas on the class summary chart—a tool used to track questions, evidence, and developing ideas for each unit in the class. On the summary chart, the teacher made boxes and arrows to represent a chain of causal mechanisms and illustrate the cause-and-effect relationships. Figure 11.2 provides the notes the class included in their class summary chart.

Figure 11.2. Notes from class summary chart

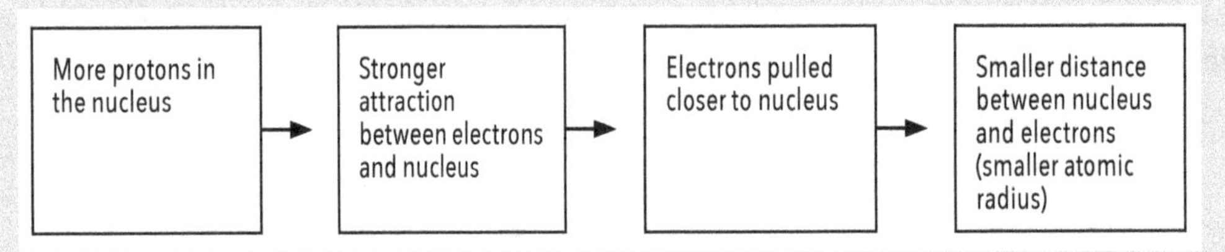

Students also focused on the increase in atomic radius as you move down a column. Again, students shared their ideas, discussed, and, as a class, came to the consensus that the additional energy levels the electrons occupy cause the increase in the atomic radius. The teacher again recorded the consensus ideas on the summary chart as a series of steps leading to the outcome described. Students studied additional patterns in atomic properties, like electronegativity, and used attractions between the protons in the nucleus and the electrons to develop cause-and-effect relationships to explain these additional patterns. Again, the idea of building causal explanations that used a series of cause-and-effect relationships was challenging for students. As a support, the teacher had students create their own charts based on the class summary chart, starting with a box about the attractions between protons and electrons, and then link a series of boxes in a chain of cause-and-effect steps to end up at the observation they were trying to explain.

Students then applied similar cause-and-effect relationships to explain the patterns in compounds. For example, the students looked for patterns in boiling point. Using mathematical reasoning, students graphed the boiling point versus molar mass of several hydrocarbon alkanes (molecules made of chains of carbon atoms surrounded

Continued

Classroom Snapshot 11.4 *(continued)*

by hydrogen atoms). Students found patterns in the relationship between the molar mass of these alkanes and the boiling points. However, students then added similar compounds, with similar molar masses, that have oxygen atoms in addition to the hydrogen and carbon atoms, and they noticed that these compounds did not fit the patterns defined earlier for hydrocarbons. Students analyzed models of electron distribution in bonds and atomic models using several simulations and activities from Unit 3 of the Interactions curriculum. Students developed cause-and-effect relationships, starting with the pull from protons in the nucleus on the electrons, to explain the pattern they found of molecules with an oxygen atom having a higher boiling point than similar molecules made of only hydrogen and carbon. The class mapped out a series of causes and effects to link the attractions between protons and electrons in order to explain why one substance had a boiling point so much higher than another. Again, boxes showing a series of cause-and-effect relationships provided a support. Figure 11.3 shows an example of how the cause-and-effect relationships could be used to explain the differences in boiling point between two compounds.

Figure 11.3. Cause-and-effect relationships used to explain differences in boiling point

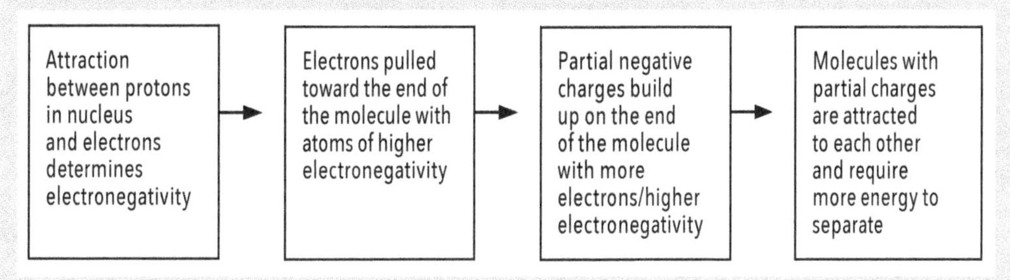

Students collected data by, for example, heating water and recording the temperature as water heated to a boiling point. Students noticed the temperature did not change while the water was boiling. Again, the class developed cause-and-effect relationships to explain why the temperature did not change. Students analyzed models of molecules and polarity. From there, students were given data for the temperature changes as two different substances were heated. When given data about the boiling points of two compounds, students could make predictions about each compound's structure (polar molecules with partial charges versus nonpolar molecules without partial charges) and build cause-and-effect relationships to explain the differences between the data.

Continued

Classroom Snapshot 11.4 *(continued)*

Toward the end of the unit, students completed a lab at home in which they worked with a parent. In this lab, students and their parents counted how many drops they could collect on a penny. Students had to explain to their parents why the water bubbles on top of the penny rather than running off it. The students did not explicitly discuss this phenomenon in class. Rather, students were using the knowledge they developed in class and the concept of building cause-and-effect relationships to figure out this new phenomenon with their parents. Students could use the cause-and-effect relationships to teach their parents about why the water does not run off the penny. For example, one student and his mother stated that the water piles up on the penny and looks like a "jellyfish." They wrote, "The reason why the water is staying on the penny is because the water molecules are attracted toward each other ... [with] the hydrogen atoms of one molecule being attracted toward the oxygen atoms of another molecule." Thus, the student used the cause of the attraction between the atoms within separate water molecules to explain his observation of the water drops forming a bubble on top of the penny. Similarly, on the assessment for this unit, one task provided data for lattice energy, a topic that was not covered in class. Figure 11.4 provides the assessment question.

Students could apply the crosscutting concepts used in the class to analyze and figure out this new data. They could find patterns in the data and match up patterns with the location of the elements on the periodic table. Students were able to construct cause-and-effect relationships using the interactions between protons and electrons to propose explanations for the new data they encountered on the assessment.

Continued

Classroom Snapshot 11.4 *(continued)*

Figure 11.4. Assessment question requiring students to apply understanding to a new topic

"Lattice enthalpy/energy" can be described as a measure of the forces of attraction between positive and negative ions in an ionic compound. It follows a periodic behavior. I have included a graph of the lattice energy of Group 1 chlorides.

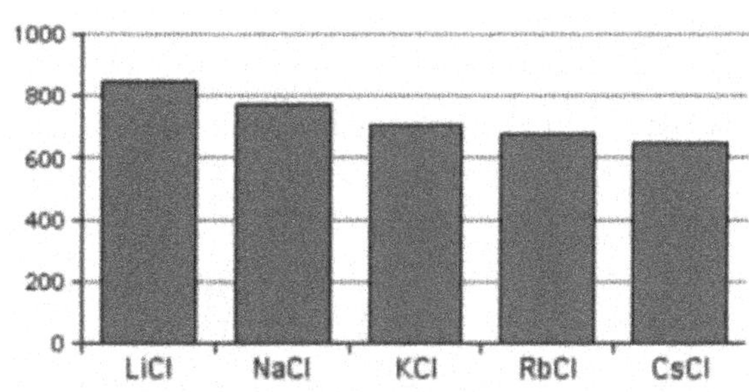

Image source: chemguide.co.uk/physical/energetics/lattice.html

Predict the ranking for the lattice energy based on the patterns in the graph for the following 3 compounds:

$BaCl_2$ $MgCl_2$ $SrCl_2$

a. Highest Energy ←——————→ Lowest Energy

b. Use evidence from the graph to support your answer.

This unit included topics of ionic bonding, covalent bonding, periodic trends, heating curves, and intermolecular forces. The chemistry textbook available for students at this school addressed each of these topics in a separate unit and chapter. However, by designing a unit around explaining observations and explicitly building the cause-and-effect relationships, students could apply their ideas to explain a new context. Students struggled to connect observable phenomena with atomic-level interactions. By repeatedly building cause-and-effect relationships that explain a range of phenomena all starting from the interactions of protons and electrons, students could connect their observations of phenomena at the macroscopic scale with the underlying cause of atomic-level interactions. Furthermore, students were able to analyze new phenomena and data that had not been covered in the class and develop rich explanations for the observations without having to learn about it in class first.

Summary

While developing ideas about the structure of matter, students often struggle to connect microscopic interactions to observable phenomena. Explicit and thoughtful use of CCCs supports students' development of structure of matter so they can use disciplinary core ideas and concepts to explain observations. Revisiting the question in Figure 11.1 (p. 248), students did not explain what happened to the bromine particles. As discussed in Chapter 15 of this book, the CCCs can be used in assessment to support students' sensemaking. The question could be revised in Part B to present students with one of the following prompts:

- Use cause-and-effect relationships to explain the changes that occurred as bromine changed from liquid to gas.
- Describe the changes in matter and energy that occurred as bromine changed from liquid to gas.

> "Explicitly incorporating CCCs in instruction and discussion supports students in developing ideas so they can use their developing particle nature of matter model to explain observations and transfer knowledge to new tasks."

Explicitly incorporating CCCs in instruction and discussion supports students in developing ideas so they can use their developing particle nature of matter model to explain observations and transfer knowledge to new tasks. Moreover, engaging students in three-dimensional learning provides them with the opportunity to understand abstract chemical concepts. Through the use of CCCs, teachers can scaffold students in describing and explaining matter concepts. Moreover, CCCs provide an opportunity for students to make sense of how what is happening on an atomic level can be used to explain what they see on a macroscale.

References

Ben-Zvi, R., B. S. Eylon, and J. Silberstein. 1987. Is an atom of copper malleable? *Journal of Chemical Education* 63 (1): 64–66.

Claesgens, J., K. Scalise, M. Wilson, and A. Stacy. 2008. Mapping student understanding in chemistry: The perspectives of chemists. *Science Education* 93 (1): 56–85.

deVos, W., and A. H. Verdonk. 1996. The particulate nature of matter in science education and in science. *Journal of Research in Science Teaching* 33 (6): 557–664.

Driver, R., E. Guesne, and A. Tiberghein. 1985. *Children's ideas in science*. Philadelphia: Open University Press.

Driver, R., A. Squires, P. Rushworth, and V. Wood-Robinson. 1994. *Making sense of secondary science: Research into children's ideas*. New York: Routledge.

Hadenfeldt, J. C., X. Liu, and K. Neumann. 2014. Framing students' progression in understanding matter: A review of previous research. *Studies in Science Education* 50 (2): 181–208.

Harrison, A. G. and Treagust, D. F., 2002. The particulate nature of matter: Challenges in understanding the submicroscopic world. In Chemical education: Towards research-based practice (pp. 189-212). Springer, Dordrecht.

Johnson, P. 1998. Progression in children's understanding of a "basic" particle theory: A longitudinal study. *International Journal of Science Education* 20 (4): 393–412.

Krajcik, J. S., R. G. Duncan, and A. E. Rivet. 2016. *Disciplinary core ideas: Reshaping teaching and learning*. Arlington, VA: NSTA Press.

Lee, O., D. C. Eichinger, C. W. Anderson, G. D. Berkheimer, and T. D. Blakeslee. 1993. Changing middle school students' conceptions of matter and molecules. *Journal of Research in Science Teaching* 30 (3): 249–270.

Mayer, K., and J. S. Krajcik. 2017. Core idea PS1: Matter and its interactions. In *Disciplinary core ideas: Reshaping teaching and learning*, eds. J. S. Krajcik, R. G. Duncan, and A. E. Rivet, 13–32. Arlington, VA: NSTA Press.

Nakhleh, M. 1992. Why some students don't learn chemistry. *Journal of Chemistry Education* 69 (3): 191–196.

National Research Council (NRC). 2012. *A framework for K–12 science education: Practices, crosscutting concepts, and core ideas*. Washington, DC: National Academies Press.

NGSS Lead States. 2013a. *Next Generation Science Standards: For states, by states*. Washington, DC: National Academies Press. *www.nextgenscience.org*.

NGSS Lead States. 2013b. *NGSS Appendix G: Crosscutting concepts*. Washington, DC: National Academies Press. *www.nextgenscience.org*.

Novick, S., and J. Nussbaum. 1978. Junior high school pupils' understanding of the particulate nature of matter: An interview study. *Science Education* 62 (3): 273–281.

Renstrom, L., B. Andersson, and F. Marton. 1990. Students' conceptions of matter. *Journal of Educational Psychology* 82 (3): 555–569.

Taber, K. S. 2003. The atom in the chemistry curriculum: Fundamental concept, teaching model or epistemological obstacle? *Foundations of Chemistry* 5 (1): 43–84.

Treagust, D. F., A. L. Chandrasegaran, L. Halim, E. T. Ong, A. N. M. Zain, and M. Karpudewan. 2013. Understanding of basic particle nature of matter concepts by secondary school students following an intervention programme. In *Concepts of matter in science education*, eds. G. Tsaparlis and H. Sevian, 125–141. Dodrecht, the Netherlands: Springer.

Chapter 12

Photosynthesis: Matter and Energy for Plant Growth

Jo Ellen Roseman, Mary Koppal, Cari Herrmann Abell, Sarah Pappalardo, and Erin Schiff

Why Is Photosynthesis Important but Difficult to Learn?

In a now classic video produced by the Private Universe Project in Science (Schneps 1997), Massachusetts Institute of Technology (MIT) graduates are shown in their caps and gowns struggling to explain how a large oak tree could grow from a tiny seed. When asked to consider whether the increased mass of the tree could have come from carbon dioxide in the air, most students dismiss the idea, calling it "surprising" and "disturbing." Despite their excellent education, they are deeply confused about photosynthesis, a process that is central to life on Earth and a topic that is taught in science classrooms starting in middle school.

What might account for the inability of these bright students to answer this fundamental question? Several factors play a role. First among them is the lack of coordination between physical and life science courses. As a result, students are not encouraged to use physical science ideas about invisible changes in matter and energy to explain life science phenomena. In characterizing problems with undergraduate biology education, Klymkowsky (2010, p. 405) includes "underestimating the need for students to come to biology with a robust grounding in physicochemical principles." Among such principles, he notes, is an understanding of "how atoms combine into molecules and how molecules behave, which implies an understanding of the factors that influence their shape, stability, interactions with other molecules, and reactivity." Interviews have shown that many undergraduate biology majors who completed a year of chemistry were "unaware or confused about how to make connections between their courses, particularly when discussing ideas such as reaction coupling, the relationship between energy and bond-breaking and bond-forming, and molecular-level collisions." (Kohn, Underwood, and Cooper 2018, p. 6). During middle and high school, where some of this cross-disciplinary foundation might be laid, the opportunity is often missed due to the lack of coordination between topics taught in life and physical science courses. For example, a typical middle school course sequence includes life science in seventh grade and physical science in eighth grade. Chemical reactions may or may not be included as a topic in eighth grade, too late for it to be useful for helping students use ideas about atom rearrangement to think about the two chemical reactions they are taught in seventh grade—photosynthesis and cellular respiration. Students commonly take high school biology before chemistry, so the only knowledge they might bring about atom rearrangement and conservation would come from their middle school courses. When students do study chemistry in high school, they are typically not asked to reflect on how knowledge of chemical reactions and conservation could help them make more sense of photosynthesis and plant growth. Nor are they asked how to use ideas about the coupling of energy-releasing to energy-requiring reactions or processes to make sense of the coupling of photosynthesis to plant growth.

Another factor making photosynthesis difficult to comprehend is the inherent level of abstraction of ideas about matter and energy at the secondary level. This can be a significant obstacle for many students. Specifically, understanding that observable growth and motion at the macroscopic level result from invisible changes in matter and energy at the molecular level requires mental models that most students have not developed. Therefore, instruction that doesn't explicitly link the physical science concepts, such as the particulate nature of matter and conservation of matter and energy, contribute to students' difficulties in understanding the biological processes of growth and movement (Anderson, Sheldon, and Dubay 1990). Even when students do have some understanding of the fundamental physical science concepts, they often have difficulty applying molecular-level ideas to living organisms (DeBoer et al. 2009; Mohan, Chen, and Anderson 2009). Many students view the basic life functions of plants as distinct and unconnected to the inert inanimate molecular phenomena they experience in a chemistry class. For example, Marmaroti and Galanopoulou (2006) found that a significant number of middle school students did not recognize photosynthesis as a chemical reaction. Assessment research has shown that students at all grade levels hold common and persistent misconceptions about photosynthesis and plant growth (AAAS n.d.), such as:

- Soil, water, and substances in the soil are food for plants.
- The mass of a gas is inconsequential and, hence, cannot contribute appreciably to plant growth.
- Plants are made of leaves, stems, and roots, not atoms and molecules.
- Plant growth does not involve chemical reactions.

- When living things grow in a closed system, the total mass of the system increases.
- When living things decompose in a closed system, the total mass of the system decreases.
- Bond-making requires energy.
- Energy is released when bonds break.

Textbooks and other kinds of curriculum materials have also contributed to the problem by not presenting students with a coherent story of photosynthesis that accounts for plant growth in terms of a few basic chemical reactions (Roseman, Stern, and Koppal 2010; Stern and Roseman 2004). For example, several middle and high school science textbooks include activities demonstrating that plants produce more starch when grown with more carbon dioxide, but the experiments are mainly used to show that plants *need* carbon dioxide rather than that carbon dioxide is a *reactant* in a chemical reaction that produces glucose. Moreover, students are not asked to explain the role this glucose plays in plant growth. This would require data showing that (a) plants are largely made up of cellulose (not glucose) and (b) plants make cellulose from glucose. Similarly, textbooks include activities demonstrating that plants *need* light but not that energy from the Sun drives an energy-requiring chemical reaction.

Roles of Crosscutting Concepts in Improving Students' Understanding of Photosynthesis

Curriculum materials and instruction that are designed to support the three-dimensional learning—called for in *A Framework for K–12 Science Education* (the *Framework*; NRC 2012) and the *Next Generation Science Standards* (*NGSS*; NGSS Lead States 2013)—have the potential to help overcome the challenges of teaching and learning about photosynthesis. Crosscutting concepts (CCCs) describe connections across disciplines that can contribute to the development of a coherent content storyline of science ideas that, along with science practices, might help students make sense of seemingly disparate phenomena across disciplines. According to the *Framework*, the value of CCCs comes from their utility as tools that students can use to (1) enrich their understanding of disciplinary core ideas (DCIs), (2) enhance their application of science and engineering practices (SEPs), and (3) extend their ability to make connections and think across disciplines (NRC 2012).

Consider, for example, the following eighth-grade student's explanation of where the mass of a dry willow tree comes from:

> *Most of the mass of a dry willow tree comes from the carbon and oxygen atoms that make up CO_2. … Carbohydrates make up plants' body structures, and carbohydrate polymers are made up of glucose monomers.… Glucose, which makes up carbohydrates, is formed by the process of photosynthesis, where plants use H_2O and CO_2 molecules. … Inside the plant's body, the*

H_2O and CO_2 molecules are broken down. The carbon from the CO_2 form glucose. … Carbohydrates, which are made up of carbon, hydrogen, and oxygen, make up 96% of the plant's dry weight[1]. Therefore, since a willow tree is a plant, carbon, hydrogen, and oxygen make up the majority of a plant's mass. When plants grow and repair, they increase in mass. … The increase in mass is caused when CO_2 and H_2O form $C_6H_{12}O_6$ (glucose) … which in turn forms carbohydrates that build the body structure of a plant such as a willow tree.

What might have contributed to this middle school student's ability to provide such a coherent response to a question that stumped the MIT graduates shown in the *Private Universe* videos? Many factors, of course, could be cited, but in this chapter, we examine the support provided by two curriculum materials that are specifically designed to help students achieve the goals of the *NGSS*—the *Toward High School Biology* (*THSB*) unit at the middle school level (AAAS/Project 2061 2017) and the *Matter and Energy for Growth and Activity* (*MEGA*) unit at the high school level (AAAS/Project 2061 2020). We focus on two CCCs that were central to the design of the units: energy and matter: flows, cycles, and conservation; and systems and system models. The abstract nature of middle and high school science ideas about the role of photosynthesis in plant growth made these two CCCs especially important for helping students use basic physical science ideas about changes in matter and energy to explain more complex processes that take place in living organisms. Whereas other chapters in this volume discuss the scientific and pedagogical foundations for the CCCs, our aim in this chapter is to illustrate the application of CCCs to the practical task of developing instructional materials to support effective teaching and learning. We explore how the design of the *THSB* and *MEGA* units took advantage of the *Framework*'s three intended roles of CCCs and provide examples of classroom activities that helped further students' understanding and use of the CCCs. In the process, we highlight how the CCCs were also used to help students overcome many of the most common difficulties in understanding the relationship between photosynthesis and plant growth.

Role 1: To Enrich Understanding of Disciplinary Core Ideas About Plant Growth

Although earlier science education reform documents (e.g., *Science for All Americans* in 1989 and *National Science Education Standards* in 1996) acknowledged the utility of certain ideas as tools for thinking across the disciplines, the *Framework* goes even further by describing CCCs as providing "an organizational framework for connecting knowledge" (NRC 2012, p. 83). However, none of these documents provides guidance for how such connections should be made in curriculum or instruction or how to support teachers in guiding their students to make them.

1. The student knew this from a data table in their textbook.

In designing the *THSB* and *MEGA* units, we began by unpacking, or deconstructing, the often highly condensed DCIs and CCCs into their essential parts. We then created smaller-grain *science ideas* that integrate elements of both *NGSS* dimensions, organized them into coherent *content storylines*, and created maps that represented the sequence of the ideas and connections among them. The storylines that follow guided the selection of phenomena that students engage with and the science practices students use to investigate and make sense of the phenomena.

In both *THSB* and *MEGA*, students encounter the science ideas at critical points in the units, usually after they have developed their own preliminary ideas based on generalizing across phenomena, data, or models they have experienced in the lessons. The science ideas are presented as generalizations scientists have constructed about how the world works based on a wider range of observations and data, and students are able to compare their own ideas to them and to find examples from their work that support the science ideas.

Toward High School Biology Content Storyline

Figure 12.1 (p. 274) shows the 17 science ideas (in numbered boxes that reference lessons where the ideas are introduced) that make up the content storyline for the middle school *THSB* unit. The storyline begins at the bottom of the map and includes three related, vertical strands, each composed of a set of science ideas.

The chemistry strand is in the middle, and its science ideas are drawn from physical science DCIs and the CCC of energy and matter: flows, cycles, and conservation. The plant growth and animal growth strands are shown on the left and right of the map, and their science ideas are drawn from middle school life science DCIs and the CCC on matter conservation. Arrows indicate connections among ideas in each strand and between the chemistry and plant and animal growth strands.

The plant strand culminates in an overarching idea (Science Idea #14) that applies the component of the CCC that says "matter is conserved because atoms are conserved" (Science Ideas #9 and #10) to plant growth:

> When plants grow or repair, they increase in mass. Atoms are conserved when plants grow. The increase in measured mass comes from the incorporation of atoms from molecules that were originally outside of the plants' bodies.

To help emphasize the connections between the plant and animal growth stories and the chemistry story, and hence, across the physical and life science disciplines, we used similar language to express the science ideas whenever possible (e.g., see Science Ideas #5, #12, and #13; Science Ideas #6, #13, #16; and Science Ideas #10, #14, and #17).

Figure 12.1. Content storyline for the middle school *THSB* unit

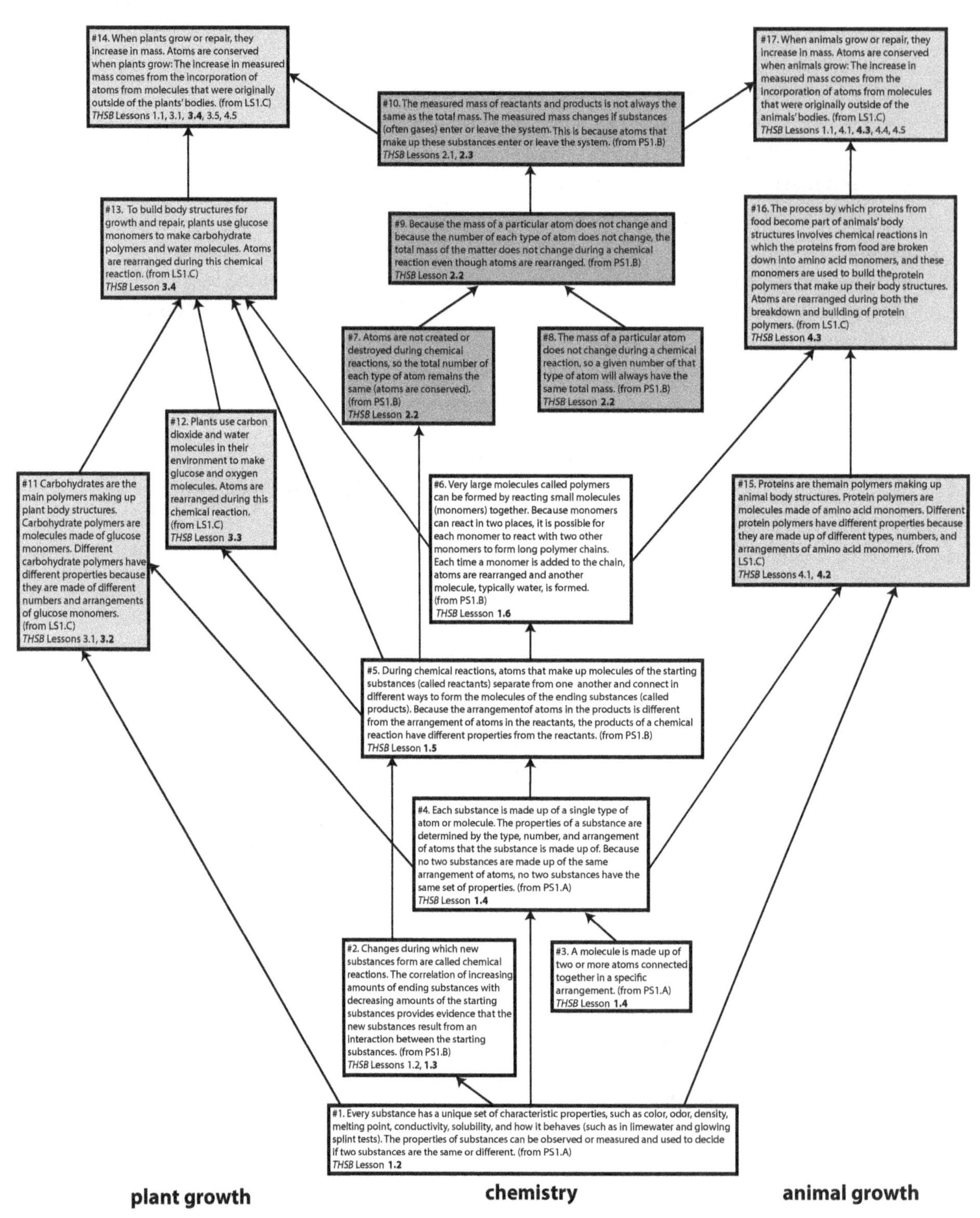

Figure 12.2. Content storyline for the high school *MEGA* unit

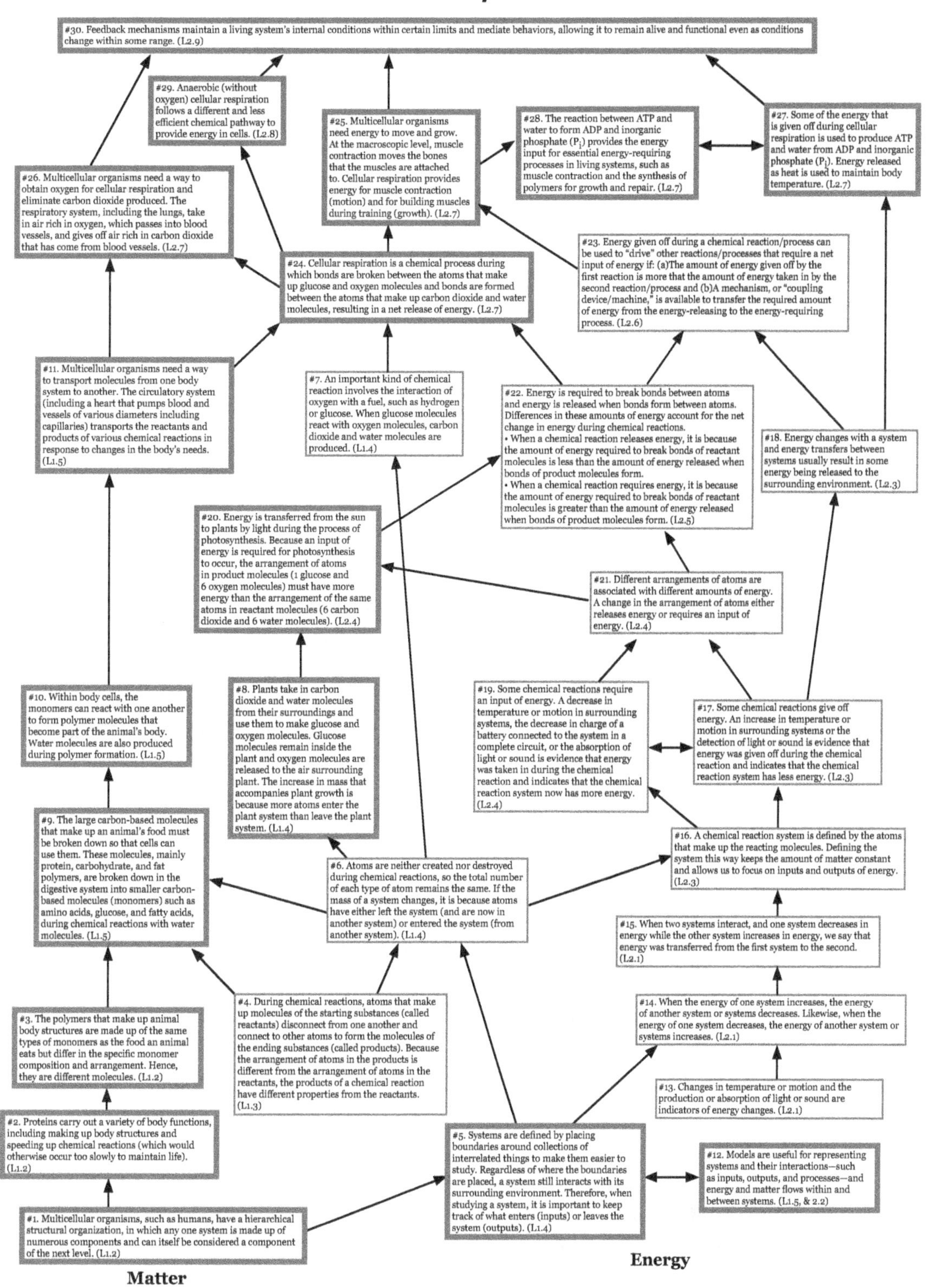

Matter and Energy for Growth and Activity Content Storyline

As with the *THSB* storyline, the 30 numbered science ideas in the *MEGA* unit's storyline draw on both DCIs and CCCs. Shown in Figure 12.2 (p. 275), the *MEGA* storyline is composed of science ideas organized into a matter strand and an energy strand. The CCC of systems and system models (Science Ideas #5, #12, and #14–16) provides a foundation for thinking about matter and energy changes in life and physical science. The matter strand shows the progression of life science ideas, as well as the contribution of matter changes at the molecular level to growth at the macroscopic level (Science Ideas #1–3 and #8–11) and their physical science precursors (Science Ideas #4 and #6). The energy strand shows the progression of life science ideas about biological growth and motion (Science Ideas #20 and #24–29) and their physical science precursors (Science Ideas #13, #17–19, and #21–23) that develop explanations of an athlete's motion in terms of bond-breaking and bond-forming at the molecular level during chemical reactions.

In developing the energy strand for the *MEGA* unit, we recognized that a quantitative treatment of energy conservation, as described in the CCC, was inappropriate for an introductory high school biology course. Instead, we took a qualitative approach to energy conservation, making use of ideas from the CCCs of energy and matter and systems and system models to address energy changes and transfers within and between a wide range of physical and living systems (Science Ideas #14 and #15). The *MEGA* unit culminates with a task that draws on ideas from both CCCs, asking students to predict changes in human body systems when an athlete engages in intense exercise.

Pulling It Together: An Example of Role 1 From the *Toward High School Biology* Unit

Lessons in both the *THSB* and *MEGA* units typically end with a set of Pulling It Together questions that provide opportunities for students to reflect on what they have learned so far and to apply that knowledge to related phenomena in a different context. For example, in the *THSB* unit, a Pulling It Together question asks students to make connections between physical and life science phenomena and to draw on ideas from both DCIs and the CCC on matter conservation to answer the following question: Where does the mass of a growing plant come from? The rest of this section describes and illustrates how the CCC on matter conservation enhances students' understanding of the DCIs in the *THSB* unit and their use in making sense of plant growth at the molecular level.

The plant growth story in *THSB* unfolds for students in Chapter 3. By this point in the unit, the class had observed and modeled a range of chemical reactions in physical systems that provide evidence for the CCC on matter conservation: While atoms rearrange during chemical reactions to create molecules of new substances, the atoms themselves are not created or destroyed; therefore, any changes in mass must have resulted from atoms entering or leaving the system.

Activities in Chapters 1 and 2 helped students appreciate the validity of the CCC and how it could be used to explain a range of phenomena, starting with those occurring in physical systems. For example, students carried out and/or observed and modeled three chemical reactions: (1) baking soda and vinegar reacting to form water, sodium acetate, and carbon dioxide (a gas that leaves the system); (2) iron and oxygen (a gas that enters the system) reacting to form rust; and (3) hexamethylenediamine reacting with adipic acid to form nylon (a solid pulled from the system) and water.

For all three chemical reactions, students had multiple opportunities to see what the reactions had in common and how the same science principles (including DCIs and the CCC on matter conservation) could be used to explain what was happening to the substances that were involved in the reactions. They observed that a substance with different properties, and hence a new substance, is produced during each reaction and that an increase in the amount of the new substance correlates with a decrease in the amounts of the starting substances. They examined data showing the atomic composition of each reactant and product, noticed that the starting and ending substances were made up of the same atoms, and then used LEGOs and ball-and-stick models to model the atom rearrangements that occur as each chemical reaction proceeds. They noticed that the measured mass increases when iron rusts in an open system and then used LEGOs to model on a balance how the measured mass could increase even though no atoms were created. Finally, students investigated the reaction that forms nylon polymers, which is like the reaction that produces the polymers that make up the body structures of plants and animals. Students were then ready to apply ideas about atom rearrangement and conservation to explain where the mass of growing plants comes from.

At the beginning of Chapter 3, students observed time-lapse photos of corn growing; identified body parts the corn was making as it grew; examined data showing that, in addition to water, the leaves, stems, flowers, and roots of various plants are made mostly of carbohydrate polymers; and compared properties of glucose to cellulose and starch. From these experiences, students concluded that plants must make carbohydrate polymers (not just glucose) if they are to grow. They also analyzed data from an isotopic labeling experiment in which scientists were able to track specific "versions" of atoms (isotopes) present in the reactants to find out where they ended up during chemical reactions. The experiment provided evidence that (a) plants make glucose and oxygen from carbon dioxide and water, with both the C and O atoms of glucose coming from CO_2 and (b) plants make cellulose and water from glucose (a reaction like nylon formation). Students then examined data showing that 96% of the dry mass of plants is due to the C, O, and H atoms.

In a Pulling It Together question in the final lesson of Chapter 3, students were asked to analyze, critique, and revise the claim made by Van Helmont some four centuries ago that the increase in the mass of a growing tree comes from water. Students who

Figure 12.3. Sample student response

Most of the mass of a dry willow tree comes from the carbon dioxide in the air. Carbohydrates are the main polymers making up plant body structures, and they are made of glucose monomers (Science idea #14). Table 1.1 shows the relative mass in grams per 100 grams of carbohydrate molecules and that carbohydrates make up most of the mass of a plant's body structure. Because a willow tree is a plant, it is made mostly of carbohydrate polymers, which are made of glucose monomers. Plants use carbon dioxide and water molecules in their environment to make glucose and oxygen molecules (Science idea #15). Table 3.3 shows that carbon dioxide and water molecules react to form glucose and oxygen molecules. Radioactively labelled carbon atoms from the air ended up in the glucose whereas the radioactively labelled oxygen atoms from the water were released into the air. Likewise, a willow tree uses carbon dioxide and water molecules to make glucose and oxygen molecules. To build body structures in plants, the atoms of glucose monomers rearrange to make carbohydrate polymers and water molecules (Science idea #16). A ball-and-stick model of the chemical reaction shows that the atoms of the glucose that are from the carbon dioxide and water molecules rearrange to form cellulose and [illegible]. Being a plant, a willow tree uses glucose monomers to make carbohydrate polymers and water molecules. The increase in measured mass of a plant comes from the incorporation of atoms from molecules that were originally outside of the plants' bodies. A table in Lesson 4.3, Activity 2 shows the mass of each atom in glucose. The carbon atoms from the carbon dioxide contribute 72 amu, the oxygen atoms from the carbon dioxide contribute 96 amu, and the hydrogen atoms from the water contribute 12 amu. Because carbon dioxide contributes the most mass to glucose and glucose is what makes up a willow tree, most of the mass of a willow tree comes from the carbon dioxide in the air.

completed this task successfully were able to integrate all three of the *NGSS* dimensions. Their explanations made use of the CCC about matter conservation, specifically the grades 6–8 idea that matter is conserved because atoms are conserved in physical and chemical processes—and its corollary that because atoms are conserved, any increase or decrease in mass in a system is because atoms have either entered or left the system. Like the student's response shown on pages 271–272, the response in Figure 12.3 cites the relevant science ideas and illustrates successful use of the crosscutting concept (highlighted in a black box) in making sense of the willow tree's increase in mass. The response also cites data from isotopic labeling experiments on plant growth and reasons quantitatively from atomic mass data to justify the claim that most of the mass of the willow tree comes from the carbon dioxide in the air. (Note that the student's response in Figure 12.3 refers to numbered science ideas based on the numbering used in the 2015 prepublished version of *THSB*, rather than on the numbering used in the final published version of *THSB* and in the content storyline map in Figure 12.1. To identify the corresponding science idea number on the story map in Figure 12.1, subtract three from each number used in the student's response.)

Teachers who participated in the study of the *THSB* unit described in Roseman, Herrmann-Abell, and Koppal (2017) monitored a representative sample of their students' explanations on the Van Helmont task. They found that 82% of their students wrote a correct claim, 74% stated the relevant science ideas (including the idea derived from the CCC about matter conservation that "plants increase in mass as they grow because atoms involved in chemical reactions enter the plant"), 50% applied the science ideas to their explanations of the willow tree, and 27% reasoned with the science ideas to link evidence to their claim.

Role 2: To Enhance Application of Science and Engineering Practices

Students using the *THSB* and *MEGA* units engage in SEPs to investigate questions about phenomena across disciplines. Whether they are analyzing data from experiments, constructing models of atom rearrangement during chemical reactions, or drawing energy transfer models, students apply the CCCs of energy and matter and systems and system models first in simple physical systems and then in more complex biological systems.

Figure 12.4. Modeling task from *THSB* focusing on the CCC idea of atom conservation

Students observe that the measured mass stays the same when iron and oxygen react in a closed container but increases when the container is opened.

Students then use LEGOs to model the iron rusting reaction to find out what happens to the mass in closed and open containers.

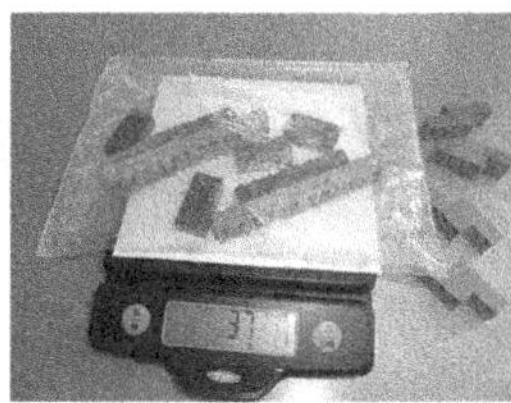

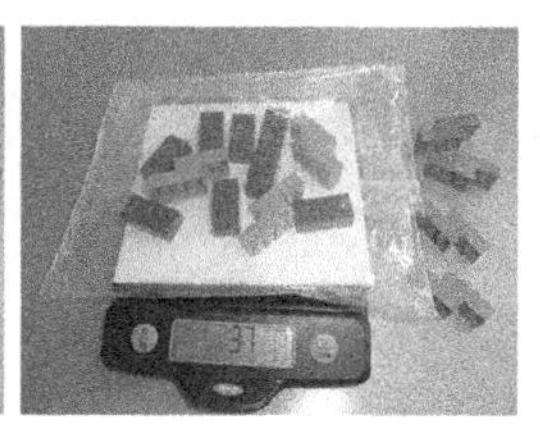

Students answer the following questions to explain why the measured mass increases even though atoms are conserved:

Explain why the measured mass was constant while the bag was sealed.

All of the atoms of reactant molecules and product molecules were trapped. No atoms could come in or go out.

Explain why the measured mass increased after the bag was opened. Where did the additional mass come from?

Oxygen atoms (from oxygen gas) could come into the bag. Two more oxygen molecules entered the bag. Since each oxygen molecule (O_2) is made up of 2 oxygen (O) atoms, that means 4 O atoms entered the bag. The four oxygen atoms that entered the bag reacted with leftover iron (Fe) atoms and formed two more iron oxide (Fe_2O_3) molecules. The measured mass increased because more atoms are in the bag. (THSB, TE, p. 214)

Applying Science and Engineering Practices in the *THSB* Unit

In the *THSB* unit, middle school students apply ideas about matter conservation as they use LEGO models to develop molecular explanations for why the mass increases when a chemical reaction involving a gaseous reactant (iron rusting) occurs in an open system. Figure 12.4 (p. 279) shows an activity in which students observe that the mass stays the same when iron rusts in a closed container but not in an open one and use LEGO models to make sense of their observations.

Figure 12.5. Explaining predictions about plant growth and ideal response

Chapter 3 – Lesson 3.3

3. The owner of a plant store wanted to see if the amount of carbon dioxide in the air would affect how her plants grow. She set up two identical rooms, Room 1 and Room 2, in her greenhouse. She put one plant in each room. Both plants were the same size and mass, and both were given plenty of water. Room 1 had normal air with the normal level of carbon dioxide in it. Room 2 was filled with air that had extra carbon dioxide in it. See Figure 3.4 below, or the color version your teacher projects.

Room 1 Room 2

Figure 3.4. Plants Exposed to Different Amounts of Carbon Dioxide

Three months later, the store owner measured each of the plants. Do you think one plant was bigger than the other? Explain your answer using science ideas, evidence, and models. Check to be sure that your explanation meets the Explanation Quality Criteria.

The plant in Room 2 would probably be bigger. Plants use carbon dioxide and water molecules in their environment to make glucose and oxygen molecules (Science Idea #12). We are told that both plants have plenty of water, so we know that the amount of water won't limit the amount of glucose that the plants can produce. However, the plant in Room 2 has more carbon dioxide than the plant in Room 1, so the plant in Room 2 can make more glucose. When we modeled the chemical reaction with ball-and-stick models, we saw that we will use 6 molecules of CO_2 and 6 molecules of H_2O for each $C_6H_{12}O_6$ molecule we make. The drawings above (another type of model) show that there are 7 CO_2 molecules in Room 1 and 18 CO_2 molecules in Room 2. So if each room has at least 18 H_2O molecules then here's what will happen according to our models:

- *The plant in Room 1 will use 6 CO_2 and 6H_2O to make 1 $C_6H_{12}O_6$ and 6 O_2. There will be 1 CO_2 and 12 H_2O left over, because there isn't enough CO_2 to make another glucose molecule.*
- *The plant in Room 2 will use 18 CO_2 and 18 H_2O to make 3 $C_6H_{12}O_6$ and 18 O_2.*

The models show how it is possible for the plant in Room 2 to produce more glucose, which it can probably use to make more polymers to build its body structures.

292

Chapter 3 – Lesson 3.3
Teacher Facilitation Notes

Teacher Talk and Actions	
Pulling It Together (cont)	
Pulling It Together Question 3 is a challenging question for students to think through. A few students may be able to reason abstractly about why the plant in Room 2 will get bigger. Most will need help. Have students work in small groups and encourage them to use LEGO or ball and stick models to help then think through their explanations. Multiple LEGO kits will have to be shared among small groups (one kit does not have enough bricks for this question). You can also let the more advanced students use ball and stick models. You can give them the following hints as needed: • Tell students that the drawings in Figure 3.4 are also models that they can use as a guide. (Remind them that real CO_2 molecules are much smaller and there are lots more in the rooms. However, the drawings give them a sense of the relative amounts of CO_2 in the two rooms). • Start by building models of CO_2 molecules in each room, using the drawing as a guide. • Then have them build models of H_2O molecules in each room. Check to be sure students have at least 18 H_2O molecules in each room. • Then have students build glucose molecules that can be formed in each room from the CO_2 and H_2O molecules they have built. When building glucose, all that matters is the atomic composition.) Once students have figured out their explanations with their models, have them construct a written explanation. Have blank explanation charts available for those who still need them, which can be found on page xviii of the introduction to the unit.	
Closure and Link	**(3 min)**
Whole class: Summarize the lesson in a few statements: We saw that in plants, carbon dioxide gas from the air and water from the ground react to form glucose monomers, which remain in the plant, and oxygen molecules, which are released into the air. Link to next lesson: In the next lesson, we will investigate how plants make the carbohydrate polymers that make up their body structures from these glucose monomers.	

293

This content appears on pages 292–293 of the *THSB Teacher Edition*.

After generalizing the idea that atom conservation explains mass conservation and changes in measured mass in simple chemical reaction systems, students use models to make sense of isotopic labeling data showing that (a) the carbon and oxygen atoms of glucose produced by a plant come from carbon dioxide in the plant's environment and (b) plants use glucose to produce the cellulose needed to build body structures as they grow. Finally, students use models to explain why their predictions that a plant will grow bigger in air enriched with carbon dioxide than in normal air make sense (see Figure 12.5).

Applying Science and Engineering Practices in the *MEGA* Unit

In the high school *MEGA* unit—after investigating, modeling, and explaining matter changes that are involved in animal growth and energy changes that are involved in animal motion and growth—students are asked to apply the process to model and explain energy changes that accompany plant growth. (Students should have already modeled matter changes associated with plant growth in *THSB*.) The *MEGA* unit develops the CCC of systems and system models to demonstrate to students the usefulness of "systems thinking" as a tool for explaining changes in energy within systems and energy transfers between systems. Students use what they learn about the importance of paying attention to inputs, outputs, and boundaries of systems to make sense of animal and plant growth (using Science Ideas #5 and #12, based on the CCC, and using Science Ideas #6, #8, and #9, based on DCIs) and then apply systems thinking to energy changes and transfers accompanying motion and growth at the macroscopic level (using Science Ideas #14–19) and then at the molecular level (using Science Ideas #20–22).

Since both ideas about and models of energy are abstract, the *MEGA* unit introduces students to them in simple physical systems. Students learn to associate changes in motion and temperature and the emission of light and sound with changes in energy. They then model energy changes and transfers in simple physical systems. Such systems can include when billiard balls collide, when a flask of hot water is placed inside a beaker of cold water, how a battery-powered toy car moves faster when connected to a new versus used 9-volt battery, and how (as shown in Figure 12.6) a solar-powered toy car moves farther under a 100-watt light bulb versus a 40-watt light bulb.

Figure 12.6. Solar-powered car setup with 100-watt and 40-watt light bulbs

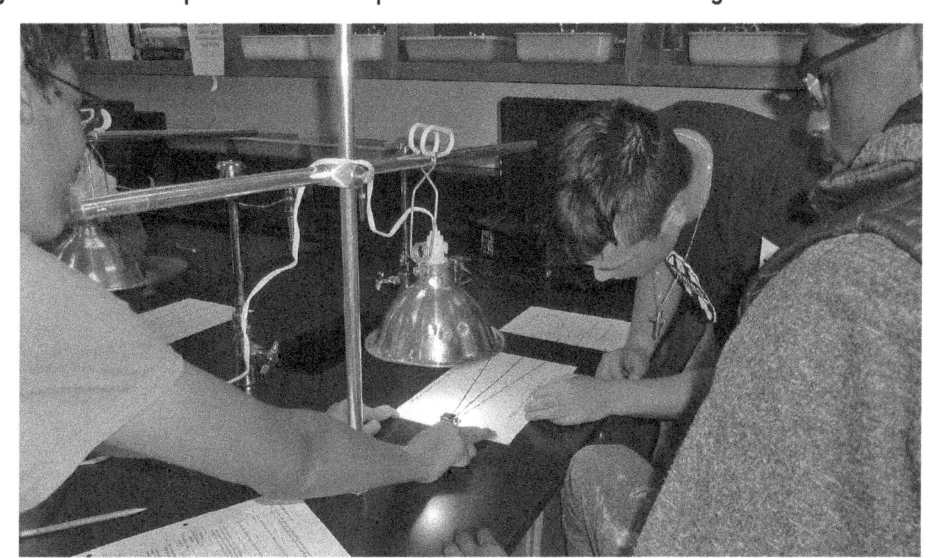

The increased motion of the solar-powered cars provides evidence that energy is being transferred from the light bulbs to the cars, which students will make use of later when constructing a model of plant growth.

Modeling Chemical Reaction Systems in Physical Science Contexts

Students next make sense of energy changes within systems and energy transfer between systems in chemical reaction systems, such as:

- Energy-releasing reactions between hydrogen (H_2) and oxygen gas (O_2) to produce H_2O and between glucose ($C_6H_{12}O_6$) and oxygen (O_2) to produce carbon dioxide (CO_2) and water (H_2O), using the emission of light and sound and increased motion and temperature as indicators that the reactions release energy, and
- Energy-requiring reactions between H_2O molecules to produce H_2 and O_2, using the increased production of gases when the system is connected in a complete circuit to a new versus used 9-volt battery as an indicator that the reaction $2\ H_2O \rightarrow O_2 + 2\ H_2$ requires energy.

Students represent energy changes *within* systems using bar graphs and system boxes in which a downward-curved arrow represents a decrease in energy of the system and an upward-curved arrow represents an increase in energy of the system. They represent energy transfer *between* systems using a thick arrow to indicate the direction of energy transfer (as shown in Figure 12.7 for the reaction $2\ H_2O \rightarrow O_2 + 2\ H_2$). Energy changes in the battery and water systems before and after the chemical reaction are represented by

Figure 12.7. Models representing energy change and energy transfer

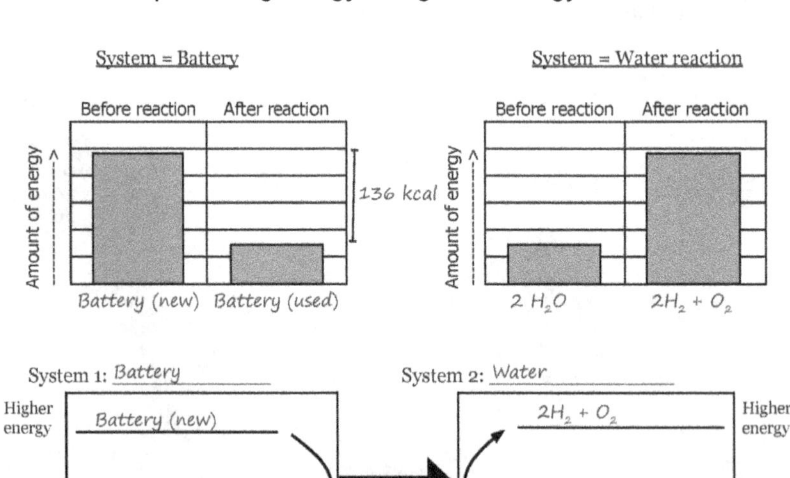

bar graphs shown in the top row of the figure. Energy transfer from the battery to the water system is represented by an energy transfer model shown in the bottom row of Figure 12.7.

For each phenomenon that students observe in both physical systems and chemical reaction systems, the following questions, based on the CCCs of systems and system models and energy and matter: flows, cycles, and conservation, guide them in constructing their models: (1) Which system is increasing in energy? (2) What evidence do you have that the system is increasing in energy? (3) What system provides the energy? (4) What evidence do you have that the system you just added to your model can transfer the required energy? These same questions then guide students in constructing models of chemical reaction systems in living organisms.

Modeling Chemical Reaction Systems in Living Organisms

In designing the *MEGA* unit, we chose each physical science phenomenon because of its applicability to biological phenomena that students will eventually encounter in the unit. For example, the solar-powered car experiment provides evidence that light transfers energy, and the water-splitting reaction shows that oxygen gas production can be used to monitor the rate of a reaction, such as photosynthesis. When students observe that aquatic plants produce more oxygen gas under a 100-watt light bulb than under a 40-watt light bulb, they conclude that the production of oxygen gas during photosynthesis requires energy. Figure 12.8 is an example from a student's notebook showing how the student defined System 2 and what was happening as energy was transferred from System 1.

Figure 12.8. Example of a student's energy transfer model for photosynthesis

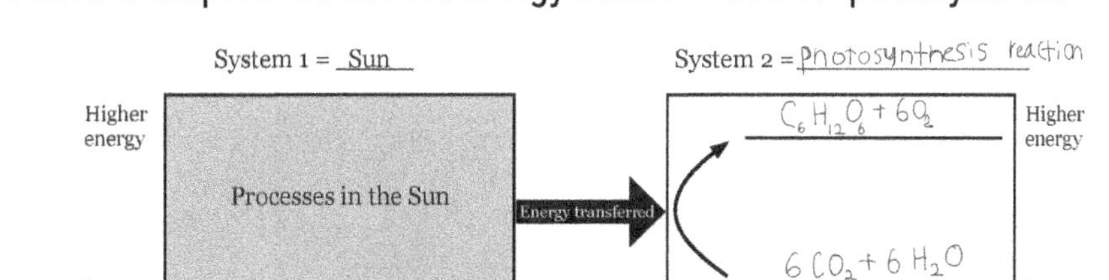

The most complex phenomena that students are asked to make sense of at the molecular level is the motion of an athlete in competition (e.g., a tennis player or an Olympic runner) and the bodybuilding of an athlete preparing for competition. Students draw inferences from multiple sets of data: a comparison of their own increased carbon dioxide (CO_2) production after exercising for five minutes to their CO_2 production after sitting quietly for five minutes, data showing how an athlete's leg muscles take up more oxygen (O_2) and more glucose ($C_6H_{12}O_6$) from his blood vessels when doing leg extensions than when at rest, and the observation that isolated muscle fibers shorten in the presence of adenosine triphosphate (ATP) but not $C_6H_{12}O_6 + O_2$—a surprising finding given the other data. Students begin to reason that motion in an animal's body requires another system that is coupled to cellular respiration and construct a three-system energy transfer model (Figure 12.9) with an arrow from System 2 to System 3. Finally, students use data collected by scientists that provide evidence that the reaction between glucose and oxygen (cellular respiration) provides energy for the synthesis of ATP from (adenosine diphosphate) ADP + Pi, which provides justification for students to draw an upward-curved arrow in System 2 and a thick arrow from System 1 to System 2.

Figure 12.9. Energy transfer model showing the ATP cycle linking cellular respiration to muscle contraction

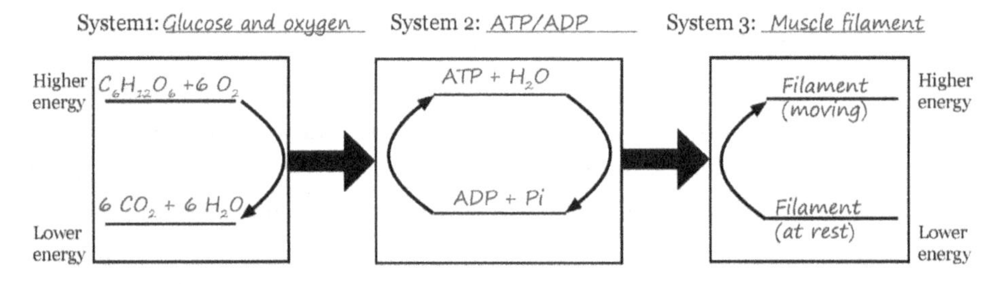

To construct an energy transfer model for animal growth, students examine data from a study of the effect of resistance training on an athlete's muscle building, which shows that the exercised arm of an athlete builds more protein in four hours than the arm not being exercised. Having learned that muscle growth involves protein synthesis and that protein synthesis requires energy (AAAS/Project 2061 2020), one student replaced the muscle filament in System 3 of Figure 12.9 with protein synthesis to create the energy transfer model shown in Figure 12.10.

Figure 12.10. Example of a student's energy transfer model for animal growth

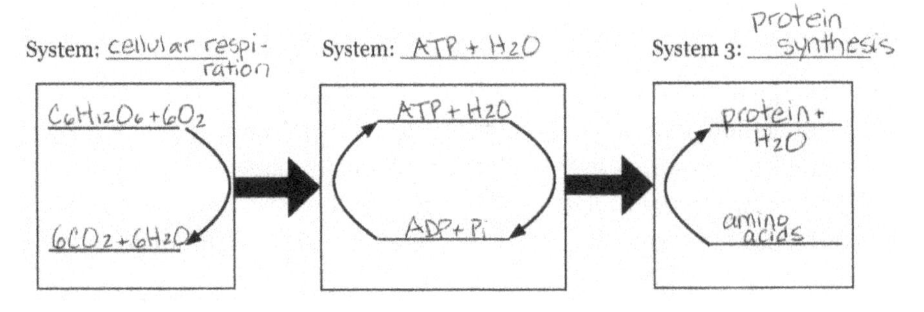

The detailed example from the *MEGA* unit that follows illustrates how the sequence of the aforementioned activities prepared students to apply DCIs and the CCCs of energy and matter and systems and system models to model and explain plant growth.

Pulling It Together: An Example of Applying Science and Engineering Practices from the *MEGA* Unit

The Pulling It Together questions in Lesson 2.7, near the end of *MEGA*'s Chapter 2, are quite challenging for students. They expect students to use ideas about cellular respiration and ATP synthesis in animals to make sense of plant growth. Many students think that plants get their energy directly from the Sun, without involving chemical reactions. Even students who have learned the photosynthesis equation don't realize that plants carry out cellular respiration just like animals do and, as with animals, cellular respiration provides the energy for growth. For the following Pulling It Together question, students worked in small groups to apply what they had learned in the *MEGA* unit to figure this out.

> Question 5. As you've just seen, models of matter and energy changes within systems and energy transfer between systems can help us make sense of phenomena involving other animals besides humans, such as hamsters. What about plants?
>
> i. As you watch the videos of plants growing, make notes about the <u>matter changes</u> that are occurring and the <u>energy changes</u> that must take place.
> ii. List all the <u>systems</u> you think should be included in your model and provide a reason for why you think each system is important to include.
> iii. Create an <u>energy transfer model</u> showing how energy from each system is transferred to the next. (AAAS/Project 2061 2020, p. 188)

Read on to see how the students responded to each of the items listed in this Pulling It Together question.

Matter and Energy Changes

To address matter changes, students reviewed the phenomena in Chapter 1 of *MEGA*, and a few recalled what they observed in their *THSB* unit that provided evidence that (a) plants make glucose (and oxygen) from carbon dioxide and water during photosynthesis and (b) plants use the glucose they make to build carbohydrate polymers such as cellulose or starch (and water) that are used to build their body structures. For energy changes, students reviewed phenomena in Chapter 2 of *MEGA* that provide evidence that (a) photosynthesis requires energy that is transferred by light and (b) polymer formation requires energy (e.g., more protein is synthesized in the arm engaged in resistance training than in the control arm).

Systems

In response to the section of the question focused on systems, students had little trouble listing the Sun system and the photosynthesis system they had previously modeled (see Figure 12.8 on p. 283). One group of students listed another system and called it "Plant Growth System" because plants would need to build polymers to grow. Another group listed "Cellular Respiration System" because they had just examined evidence that cellular respiration provides energy for human motion and growth. When classmates asked why they listed cellular respiration for plants, the group noted that science ideas about cellular respiration stated that the process provided energy for all multicellular organisms, which included plants.

Energy Transfer Model

Figure 12.11 shows the energy transfer model that students are expected to construct.

Figure 12.11. Model of photosynthesis that students are expected to construct

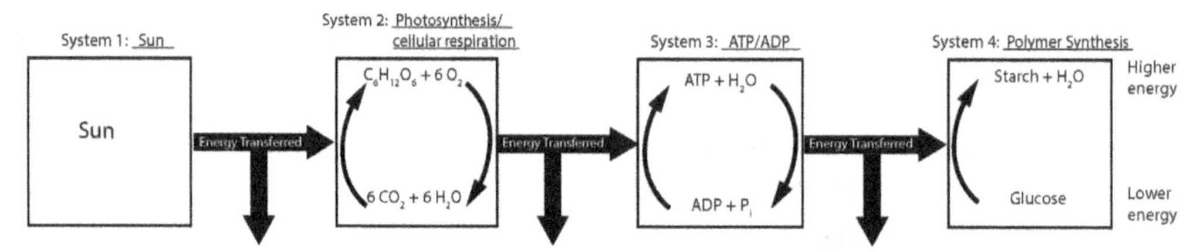

To get started on their energy transfer models, students referred to the guiding questions based on the CCCs that had helped them construct energy transfer models of physical systems and of animal motion and growth: (1) Which system is increasing in energy? (2) What evidence do you have that the system is increasing in energy? (3) What system provides the energy? (4) What evidence do you have that the system you just added to your model can transfer the required energy?

They reasoned that the system increasing in energy is the plant growth system because (1) a table in the unit titled Energy Requiring Chemical Reactions (AAAS/Project 2061 2020, p. 136) showed that the chemical reactions plants need to build body structures for growth (e.g., making protein, carbohydrate, or fat polymers from their respective monomers) all require energy; (2) the energy must come from cellular respiration, the energy-releasing chemical reaction between glucose ($C_6H_{12}O_6$) and oxygen (O_2), which produces the carbon dioxide (CO_2) and water (H_2O) that all multicellular organisms—including plants—use to obtain energy (as stated in the science ideas); and (3) the table Energy Released During Reactions of Different Fuels With Oxygen (AAAS/Project 2061 2020, p. 121) shows that the reaction between glucose and oxygen (which students now know is the process of cellular respiration) releases sufficient energy for building polymers that make up plant body structures.

Several groups of students placed the plant growth system at the far right of the model and added the cellular respiration and ATP/ADP systems to its left, so it resembled the energy transfer model they had drawn for animal growth (Figure 12.10, p. 284). They realized that the far-left side of the model should start with the Sun and photosynthesis systems shown in Figure 12.8 (p. 283), but they didn't know how to connect the photosynthesis system to the cellular respiration system. After letting students wrestle with the problem for a while, the teacher asked if another system in their model could give them ideas. One student asked if it would be OK to combine the photosynthesis and cellular respiration systems into a single system, like the ATP/ADP system. Another student thought that wasn't a good idea because the processes occur in different organisms. The first student argued that both processes do occur in plants, and most students agreed that the two processes could be combined into a single system to show a cycle as with the ATP/ADP system.

One curious student then wondered how to represent the fact that glucose serves two purposes in plants—as a fuel that reacts with oxygen to release energy and as a building block of polymers that make up the plant's body structures. The student realized that if the plant burns all its glucose as fuel, then none will be left for building its body structures. The student's group set about trying to represent that in their model.

Role 3: To Make Connections Across Disciplines

In contrast to traditional middle and high school curriculum materials, *THSB* and *MEGA* both target ideas about matter and energy changes in chemical reactions in both simple physical systems and complex biological systems in the same unit. Furthermore, both *THSB* and *MEGA* first introduce abstract ideas in simple physical systems that allow firsthand observations of the phenomena before having students apply them to complex biological systems that require inferences from data. Unit activities use similar models to represent matter and energy changes within systems and energy transfers between systems in phenomena in both simple physical and complex biological systems. This integration of physical and life science content enhances the explanatory power of the science ideas, both DCIs and CCCs related to energy and matter and systems and system models.

Both DCIs and CCCs played a critical role in our selection and sequencing of the phenomena students would experience in the *THSB* and *MEGA* units. Guided by the sequence of science ideas in each unit's content storyline (see Figure 12.2 [p. 275] and Figure 12.3 [p. 278]), we selected phenomena from both physical and life science contexts that clearly aligned with the science ideas, had common features (e.g., produced a gas, used energy from a battery or light), and could be experienced by students either directly or through inferences from data. Our strategy was to use directly observable physical science phenomena to help students generate science ideas they could then apply to complex biological phenomena that required inferences from data. Seeing the applicability of the science ideas—including ideas from CCCs about conservation and systems—across disciplines helped students appreciate their explanatory power.

In the *THSB* middle school unit, for example, students are challenged to make sense of phenomena involving mass changes during chemical reactions that contribute to plant growth (Science Idea #14 in Figure 12.2, p. 275). To do this, they apply ideas about conservation that were used to explain mass changes when iron rusts. As shown in Table 12.1, in both life and physical science phenomena, atoms from a gaseous reactant enter the system and react to form a solid product that remains in the system, thereby increasing its mass. Without the ability to make this kind of conceptual connection between disciplines, students would be at a loss to explain the molecular mechanisms of plant growth. Although students might be able to observe that the pores in plant leaves open and close and that plants produce starch in the light, they would not be able to link the carbon dioxide that is taken in by plant leaves to the sugar and cellulose that plants produce.

Similarly, students might learn that scientists are able to track specific "versions" of atoms (isotopes) to find out where they end up during chemical reactions, but students might rightly wonder why they would be learning about isotopic labeling experiments unless they had *accepted the idea* that the products of a chemical reaction come from the reactants. It makes more sense for students to observe directly for themselves that products come from reactants in simple physical systems and only then consider whether the same thing happens in plant leaves. Isotopic labeling, then, solves the problem of tracking atoms from reactants to products that are mixed with lots of other molecules that make up plants.

Table 12.1. Physical and life science phenomena in *THSB* selected to make matter changes during plant growth credible to students

Science Idea #12: Plants use carbon dioxide and water molecules in their environment to make glucose and oxygen molecules. Atoms are rearranged during this chemical reaction.	
Life Science Phenomena (Lesson)	Related Physical Science Phenomena (Lesson)
• Data show substances plants take in from their environment (water, carbon dioxide, and minerals) (L3.3). • Data from isotopic labeling experiments show where carbon from carbon dioxide ends up and where oxygen from water ends up during the photosynthesis reaction (L3.3).	• Rust forms when iron and oxygen in the air interact, and the more iron and oxygen that are used, the more rust forms (L1.2 and L1.3). • Carbon dioxide (and other substances) forms when baking soda interacts with vinegar, and the increase in carbon dioxide correlates with the decrease in vinegar (L1.2 and L1.3).
Science Idea #13: To build body structures for growth and repair, plants use glucose monomers to make carbohydrate polymers and water molecules. Atoms are rearranged during this chemical reaction.	
Life Science Phenomena (Lesson)	Related Physical Science Phenomena (Lesson)
• Data from a radioactive labeling experiment show the percent of labeled carbon atoms from glucose that end up in cellulose in plants grown with and without herbicide (L3.4).	• Nylon (a polymer) forms when hexamethylenediamine and adipic acid interact, and the increase in the amount of nylon correlates with the decrease in the reacting substances (L1.2 and L1.3).
Science Idea #14: When plants grow or repair, they increase in mass. Atoms are conserved when plants grow; the increase in measured mass comes from the incorporation of atoms from molecules that were originally outside of the plants' bodies.	
Life Science Phenomena (Lesson)	Related Physical Science Phenomena (Lesson)
• Data showing percent of dry weight of types of atoms making up an evergreen tree show that C, H, and O atoms make up most of the tree's dry mass (L3.5). • Calculations using data about the mass of C, H, and O atoms show the percent that each atom contributes to the mass of glucose $C_6H_{12}O_6$ (L3.5). • Data from Van Helmont's willow tree experiment show that the increase in the tree's mass cannot be accounted for by the decrease in the mass of the soil (L3.5).	• When baking soda and vinegar interact in a sealed container, the measured mass does not change; but when the container is opened a hissing sound is heard and the measured mass decreases (L2.1). • When iron and oxygen interact in a sealed container, the measured mass does not change; but when iron and oxygen interact in an open container, the measured mass increases (L2.1).

For the *MEGA* high school unit, we took a similar approach to helping students first make sense of phenomena involving energy changes during chemical reactions in simple physical systems. Consider the idea from the *Framework* that "energy from light is needed for plants because the chemical reaction that produces plant matter from air and water requires an energy input to occur" (NRC 2012, p. 153). Elementary school students are expected to know only that plants need light, but the *Framework* expects high school students to know why. This requires knowing that some chemical reactions, including photosynthesis, require an energy input to occur (see Science Idea #20 in Figure 12.3, p. 278), which requires that students have accepted earlier ideas in the storyline, specifically that (a) changes in temperature or motion and the production or absorption of light or sound are indicators of energy changes; (b) when the energy of one system increases (or decreases), the energy of another system decreases (or increases); (c) when two systems interact, and one system decreases in energy while the other system increases in energy, we say that the energy was transferred from the first system to the second; and (d) because different arrangements of atoms are associated with different amounts of energy, a change in the arrangement of atoms during a chemical reaction either releases energy or requires an input of energy.

Without the foundational ideas developed first in a physical science context, students could observe that aquatic plants produce more oxygen gas with more light and less oxygen gas with less light, but they would be unlikely to link the input of energy from light to the production of higher energy products (that can react to release energy needed for growth). It makes sense for students to directly observe indicators of energy changes in simple physical systems and model energy changes within systems and energy transfer between systems in those simple physical systems before being asked to apply the indicators and models to make sense of energy changes and energy transfers during photosynthesis and cellulose production. Physical science phenomena that could be used to develop Science Idea #20 are shown in Table 12.2.

Table 12.2. Physical and life science phenomena in *MEGA* selected to make energy changes during photosynthesis credible to students

Science Idea #20: Energy is transferred from the Sun to plants by light during the process of photosynthesis. Because an input of energy is required for photosynthesis to occur, the arrangement of atoms in product molecules (1 glucose and 6 oxygen molecules) must have more energy than the arrangement of the same atoms in reactant molecules (6 carbon dioxide and 6 water molecules).	
Life Science Phenomena (Lesson)	Related Physical Science Phenomena (Lesson)
• Aquatic plants produce more oxygen gas when a 100-watt light bulb shines on them than when a 40-watt light bulb shines on them (L2.4).	• The wheels of a battery-powered toy car spin and the car moves faster and hums at a higher pitch when connected in a circuit to a new versus used 9-volt battery (L2.1). • A solar-powered toy car moves farther when a 100-watt light bulb shines on it than when a 40-watt light bulb shines on it (L2.1). • Water produces more oxygen gas when connected in a circuit to a new versus used 9-volt battery (L2.4).

Summary

This chapter describes how two crosscutting concepts—matter and energy and systems and system models—can be used with life and physical science disciplinary core ideas about matter and energy to help students make sense of plant growth in terms of chemical reactions. We were able to use the crosscutting concepts in the design of our curriculum materials to (a) enrich students' understanding of the life and physical science ideas and connections among them that explain plant growth, (b) use appropriate science practices to make sense of the phenomena, and (c) use paired phenomena in life and physical science that illustrate the science ideas and their explanatory power to help students appreciate connections across life and physical science. The choice of specific crosscutting concepts was strongly influenced by a clear understanding of the disciplinary core ideas at each grade level and the common misconceptions and preconceptions students were likely to bring.

It is not possible, of course, to attribute student learning gains directly to the three-dimensional design of the units or their use of crosscutting concepts (or any other specific design feature), but both the *THSB* and *MEGA* units have shown promise. In separate randomized control trials, students who used each unit showed greater learning gains than students in the comparison groups who used district- and/or teacher-developed materials they thought would target the same learning goals. Students using the middle school *THSB* unit also experienced a substantial decrease in misconceptions about where the mass of a growing plant comes from (Herrmann-Abell, Koppal, and Roseman 2016), and teachers benefited from the supports provided for teaching the unit and for implementing *NGSS* recommendations (Roseman, Herrmann-Abell, and Koppal 2017). Moreover, high school students who experienced the *MEGA* unit gained a better understanding of the role that matter and energy play in chemical reactions occurring in simple physical systems and complex biological systems than students in the comparison group (Herrmann-Abell, Hardcastle, and Roseman 2019; Roseman et al. 2019).

Despite overall success in improving student learning with the two units, several challenges remain. The content storylines at the middle and high school level are cognitively complex. Moreover, given the relative newness of the *NGSS* and the time it takes to create new materials, few students will be entering middle and especially high school with the prerequisite knowledge and skills to complete the units in the recommended time frames. Teachers play a critical role in advocating for and selecting materials that meet *NGSS* design specifications, developing expertise in using those materials that are available, and eagerly participating in curriculum research and development projects.

References

American Association for the Advancement of Science (AAAS)/Project 2061. n.d. *http://assessment.aaas.org*.

American Association for the Advancement of Science (AAAS)/Project 2061. 2017. *Toward high school biology: Understanding growth in living things*. Arlington, VA: NSTA Press.

American Association for the Advancement of Science (AAAS)/Project 2061. 2020. *Matter and energy for growth and activity*. Arlington, VA: NSTA Press.

Anderson, C. W., T. Sheldon, and J. Dubay. 1990. The effects of instruction on college nonmajors' conceptions of respiration and photosynthesis. *Journal of Research in Science Teaching* 27 (8): 761–776.

DeBoer, G. E., C. F. Herrmann-Abell, J. A. Wertheim, and J. E. Roseman. 2009. Assessment linked to middle school science learning goals: A report on field test results for four middle school science topics. Paper presented at the National Association for Research in Science Teaching (NARST) Annual Conference, Garden Grove, CA.

Herrmann-Abell, C. F., J. Hardcastle, and J. E. Roseman. 2019. Evaluating a unit aimed at helping students understand matter and energy for growth and activity. Paper presented at the American Educational Research Association Annual Conference, Toronto, Canada.

Herrmann-Abell, C. F., M. Koppal, and J. E. Roseman. 2016. *Toward High School Biology:* Helping middle school students understand chemical reactions and conservation of mass in nonliving and living systems. *CBE-Life Sciences Education,* 15 (4): ar74, 1–21

Klymkowsky, M. W. 2010. Thinking about the conceptual foundations of the biological sciences. *CBE-Life Sciences Education,* 9 (4): 405–407.

Kohn, K. P., S. M. Underwood, and M. M. Cooper. 2018. Energy connections and misconnections across chemistry and biology. *CBE—Life Sciences Education* 17 (1): ar3. DOI: 10.1187/cbe.17-08-0169.

Marmaroti, P., and D. Galanopoulou. 2006. Pupils' understanding of photosynthesis: A questionnaire for the simultaneous assessment of all aspects. *International Journal of Science Education* 28 (4): 383–403.

Mohan, L., J. Chen, and C. W. Anderson. 2009. Developing a multi-year learning progression for carbon cycling in socio-ecological systems. *Journal of Research in Science Teaching* 46 (6): 675–698.

National Research Council (NRC). 2012. *A framework for K–12 science education.* Washington, DC: National Academies Press.

NGSS Lead States. 2013. *Next Generation Science Standards: For states, by states.* Washington, DC: National Academies Press. *www.nextgenscience.org.*

Roseman, J. E., C. F. Herrmann-Abell, and M. Koppal. 2017. Designing for the *Next Generation Science Standards*: Educative curriculum materials and measures of teacher knowledge. *Journal of Science Teacher Education* 28 (1): 111–141.

Roseman, J. E., L. A. Stark, C. F. Herrmann-Abell, K. M. Bass, G. DeBoer, D. Drits-Esser, J. Hardcastle, S. Homburger, M. Malone, and R. Nehm. 2019. *Developing high school biology curriculum materials that support* NGSS *teaching and learning: Opportunities and challenges.* Symposium presented at the Annual International Conference of NARST, Baltimore, MD.

Roseman, J. E., L. Stern, and M. Koppal. 2010. A method for analyzing the coherence of high school biology textbooks. *Journal of Research in Science Teaching* 47 (1): 47–70.

Schneps, M. 1997. Minds of our own: Lessons from thin air [Video]. Cambridge, MA: Science Media Group, Harvard University. *www.youtube.com/watch?v=z7qH-GhabiA.*

Stern, L., and J. E. Roseman. 2004. Can middle school science textbooks help students learn important ideas? Findings from Project 2061's curriculum evaluation study: Life science. *Journal of Research in Science Teaching* 41 (6): 538–568.

Chapter 13

Re-Envisioning Instruction With Crosscutting Concepts: Weather and Climate

Ann E. Rivet and Audrey Rabi Whitaker

Every day we walk outside, and consciously or not, we are aware of the weather around us. We make daily lifestyle choices based on this awareness, including what we will wear and what activities we will engage in during the day. Many longer-term choices, including where we live and the kinds of food we eat, are influenced by the climates we live in. Have you ever wondered what makes the weather act the way it does or why different locations have different climates? These questions are at the heart of understanding weather and climate and why they change over time.

This chapter aims to illustrate what instruction looks like when crosscutting concepts are an explicit and intentional part of teaching and learning about weather and climate, in contrast to traditional instruction that focuses on teaching only the ideas. We exemplify how three-dimensional learning in this area differs in both substance and style when compared to a traditional presentation, such as what would have been expected in instruction aligned with the *National Science Education Standards* (NRC 1996). We will present comparative examples of teaching weather and climate at the elementary, middle, and high school levels to illustrate these differences, highlighting how a focus on crosscutting concepts affects student learning. Finally, we will emphasize how a three-dimensional perspective on teaching climate and weather that foregrounds crosscutting concepts changes the learning goals and expected outcomes at each grade level.

What Do We Mean by Weather and Climate?

Before diving into instructional examples, we first step back to look at the aims for understanding weather and climate overall. To start, it is important to understand that *A Framework for K–12 Science Education* (the *Framework*; NRC 2012) and the *Next Generation Science Standards* (*NGSS*; NGSS Lead States 2013) have reframed disciplinary Earth science concepts in ways that are very different from past standards. The ESS2: Earth's Systems disciplinary core idea (DCI) is a broad umbrella that includes five component key ideas. Together, these key ideas describe and explain the processes and mechanisms that drive the phenomena we observe on Earth's surface at a variety of different scales. In other words, this disciplinary core idea is focused on students understanding how Earth "works" to create the world we see around us every day (Rivet 2016). Across K–12, the Earth's systems DCI aims for students to come to understand two broad features of the planet: (1) Earth is constantly changing, and (2) different aspects of Earth are interconnected and influence one another.

The topic of weather and climate is one of the key components to the Earth's systems disciplinary core idea. The guiding question posed in the *Framework* for this topic is as follows: "What regulates weather and climate?" The focus on *regulation* as a mechanism is a noteworthy shift from previous standards and guidelines. In prior standards, the expectation for teaching about weather and climate was that by the end of high school, students could name the factors of local weather, describe the features of local- and regional-scale weather systems, define the concept of climate, and compare the characteristics of different climates. Rarely did Earth science instruction provide students with opportunities to explore the dynamics of weather and climate at greater levels of sophistication. However, the new focus in the *Framework* and the *NGSS* emphasize that now, students are expected to understand and explain *what regulates* weather and climate, including both stability and change perspectives. The *NGSS* expectation is for students to understand not just what weather factors are but how they interact and what causes them to change. This means understanding *why* weather systems operate at different time and spatial scales, as well as what causes these weather systems to develop in complex yet predictable ways. Additionally, whereas weather changes frequently, climate systems often remain relatively stable. To understand the regulation of climate systems, students must understand the physical, biological, and chemical processes that maintain climates at regional and global scales, as well as the perturbations to these conditions that would cause climates to change over time. In concordance with the *Framework*, the emphasis is on modeling, explaining, and predicting the constant changes that are occurring in the atmosphere across time and spatial scales. Crosscutting concepts provide the fundamental tools to achieve these learning goals.

Weather and climate are complex ideas that are not learned once but developed over time. This is necessary in part because in order to understand how Earth's atmospheric system works, students need to draw on core ideas from other disciplines. For example, in order to explain a thunderstorm cell or a tornado, students need to understand convection. In order to understand and predict climate variations, students need to apply ideas about the heat capacity of different gases. As a consequence, the teaching of weather and climate needs to occur across grade levels and build concurrently with the development of core ideas in the other disciplines.

The Role of Crosscutting Concepts in Understanding Weather and Climate

The key focus across K–12 for understanding Earth's weather and climate systems is on *how they work* and *how they influence other aspects of the environment,* both natural and human-built. Thus, it is not sufficient to be able to list types and name facts when it comes to weather and climate. The goal is now explaining mechanisms and processes at grade-appropriate levels that develop over time. This is where crosscutting concepts become central for understanding this important core idea. According to the *Framework* (NRC 2012), crosscutting concepts are "concepts that bridge disciplinary boundaries, having explanatory value through much of science and engineering. … These concepts help provide students with an organizational framework for connecting knowledge from various disciplines into a coherent and scientifically based view of the world" (p. 83). Crosscutting concepts play a unique role in three-dimensional learning in that they frame how to think about the purpose or goal of explaining phenomena in ways that connect to the disciplinary concepts. As such, they serve as a set of heuristic approaches that can be used in different ways to develop and reflect on explanations of complex weather and climate phenomena (Rivet et al. 2016).

One of the key ways that crosscutting concepts can be used in learning about weather and climate is as lenses or frames through which students can observe, analyze, and make sense of real-world instances of phenomena. For example, students can observe patterns in data and note that different kinds of patterns can be described at different spatial and temporal scales (e.g., local patterns versus regional patterns; patterns over days and weeks versus patterns across years and decades). Alternatively, students can examine a phenomenon by looking specifically for its causes and effects—for example, the causes of a big July thunderstorm over a city or the potential causes and effects of sea level rise.

A second important role of crosscutting concepts is to serve as bridges to connect disciplinary concepts from a range of scientific areas (e.g., physics, chemistry, biology) to atmospheric phenomena. For example, to explain how water vapor in the atmosphere condenses to create clouds and precipitation, students can apply the cause-and-effect explanation for the phase change in condensation on the outside of a glass of ice, which they learned about in chemistry. Similarly, characteristics of interacting systems that are first learned in biology in relation to cells and body systems can be connected to understand the nature of stability and change in air masses and climate systems.

A third approach is for students to use the crosscutting concepts as tools to problem solve how to develop an explanation or model for weather or climate phenomena. For example, when asked to explain why a snowstorm formed over the White Mountains or why it rains so much in Seattle, students can start the task by choosing a crosscutting concept (such as patterns, cause and effect, or scale and proportion) to frame an initial model or explanation. A fourth, and related, way that crosscutting concepts can be used in learning about weather and climate is to consider them as rules for how to explain processes underlying observable phenomena. For example, if students are trying to understand how the flow of energy in and out of the atmosphere affects climates, they could use the rule of distinguishing *causes* of change versus *effects* of change to more clearly specify the influence of incoming and outgoing energy to the system.

"Thus, the power in foregrounding the crosscutting concepts in instruction on this topic is in how the different crosscutting concepts enable students to "spin" or "pivot" their perspective on the specific phenomenon in ways that can effectively support their learning."

A key perspective to keep in mind about weather and climate ideas is that any specific phenomenon related to weather and climate is complex and multifaceted. Thus, the power in foregrounding the crosscutting concepts in instruction on this topic is in how the different crosscutting concepts enable students to "spin" or "pivot" their perspective on the specific phenomenon in ways that can effectively support their learning. By doing so, it prevents students from developing the belief that Earth phenomena have one simple explanation or should only be thought of in one correct way. It opens the space for new, competing models and explanations to be considered and debated between peers, fostering the opportunity for scientific argumentation, critique, and reflection in classroom discussions. For example, instead of simply applying a definitional relationship to why rainstorms form along cold fronts, students can first begin to look at patterns in precipitation and temperature relative to the location of a cold front; then reason about what is happening to the water vapor at different vertical and horizontal scales; then use a model to identify the causes and effects of cold air converging with warm air. Each of these different perspectives addresses the phenomenon in different ways, and each gives students a different understanding of rainstorms along a cold front that *together* add up to the robust, applicable understanding of weather and climate that the *NGSS* is aiming for.

In the *NGSS*, the performance expectations (PEs) for weather and climate in elementary school often refer to the crosscutting concept of patterns; in middle school, the emphasis is on the crosscutting concept of cause and effect; and in high school, the PEs include both the crosscutting concepts of cause and effect and systems and system models. However, it is important to note that there is not a one-to-one mapping of crosscutting concepts and grade level or DCI. The crosscutting concepts all together are needed to fully understand and build sophistication of weather and climate concepts. Although specific performance expectations in the *NGSS* highlight patterns or systems as the learning goal, the reality is that the full set of crosscutting concepts is necessary to explain how weather and climate work.

Contrasting Examples of Instruction

For each grade band, we present two Instructional Applications synthesized across lessons observed from multiple classrooms, which contrast traditional teaching of weather and climate with learning that focuses on crosscutting concepts. The crosscutting concept emphasized in each Instructional Application varies across grades. Each example represents an instructional sequence that would likely take place over multiple class periods. Note that each instructional sequence is aligned with a performance expectation at grade level, although the lesson itself may emphasize a different crosscutting concept than the one embedded in that particular performance expectation. After each set of contrasting Instructional Applications, we present a brief analysis that highlights the differences and emphasizes how students' resulting understanding of weather and climate would be different in each case.

Elementary School

Instructional Applications 13.1 and 13.2 (both on p. 300) are both set in third-grade classrooms. The former details a traditional lesson sequence, whereas the latter features a sequence with a focus on the CCC of patterns.

INSTRUCTIONAL APPLICATION 13.1
Traditional Lesson Sequence

Ms. Karen introduces her weather unit to her third-grade class by asking them to read and respond to a short article about the seasons. As they discuss the article, the students agree that different seasons tend to have different types of weather. Ms. Karen then introduces them to their weather journals. Each day for two weeks, the students record the weather, including temperature, cloud cover, and wind. They compile their observations and create bar charts for the number of sunny days and cloudy days, the number of colder days and warmer days, and the number of windy days and calm days. These graphs are displayed on the bulletin board outside of the classroom. Then Ms. Karen asks the students to think about how the graphs would look different for a different season. The students share their ideas and agree that the temperature graph would be lower in the winter since they know from experience that winter is a colder season than spring.

INSTRUCTIONAL APPLICATION 13.2
NGSS Lesson Sequence With CCC Focus on Patterns[1]

Ms. Abby's third graders have collected a class set of weather data for a month. Today, Ms. Abby passes out the tables of daily temperature, cloud cover, precipitation, and wind speed data. Students work in groups to create bar charts for each data set. As a class, they look for patterns in the relationship between temperature and other weather conditions.

During the discussion, they share observations that rainy days are more often accompanied by cooler temperatures than sunny days, and the temperature is also lower on windy days. Ms. Abby points out that the data was collected in April in Chicago and leads a class discussion about how it might look different if they had collected the data in January. Students discuss how the temperature data would probably be colder and the precipitation would probably be in the form of snow, but the *relationship* between sunny days and changes in temperature would stay the same.

Ms. Abby then shares a local 10-day forecast that calls for the temperature to cool down in the next three days. She asks students to predict what they think the weather conditions will be like during these days, including precipitation, wind, and cloud

1. Aligned with PE 3-ESS2-1: Represent data in tables and graphical displays to describe typical weather conditions expected during a particular season.

Continued

Instructional Application 13.2 *(continued)*

cover. Students write individual predictions, along with reasons for each prediction, and they share with the class. Most agree that it will likely be cloudy and rainy with an increase in wind because the patterns in the data show these relationships.

Ms. Abby concludes the lesson by asking students to reflect on the nature of the patterns they observed and how it helped them make predictions. She draws attention to the different aspects of the data pattern: Not only did they identify relationships between variables, but they also observed a pattern of change. She asks, "Why did you need to know these patterns to make a good prediction about the weather?" Students say that the weather is going to change in the future, and therefore, it helps them to see a pattern of weather variables changing so they aren't just making a random guess.

Comparative Analysis

In both Instructional Applications 13.1 and 13.2, students are working with local weather data they gathered themselves, and they create graphs of the data. However, there are distinct differences between the lessons in how students use the data and the kinds of explanations they generate.

In the traditional case, students observe the variations in temperature, cloud cover, and wind in the data they collect, but they are not asked to explain them. The focus on variations in weather in this lesson is driven by students' own experiences comparing summer and winter, not from explaining relationships in the data. In the *NGSS* lesson, students use the crosscutting concept of patterns as a lens to examine the student-collected data. Ms. Abby focuses the students' attention on both identifying the patterns in the data and developing explanations for those patterns. Furthermore, in this lesson, students not only retrospectively consider how these patterns in the data change by season, but they also apply the explanations of these patterns as evidence to make predictions for future weather events. By observing and explaining a pattern in the relationships between weather variables, these third-grade students are laying a foundation for increasingly sophisticated understandings of weather in the future, when they will begin to consider interactions between related components of the weather system.

Middle School

Instructional Applications 13.3 (p. 302) and 13.4 (p. 303) are set in sixth-grade classrooms. The former details a traditional lesson sequence, whereas the latter features a sequence with a focus on the CCC of cause and effect.

INSTRUCTIONAL APPLICATION 13.3
Traditional Lesson Sequence

Mr. Williams starts his usual sixth-grade unit on weather by displaying a weather map of the eastern United States. He asks the students to guess what each symbol on the map means, filling in any information missing from their responses. The *L* indicates a low pressure system, the *H* indicates a high pressure system, the line with the blue triangles is a cold front, and so on. He then displays a diagram showing a cross section of the air masses on either side of a cold front and asks the class how the different temperatures of the air would affect the weather conditions. One student offers that a cold front is where it often rains or snows because the cold air on one side of the front is colliding with the warm air on the other side. Mr. Williams points out that the cold front is associated with a low pressure system. He then plays a video that shows an animated weather map and asks the students to observe how the pressure centers, air masses, and fronts move. When the class shares the observation that the characteristic movement of fronts is from west to east, he explains that this is because of the jet stream that moves air around the North Pole, which is caused by Earth's rotation.

Mr. Williams then gives the students a current weather map with a cold front over St. Louis and asks them to create a weather forecast for Nashville. Students work in groups to prepare their forecasts. Each group gets a blank map of the region. They draw the cold front, shifted farther to the east than it was on the current weather map, and they add the *L* symbol to indicate low pressure. The groups then prepare scripts in which they take on the role of a TV meteorologist describing the weather forecast. The next day, each group presents its forecast to the rest of the class. They are generally in agreement that the cold front moving eastward will bring rain to Nashville over the next 24 hours.

INSTRUCTIONAL APPLICATION 13.4

NGSS Lesson Sequence With CCC Focus on Cause and Effect[2]

To start a unit on weather, Mr. Sanders shows the class a recent local TV weather forecast. He asks students to write down their observations and questions while watching the three-minute clip. Afterward, students share their observations. They make a list of symbols they saw, such as the big letters *L* and *H*, small arrows moving around the letters, and lines with blue triangles or red semicircles. The students note that the areas of rain shown by the radar move from west to east across the map, whereas the small arrows appear to be moving in a spiral.

Mr. Sanders then shows the class diagram models of a high pressure system and a low pressure system and asks students to work in small groups to describe what is going on in each diagram using the ideas of convection and energy introduced in the prior unit (when they made models of boiling water in a pot and condensation outside of a cold soda can). The students share their observations of these diagrams, noting that in the low pressure system the air is rising from the surface to higher altitudes in the atmosphere, whereas the opposite is happening at the high pressure system. They hypothesize that, since warm stuff rises when it is a lower density than the material above it, the air around the low pressure system might be warmer at the surface and cooling as it rises.

After discussing some of these initial explanations, Mr. Sanders revisits concepts from earlier instruction, including temperature differences, convection, and phase change. Students use these ideas to explain how in the atmosphere, the rising warm air mass will cool and cause condensation of water vapor in that air, leading to the formation of clouds and rain.

Next, Mr. Sanders asks the class to extend their cause-and-effect reasoning by considering what happens when two adjacent air masses interact, as students build explanations for why it rains along cold and warm fronts using their convection model of air masses as evidence. To start, students construct a physical model that uses warm water with red dye to represent warm air and cold water with blue dye to represent cold air. Students observe a phenomenon in the model where the cold blue water moves along the bottom of a clear tank, causing the warm red water to move up and over it. Mr. Sanders then passes out cross-section diagrams of warm and cold fronts that show the shape of the frontal zone and the movement of surface winds

2. Aligned with PE MS-ESS2-5: Collect data to provide evidence for how the motions and complex interactions of air masses result in changes in weather conditions.

Continued

Instructional Application 13.4 *(continued)*

indicated by arrows. Working in groups, students use their observations of the water model as evidence to annotate the diagrams; they show that in both cases, warm air is forced above cooler air, causing it to rise. They write explanations that connect the rising movement of warm air to the cooling and condensation along the boundary between a warmer air mass and a cooler air mass, which causes clouds and rain to form along the front.

Students then apply these explanations to predict what the future weather might be in their local area, using evidence from current weather reports and seasonal weather trend data. They justify their predictions in terms of the causes and effects of the moving air that will create the predicted weather. In a summative discussion, Mr. Sanders asks individual students to share a single aspect of their prediction, such as "clouds will form," along with the cause of that prediction, such as "the water in the air is cooling so much that it forms condensation." He elicits additional steps in each cause-and-effect relationship, asking the students, "What caused the air to cool so much?" He charts the cause-and-effect relationships described by the students so they will be able to refer to them in subsequent lessons, pointing out that all these connected relationships work together as part of Earth's weather system.

Comparative Analysis

The two lessons both engage students with common representations of weather that they see on TV or in the newspaper—that of a weather map with high pressure, low pressure, and fronts indicated by symbols. However, although the traditional lesson teaches students to identify the symbols on a weather map and how they correspond with weather events, it does not engage students with understanding *why* changes to weather would occur at the featured locations. In other words, the traditional lesson does not dive into the mechanisms for what creates rain or snowstorms along frontal boundaries or the conditions that influence those observable weather outcomes; it merely points out that there are temperature differences between high and low pressure fronts.

The *NGSS* lesson also begins by introducing students to the symbols on a weather map, but Mr. Sanders then moves into engaging students with cause-and-effect explanatory models that build on prior science understanding. By explicitly asking students to consider weather conditions like rain as the *effect* of an underlying *cause* (the rising of warm air due to convection and density differences) and applying these models to create future weather predictions, this lesson builds students' capacity to use the crosscutting concept of cause and effect as a tool for understanding how Earth systems function and change over time.

High School

Instructional Applications 13.5 and 13.6 (p. 306) are set in ninth-grade classrooms. The former details a traditional lesson sequence, whereas the latter features a sequence with a focus on the CCC of systems and system models.

INSTRUCTIONAL APPLICATION 13.5
Traditional Lesson Sequence

To begin her unit on climate, Ms. Smith assigns students to different climate regions and directs them to complete a research project describing the region's location on Earth, the area's typical weather in summer and winter, and common plants and animals that live in the region. The following week, each group presents its report, while Ms. Smith records a summary of their findings on a large world map. As a class, the students discuss the variations in climate around the world, and Ms. Smith poses two questions to the class: "Why are there different climates in different areas?" and "Has it always been like this?"

Ms. Smith then shows a diagram of Earth receiving incoming radiation from the Sun and describes how the curve and tilt of Earth mean that different parts of the surface receive different amounts of the Sun's energy. She helps students connect this representation to the definition of climate as the average description of weather over a long period of time. Each student writes a short explanation of the climate patterns in his or her assigned location. The explanation includes how the region's position on the globe determines both the average temperature and the annual temperature range based on how sunlight hits that part of the planet throughout the year. The students share their explanations and discuss what additional factors besides sunlight could account for differences in climate. Using the world map, they make a list of possibilities, including proximity to mountains and large bodies of water. They infer that the local geography affects the temperature of the atmosphere and the way the Sun's energy is absorbed in different areas.

Next, students read a popular science text on how Earth's climate has changed over past geologic eras. The text describes how it was mostly warm and tropical when the dinosaurs existed and how it became cold, with large areas covered in ice and snow, during the ice ages. Finally, students discuss how Earth's climate might continue to change in the future.

INSTRUCTIONAL APPLICATION 13.6

NGSS Lesson Sequence With CCC Focus on Systems and System Models[3]

Ms. Lance starts the climate unit by reminding students about what they have learned about weather previously (in middle school): Weather is a description of temperature, precipitation, wind speed and direction, and air pressure and humidity; and weather changes over time depending on how air masses move and interact with each other. She asks students to discuss what weather is typically like in their town and how it is different from what it is typically like in their favorite place to visit (or where they want to visit). After brief small-group discussions, Ms. Lance emphasizes that this "typical" description of weather is really what scientists refer to as climate—descriptions of the average weather conditions and variations over long periods of time (i.e., many years or decades). She reminds students that weather changes are caused by energy flowing in the atmosphere, which moves the air and water vapor both horizontally and vertically, causing the water to change state from gas to liquid or solid, which creates clouds and precipitation. Ms. Lance poses the following to the class: "So let's think about this at a different scale. How can we describe energy flow when we are talking about *climates* instead of *weather*? Does energy flow in the atmosphere change over longer periods of time?"

Ms. Lance then directs the class on how to use computer-based climate models to examine the ways that energy enters and leaves the atmosphere (looking at inputs from the Sun and its subsequent absorption, reflection, or reradiation) and where it is stored in the atmospheric system. She assigns a different location on Earth to each pair of students. The students work in these pairs to create conceptual energy-flow diagrams of how the energy enters and moves between the land, ocean, and atmosphere in their assigned location. Next, they use an online climate database to gather information about the typical temperature and precipitation patterns for that location and describe the connection between the climate data and their energy-flow models of the climate systems governing that region.

After student groups present their explanations, Ms. Lance leads students into class discussions about whether and how climate systems could change over time, building on their prior discussions of inputs, outputs, and feedback loops in other Earth systems. The class brainstorms different factors that could change the way energy flows through Earth's climate system, such as the Sun's radiation output,

3. Aligns with PE HS-ESS2-4: Use a model to describe how variations in the flow of energy into and out of the Earth's systems result in changes in climate.

Continued

Instructional Application 13.6 *(continued)*

variations in Earth's orbital eccentricity or axial precession, volcanic eruptions and other tectonic events, and changes to the atmospheric composition caused by life-forms on Earth. Student pairs then return to their laptops to test these hypotheses by varying these energy factors in the global climate models. The models predict that these variations in energy flows in and out of the climate system would change the characteristics of the climate over different time periods. The students write out descriptions of these findings. Their written descriptions include several parts: They define the inputs and outputs of the climate system captured by the model; describe the predicted energy flow between model components; identify the timescale over which both the energy flow and the change to the system might occur; and discuss limitations of their predictions due to any approximations or simplifications inherent to the model.

Comparative Analysis

In both of these lessons, students explore climate factors in a specific location on Earth, and teachers use these local investigations to drive the development of students' understanding of the global climate system. However, in the *NGSS* lesson sequence, Ms. Lance's prominent use of the crosscutting concept of systems and system models supports students in both explaining and predicting the behavior of Earth's climate system on different time and spatial scales. In contrast, the traditional lesson addresses components of the system in isolation rather than asking students to consider the complexities of how they interact. The focus on interaction and feedback of system components and on students' engagement with manipulating system models in the *NGSS* lesson is particularly important for this topic because the vast majority of scientific work around climate change due to human activity is grounded in the development and use of increasingly complex climate system models by scientists. Being able to reason about the components, interactions, limitations, and predictive powers of different climate system models prepares students for subsequent learning goals associated with this performance expectation.

Summary

The aim of this chapter is to illustrate the importance of crosscutting concepts in teaching about the topic of weather and climate. Like many other concepts in Earth science, weather and climate are central to building students' understanding of how the world around them "works," yet these concepts are conceptually challenging to explain due to the scale and complexity of the phenomena. Crosscutting concepts are a crucial component of three-dimensional teaching and learning of Earth science concepts, particularly

> A key shift made by the *NGSS* is the expectation that, rather than simply describing weather and climate, students will make explanations and predictions based on understanding mechanisms for how weather and climate *systems* work. This shift is facilitated by engaging students with the full range of crosscutting concepts to support their meaning-making around this topic.

in how they serve as unifying frameworks for understanding and making sense of the world.

A key shift made by the *NGSS* is the expectation that, rather than simply describing weather and climate, students will make explanations and predictions based on understanding mechanisms for how weather and climate *systems* work. This shift is facilitated by engaging students with the full range of crosscutting concepts to support their meaning-making around this topic. In particular, crosscutting concepts can act as bridges to help students transfer useful ideas from prior learning and other content areas, can be used as tools and guiding rules to help students explain complex phenomena, and can serve as lenses to help students make connections between different parts of the Earth system and its interconnected subsystems. Together, the crosscutting concepts are central to helping students build robust, sophisticated, and nuanced understandings of Earth and its behavior.

Using crosscutting concepts when teaching about weather and climate builds the progression of learning goals from the elementary to middle to high school grade bands and strengthens the connections between ideas developed at each of these levels. To reinforce this development of sophistication over time, teachers and students should intentionally use the language of the crosscutting concepts. Teachers can facilitate this by referring to the crosscutting concepts by name in descriptions of student tasks, planned questions, or discussion topics and written prompts to which students respond.

Acknowledgments

We would like to thank Michael Ford for his helpful feedback on this chapter. A portion of Ann Rivet's contribution to this work was completed while employed at the National Science Foundation and was supported through the Independent Research and Development (IR/D) program at the NSF. The views expressed in this work are solely those of the authors and do not reflect the views of the National Science Foundation.

References

National Research Council (NRC). 1996. *National science education standards*. Washington, DC: National Academies Press.

National Research Council (NRC). 2012. *A framework for K–12 science education: Practices, crosscutting concepts, and core ideas*. Washington, DC: National Academies Press.

NGSS Lead States. 2013. *Next Generation Science Standards: For states, by states*. Washington, DC: National Academies Press. *www.nextgenscience.org*.

Rivet, A. 2016. Core idea ESS2: Earth's systems. In *Disciplinary core ideas: Reshaping teaching and learning*, eds. R. G. Duncan, J. S. Krajcik, and A. E. Rivet, 205–224. Arlington, VA: NSTA Press.

Rivet, A. E., G. Weiser, X. Lyu, Y. Li, and D. Rojas-Perilla. 2016. What are crosscutting concepts in science? Four metaphorical perspectives. In *Transforming learning, empowering learners: The international conference of the learning sciences (ICLS) 2016, Volume 2*, eds. C. K. Looi, J. L. Polman, U. Cress, and P. Reimann, 970–973. Singapore: International Society of the Learning Sciences.

Chapter 14

Crosscutting Concepts in Engineering

Christine M. Cunningham, Kristen B. Wendell, and Deirdre Bauer

A Glimpse Into an Engineering Classroom

This chapter details the use of crosscutting concepts (CCCs) in engineering instruction. The example that follows offers a brief look into an engineering classroom, describing how concepts related to patterns and scale and proportion arise during the activity.

Engineering Parachutes

A class of fifth-grade students has been studying the physics of falling bodies. Last week, they learned about the force of gravity on Earth and compared it to gravity on other planets. To observe and measure the force of gravity, students conducted science experiments such as dropping a number of small items, like binder clips, and timing their descent. Then the teacher asked students to think about some forces and technologies that counteract the force of gravity. A lively discussion of friction, drag, and air resistance ensued. One of the technologies students named that slows rate of descent was the parachute. The teacher then revealed the class's next challenge—to engineer parachutes that decrease the rate of fall on another planet enough so that a valuable load (modeled with a piece of bowtie pasta) is undamaged. She specified a range of safe descent rates for the parachutes—on Earth, they should fall slower than five feet per second.

Students jump into the problem, first identifying key variables that might affect the parachute design, such as the canopy size, material, and suspension line length. They conduct controlled experiments—testing one variable while holding all others constant—to learn more about how these variables behave. They investigate which canopy materials (netting, sheer fabric, coffee filters, or plastic) fall the slowest. They test and collect data about various canopy sizes. Similarly, they change suspension line length and observe the impact on the parachutes' performance. Through extensive testing and observation, students come to recognize *patterns* in how parachutes fall. As a class, they discuss and document the following patterns: (1) Canopies made of materials that let air through, such as netting or sheer fabric, fall more quickly than those that better trap the air, such as plastic or coffee filters. (2) Parachutes with larger canopies generally work better to slow the descent than those with smaller canopies. (3) Canopies with longer suspension lines also generally work better. As they generate such statements, the teacher asks students to explain *why* this might be the case. How do these variables affect the upward forces (drag) that act against gravity?

After exploring the basics of parachute design, the students engineer a parachute of their own. Now they need to consider all three variables (material, canopy size, and suspension line length) together. The teacher also introduces another constraint: The parachutes will be deployed to another planet. They will be transported by spacecraft that have a limited amount of space (or payload) available. Thus, their parachute can only take up a certain amount of space (calculated by considering the size of the canopy and line length). Students design original solutions, test them, and collect data for each trial. Students share their data with the class in data charts that display the information from all the groups. As students make sense of the data, they distill important relationships between the *scale and proportion* of the parts of their parachutes. One student realizes that a small change in radius means a significant change in canopy area. Another student surfaces an important relationship between suspension line lengths and canopy size, pointing out that the suspension lines need to be long enough to allow the canopy to fully open so it can trap the air that creates the drag.

Real-world testing of these ideas through subsequent, iterative parachute redesign encourages students to reconsider and refine their ideas. By engaging with disciplinary core ideas (DCIs) and science and engineering practices (SEPs), the students are able to present evidence to support the redesign of their engineered parachutes. Teachers can support students' engagement in crosscutting concepts (CCCs) with assessment prompts such as those presented in Table 15.4 (p. 345). Structuring activities and conversations using these prompts allows students to seek and discuss patterns in their data. Compiling class data into shared tables and leading discussions that ask students to consider the relationships between the scale of various parts encourages students to use the CCCs and make connections between engineering and science. Explicitly naming the CCCs can help students recognize these tools, but such naming should always be anchored in use.

Engineering, Science, and the *Next Generation Science Standards*

For the first time in the United States, the recommended science standards—the *Next Generation Science Standards* (*NGSS*; NGSS Lead States 2013)—include learning engineering as an explicit educational goal. This chapter focuses on the uses of three-dimensional learning with engineering. This chapter is a little different from the previous chapters in Part 3, each of which focused on a discrete scientific disciplinary core idea. Engineering is a broad discipline that encompasses many different fields—for example, mechanical, biomedical, or environmental engineering. Why introduce something new to an already-packed curriculum? There are a number of reasons why engaging children in engineering makes sense (Cunningham 2018).

- Engineering helps children understand, and improve, the world they live in. Almost all of our lives are spent interacting with the human-made (or engineered) world. We are surrounded by engineered objects—pens, milk cartons, bicycles, thermostats, smartphones. To understand the world they live in, children should understand how these objects come to be. They should also recognize that humans shape the world around us and that this ability comes with responsibilities and consequences.
- Engineering fosters problem-solving skills and dispositions. Children naturally solve problems as they explore the world around them. Engineering taps these proclivities, providing supports and structures that allow children to tackle increasingly complicated problems, while building their strategies and confidence in their problem-solving abilities.
- Engineering increases motivation for and engagement in learning. An authentic, real-world problem motivates students, especially if it is open-ended, permitting many possible, innovative solutions. Seeing how concepts connect and why knowledge matters encourages students to engage with content in deeper ways (Cunningham 2018).
- Engineering improves math and science achievement. Applying knowledge helps students learn better—engineering projects invite students to use and manipulate knowledge in meaningful ways (Wendell and Kolodner 2014).
- Engineering increases access to careers. Engaging in engineering projects helps students understand the kind of work that engineers do and build their agency and identity as engineers.

Whereas engineering and science rely on and inform each other, engineering has a unique set of attributes that distinguishes it from science. One way of distinguishing the disciplines is to think about their purposes. Science aims to describe, explain, and predict the natural world and its physical properties. Science explains why something gets hot or how a muscle works to move a limb. Engineering is the systematic application of knowledge to design objects, systems, and processes that meet human needs and desires. Engineering aims to design more effective solar ovens or create prosthetic limbs. The goal or end-product of the overarching activity in each discipline—explanation, generation, or solution design—is different for science and engineering. However, in the real world, scientists and engineers move back and forth between the disciplines as they pursue their goals. For example, scientists might need to redesign an instrument (an engineering task) to more accurately measure temperature for their experiments. Engineers working on artificial limb design need to thoroughly investigate an anatomical system before they design a functional and comfortable device. Science and engineering are integrally linked.

The core idea of engineering is that engineers use a systematic, iterative process to design solutions to problems. This is reflected in the *NGSS*'s DCI for engineering, ETS1: Engineering Design, which identifies three components to this process—defining and delimiting an engineering problem, developing possible solutions, and optimizing the design solution. As they design solutions to problems, engineers engage in practices. Some of these are similar to practices used by scientists—for example, analyzing and interpreting data. Other practices, such as defining problems and identifying constraints and criteria, are unique to engineering (Cunningham and Kelly 2017). As the three dimensions of the *NGSS* illustrate, CCCs also traverse science and engineering.

Crosscutting Concepts Bridge Science, Engineering, and Mathematics

> "Because they are crosscutting, these concepts offer tools that can help students bridge science and engineering, as well as math, by connecting what they learn in one discipline to another."

As they are in science, CCCs are critical in engineering for creating high-quality engineering solutions. Engineering activities offer many opportunities for students to engage with CCCs. Because they are crosscutting, these concepts offer tools that can help students bridge science and engineering, as well as math, by connecting what they learn in one discipline to another.

For example, engineering designs often invite students to wrestle with how a change in a specific parameter affects the performance of a design to satisfy a criterion. Such understanding often requires identifying patterns. Relationships and patterns become more meaningful when students observe a physical phenomenon and want to make sense of it because doing so can improve their design's performance. During the parachute testing, for example, students wanted to design parachutes that met and surpassed the criteria. They closely observed the performance of various parachute designs during ad hoc and systematic tests. Through these physical tests and the subsequent analysis of data tables that displayed the numerical data they collected, students discerned patterns in how the designs performed.

During the parachute challenge, students conducted controlled experiments, changing only one variable (canopy material or size or suspension line length) to investigate how they affected parachute descent rate. They observed performance and collected data from multiple trials and then analyzed their data by looking for the patterns in rates of descent. Students looked across their designs and test results to notice patterns in how the designs behaved with respect to the engineering goal.

While constructing parachutes, students asked themselves which canopy size worked best to slow parachute descent. When students tested canopies of different sizes, they again noticed patterns in terms of which fell the slowest. By testing many parachutes, students noticed that smaller canopies fell more quickly and that those with longer suspension lines worked better—up to a point, when they no longer enabled enhanced performance. This led them to understand that parachutes with larger canopies tended to fall more slowly. Thus, canopy size emerged as a relevant variable that affects the ways objects fall.

Lessons are strengthened when students are encouraged to consider why objects and designs behave as they do. In the parachute example, the teacher prompted students to consider what caused the patterns they saw to occur. In this challenge, searching for patterns in their data helped students understand the physics of falling bodies. The explanation of the phenomenon they observed invited science thinking and reasoning. Ultimately, students learned that larger canopies work well because they provide increased surface area and "catch more air," which increases drag, a force that counteracts gravity.

During the parachute design, students also observed another pattern—that seemingly small changes in radius length can really increase parachute area and performance. This provided a natural opportunity to explore and understand firsthand the importance of scale and proportion for the creation and functioning of devices. Sharing the mathematical formula for area, exploring the relationship between radius and area, and discussing why this happens (the power of squaring a variable) also connects mathematics with science and engineering. Thus, we can see pathways to interdisciplinarity, as the use of patterns allows students and teachers to move back and forth across understanding (science and mathematics) and designing (engineering).

We turn now to what CCC-infused classrooms might look like. Instructional Application 14.1 (set at the elementary level; p. 316) and Instructional Application 14.2 (set at the middle school level; p. 321) illustrate how careful thought and explicit attention to CCCs can support three-dimensional learning in engineering.

INSTRUCTIONAL APPLICATION 14.1

Designing a Solar Oven: Engineering in an Elementary Classroom

Engaging children in meaningful engineering problems can be a promising way to help students build science and engineering knowledge, familiarity with practices in these fields, and understanding of crosscutting concepts. In this Instructional Application, as students engineer insulation for a solar oven, they draw on what they have learned about energy flow and transfer to design a solution that retains maximal heat.

Ms. Alvarez's fourth-grade class has just completed a science unit on energy transfer, guided by *NGSS* performance expectation 4-PS3-2: Make observations to provide evidence that energy can be transferred from place to place by sound, light, heat, and electrical currents. Ms. Alvarez is now challenging students to use what they learned about energy and its transfer as they engineer a solar oven. The task is a bit unconventional; instead of designing the external components of the oven (box and window), Ms. Alvarez challenges students to think about how they might engineer *insulation* for the oven system so it retains heat. Thus, the students are challenged to think about energy flow and how this might be affected to conserve heat. As they engage with this challenge, they will meet *NGSS* performance expectation 4-PS3-4: Apply scientific ideas to design, test, and refine a device that converts energy from one form to another. Students will be constructing a solar oven that converts light energy into heat energy. In the unit, Ms. Alvarez plans to highlight two crosscutting concepts: structure and function and energy and matter.

Ms. Alvarez introduces the engineering problem to the class by situating it in a larger context. She explains to her students that one of the chores that children in a number of countries are responsible for is collecting firewood. She brings this task to life by showing her students a few short videos, pictures, and statistics about collecting firewood. She asks students to imagine their lives if they had this responsibility each day and introduces them to the environmental impacts of using wood-fired stoves. Explaining that some communities are starting to explore other options for cooking, she introduces solar power and solar ovens. Ms. Alvarez shows students a model solar oven made of a shoebox and explains they are going to work on figuring out how to design insulation for the oven so it effectively traps the heat.

Solar Oven Parts and Purposes

To encourage students to adopt a knowledge-based approach to the design of their solar oven insulation, Ms. Alvarez asks her students, prior to tackling the full

Continued

Instructional Application 14.1 *(continued)*

challenge, to think about the *structure and function* of a solar oven. First, she has students think about the purpose of a solar oven, asking, "What is its goal? How do you know it is functioning well?" Students identify that the oven needs to get hot—hot enough to heat or cook the desired food or drink. Then Ms. Alvarez has students think about the oven's various parts and why they are designed as they are, asking, "How do the parts need to be structured to successfully contribute to the functioning of the oven system?" The students learn that the solar oven will be made of a box that has a reflector panel, covered in aluminum foil. They discuss why the reflector is covered in foil and how it functions (to direct sunlight—light energy—into the oven). The oven has a "window" on the top that is cut from the larger box. Clear plastic covers the hole and is affixed tightly to the box. Previous investigations have helped students understand why clear plastic is used for this element—it allows light in but traps heat from leaving the oven. This discussion elicits concepts related to *energy flows* as students recognize that what the oven technology needs to do is direct and trap light and heat energy from the Sun and transfer it to heat energy to cook food. During this conversation, students further define the problem they will solve.

Exploring Materials' Insulative Properties

Before students begin to design their insulation solutions, Ms. Alvarez has them engage in a scientific exploration to develop their understanding of how and why insulators work and to examine the properties of the materials that work well as insulators. Students begin by predicting which materials—from a selection of foam sheets, felt, aluminum foil, newspaper, and plastic, in either flat and shredded form—will work best as insulators for their oven designs and why. Student groups then conduct controlled tests to collect data about how well the materials perform as thermal insulators. The five materials in two forms are each put into a plastic cup (for 10 cups total). These will be compared to a control—a cup with only air in it. A thermometer is added to all 11 cups, which are then placed in an ice bath. The temperature of each cup is recorded every 30 seconds for five minutes.

Once they have collected their data, the student groups calculate the change in temperature in each cup and graph the temperature data over time to see how well each material functions as an insulator. In their groups, they analyze how the materials performed and share their data and observations with the class. The class is challenged to explain what they observed in terms of the flow of heat energy and to make observations about how the properties and *structure* of each material determines how well it *functions* as an insulator. During this conversation, students begin to recognize that materials that are more "fluffy" (filled with air), like foam and felt,

Continued

Instructional Application 14.1 *(continued)*

work better as insulators and that the crumpled version of materials work better than flat sheets.

Ms. Alvarez draws sketches of a few of the cups with various types of materials on the board and then uses physical gestures to model where heat energy moves more rapidly (through compact, solid materials) and where it moves more slowly (through materials that have air trapped within solids). Ms. Alvarez writes "air, especially trapped air, is an insulator" on the board to help students remember this important point. She directs students to talk in their small groups to "explain how each material in the cups functioned in terms of *energy flow and heat conservation.*" The explicit use of CCC terminology in Ms. Alvarez's questions invites relevant and meaningful use of these concepts during student discussions. She helps students see how careful thinking about the properties of a material is an important part of using the CCCs in engineering design. The properties of a material can determine what *structures* it can form and therefore what range of *functions* it can carry out, and knowledge about material properties can help engineers determine the cause of a particular design performance.

Designing Oven Insulation

Having developed some insights into the functioning of materials they will use, students are better prepared to make informed decisions as they tackle the design challenge of creating insulation for their oven. Ms. Alvarez reminds the class of their goal—designing an oven that will trap as much heat for as long as possible. However, she introduces another constraint—the teams will need to work within a budget and try to create an inexpensive solution. Each material has a cost. Those that work better as insulators are generally more expensive.

With the problem defined, students are ready to develop possible solutions and optimize them. They revisit the data from their previous experiment as they make design choices about which materials to use, how much of each material to use, and the configuration of the materials inside the oven. The students are motivated to design a high-performing oven and authentically use CCCs. In groups, they propose and discuss possible design solutions, applying understandings related to *structure and function* and *energy flow and heat conservation*. As they select materials, students recall which functioned best as insulator and in what form. They also justify their design decisions by referencing their previous experiments, diagrams, and conversations about how energy flows. Students log and document their design decisions in their Engineering Journals.

Continued

Instructional Application 14.1 *(continued)*

Students create an initial design. Outside in the bright Sun, they test how quickly the oven heats up and how well it retains heat when removed from the sunlight, logging temperature data in 30-second intervals for five minutes in the Sun followed by five minutes out of the light. They also calculate the cost of their insulation.

Mapping Energy Flows

To focus students on thinking about energy flows, Ms. Alvarez tasks each group with looking at the diagram of the oven they sketched. She first asks them to "draw blue arrows that trace the path of energy as it goes from light energy from the Sun to the heat energy moving in and through the oven." Then she asks them to "identify where in their oven system energy moves more rapidly by coloring those parts green and also to identify where air functions as an insulator to slow heat flow and conserve the heat energy by coloring those parts red." During their group deliberations, students point and gesture to parts of the physical and sketched model. Moreover, their discussions include words such as *energy, insulator, flow, rapid, slow, heat, solid, air, function, fluffy*, and *trapped*. Because every group's design is unique, each one needs to figure out how their particular oven functions.

Sharing Results to Construct Understanding

After each group analyzes its own data, the students share the design and testing results with classmates. Ms. Alvarez has the class look across all groups' heating and cooling data and generate explanations about why some ovens functioned better than others. She encourages students to reference the oven structures and use concepts related to energy transfer. When students offer incomplete explanations, such as "the foam made [the oven] hot," she asks additional, probing questions to promote scientific explanations. (A more robust student explanation might include a more robust description that links the structural and scientific elements—for example, "The energy of the light became heat energy. It was trapped best by the shredded foam because the pockets of air in the foam and between the foam strips made it more difficult for the energy to move through the space and out of the box.") Ms. Alvarez also asks each group to reflect on what has worked in their designs, what has not, and why. Having students share their thoughts and reflections with the whole class allows all students to access ideas and understandings related to how solar ovens function. Then students can use such knowledge to inform their oven redesign. Ms. Alvarez's students redesign their insulation based on what they have learned about oven structure, functions, and energy flows. They again collect data and analyze it to determine whether their second designs perform better than their first.

Continued

Instructional Application 14.1 *(continued)*

Engineering With SEPs, DCIs, and CCCs

The open-ended engineering problem invites students to apply and refine their knowledge of heat transfer, energy, and insulators in an authentic, concrete context. The unit reinforces science ideas including: (a) energy can be transformed from one form (light) to another (heat), (b) some materials transfer heat energy more slowly than others—and thus function better as thermal insulators, and (c) heat energy moves from warmer places to cooler places. Thus, as students participate in the engineering challenge, they develop or deepen knowledge that is relevant to the problem at hand. As they do the challenge, they also must engage in SEPs. For example, they need to define the problem they will solve, which includes articulating how they will know whether their technology is functioning well. They design and create a solar oven and use data they collect to assess how well their design functions and identify how it might be further improved.

Finally, the engineering design challenge invites students to engage continually and in meaningful ways with CCCs. Students think about their experiments and why the system works. They ask what causes their oven system to behave at it does and how a change in one variable affects the performance of their oven. They use such knowledge to inform their next steps, design choices, and explanations. They consider the *structure* of the materials they use and the structure of the arrangement of materials in the oven as they seek to optimize the *function*, or performance, of their oven. Throughout the unit, as students design a technology that relies on *energy flow and heat conservation*, they grapple with how energy moves into and through the system and work to maximize heat being trapped within the oven. The teacher supports students' use of CCCs by encouraging them to consider these concepts during discussions, decision making, and sensemaking. The CCCs in this case make visible the ways that DCIs and SEPs inform engineering designs.

In this lesson and in many other engineering and science activities, a number of CCCs are useful for creating workable solutions to design challenges and figuring out how and why phenomena occur. A number of CCCs *could* be applied to most engineering or science activities; the key is to think carefully about which two or three are *most* important to emphasize for the lesson's objectives. This selection will depend on the context and the task and will alter the flow and focus of the activities. For example, CCCs related to patterns, cause and effect, and systems and system models also underlie the design of effective solar ovens. If Ms. Alvarez had chosen to focus on patterns, she might have concentrated on data analysis and graphing elements by having students (a) share and compare the graphs they had created related to their oven's heating and cooling and (b) compare the shapes of the graphs, making observations about them. A focus on systems and system modeling might

Continued

Instructional Application 14.1 *(continued)*

have led her to focus on the insulation's role as part of a larger solar oven system. Discussions of the role and interactions between different subsystems, such as the box, the insulation, the container for the food, and the food itself, might have all been explored. Additional experiments that probe the effects of the other elements (like the materials for the box that make up its reflector and window) and how modifications to them affect the overall system could have been articulated.

This example demonstrates how a traditional engineering activity—designing a solar oven—can be strengthened by explicitly incorporating CCCs. Traditional approaches to solar oven design have students experiment with which materials to use for the reflector and what color to paint the inside of the oven. Carefully thinking about how to emphasize the CCC of energy and matter suggests a new approach that focuses on heat flow and heat conservation using insulation. Having students explicitly consider the structure and function of the oven, the materials, and the arrangement of the materials within the oven and how these might affect energy flow invites students to think more deeply about the concepts underlying the activity.

INSTRUCTIONAL APPLICATION 14.2

Designing Biomimetics: Engineering in a Middle School Classroom

For middle school students, engineering design problems can provide meaningful contexts for investigating phenomena in the world, learning to use new tools like coding languages and electronics, and developing strategies for collaborating with peers. In this Instructional Application, we consider how a seventh-grade teacher intentionally incorporates crosscutting concepts to support her students' engineering design learning experiences.

Researching Digging Animals to Design Better Rescue Robots

To transition from their study of life science to physical science, Ms. Ren's seventh-grade students are taking on the challenge of designing and prototyping biomimetic robots. Biomimicry is the process of investigating and applying knowledge about living systems to create more effective, elegant, and environmentally friendly technological solutions. In biomimetics, the natural world inspires the human-made one. The goal for Ms. Ren's students is to learn more about animals that are excellent diggers, and then take what they learn about digging animals and use it to develop

Continued

Instructional Application 14.2 *(continued)*

a rescue robot that can dig through earthquake rubble. As they carry out this task, they will address the first part of two *NGSS* performance expectations:

- MS-LS1-3: Use argument supported by evidence for how the body is a system of interacting subsystems composed of groups of cells.
- MS-ETS1-2: Evaluate competing design solutions using a systematic process to determine how well they meet the criteria and constraints of the problem.

Ms. Ren knows that this biomimetic design challenge can help students consider how CCCs sometimes help scientists and engineers pursue different goals. In the unit, they will consider the CCCs of structure and function, stability and change, and patterns, and they will take note of how these concepts traverse life science and robotics engineering.

Ms. Ren launches the biomimetic digging robots challenge by having students consider a fictional invitation from an engineering organization working on robots for natural disaster rescue and relief scenarios. The organization is seeking new ideas for robots that can assist with search and rescue in areas where earthquakes have caused building collapse and piles of rubble. They'd like to see a range of robotic prototypes and hear students' recommendations for digging rescue bots. After designing, building, and testing prototypes of digging robots, the students will need to be ready to explain the biological inspirations for their robots and the ways in which the living systems could and could not be translated into technological ones. This unit's interweaving of life science and robotics design allows Ms. Ren to highlight how crosscutting concepts bridge science and engineering.

After revealing the overall design challenge, Ms. Ren provides students with texts, diagrams, and videos about four particularly effective and interesting digging animals. They showcase four different approaches to digging: One scratches the ground (pocket gopher), one hooks and pulls dirt (pangolin), one rotates its shoulders to swim through material (mole), and one uses its teeth like chisels (rat). For the next part of the unit, the students will explore the structure and function of the digging tools and behaviors of these animals, as biologists do: They will decompose a biological system into its component structures and their individual functions, with the purpose of better understanding and explaining the overall functioning of the organism.

Analyzing Parts, Properties, and Actions

To help the students get started, Ms. Ren engages the whole class in analyzing the structures and functions of a house cat. She chooses the cat for this "practice run" with structure-function analysis because many students are familiar with the animal;

Continued

Instructional Application 14.2 *(continued)*

however, it is not particularly well-adapted to digging, so students will still have more to learn when they later repeat this kind of analysis on good diggers. The whole class enjoys repeated viewings of a close-up video of a cat scratching in a litter box, and then they answer the questions Ms. Ren poses to scaffold the process of identifying structure-function relationships. She asks students to identify which body parts the cat uses to move the litter and which body parts provide support for those that are directly involved in digging. Next, she helps the students articulate what they notice about the properties of the structures that play primary and secondary roles in the cat's attempted digging action (e.g., the structures' shape, surface texture, relative size, flexibility). They discuss how each of these properties helps the cat move litter. Finally, they find verbs to describe the actions that each digging part or supporting structure carries out. They talk about the motion both of external structures, like claws and paws, and internal structures, like bones and leg muscles. Finally, Ms. Ren demonstrates making a labeled sketch of the cat's foreleg, paw, and claw. Ms. Ren hopes that this whole-class work with the cat will enable the students to put the CCC of structure and function into action for the first stage of their biomimetic design process. She wants them to experience how structure and function in science help pursue the goal of understanding and explaining the mechanisms for an action or behavior.

The students work in teams to analyze their assigned digging animal, chart its digging structures and functions (Figure 14.1), and sketch a multipanel "storyboard" that

Figure 14.1. Chart of digging structures and functions

Structures	Properties		Functions	
What *body parts* help the animal dig?	What *adjectives* describe those parts?	How do those properties help with digging?	What *verbs* tell how the body parts act?	How do those actions help with digging?
Front legs	-Short and stocky -Made of short, strong bones and large strong muscles -Flexible	-Provides force for powerful pushing -Helps the legs move fast	- Extend forward and push back	- Move the claws to different areas of soil
Claws	- Stiff · Hard · Curved · Sharp · Thin · Made of keratin	- Deal with constant back and forth motion - Cut through hard soil	- Slice and scrape - Scoop	- Breaks up hard soil into powder that gets pushed away
Back legs	- Short and stocky - Large strong muscles	- Keeps the whole gopher in place while front claws scratch	- Push against ground	- Braces the gopher - Keeps it from tipping or sliding

This chart features scaffolding questions and student ideas about the structure-function relationships that enable a pocket gopher to dig through soil.

Continued

Instructional Application 14.2 *(continued)*

shows the position of its structures at three different moments in its digging action (Figure 14.2). To do this work, the students not only view close-up videos multiple times but also study anatomical drawings of the animals' jaws and feet and read texts that explain how the shape and location of muscles and bones aid in digging.

Figure 14.2. Student storyboard

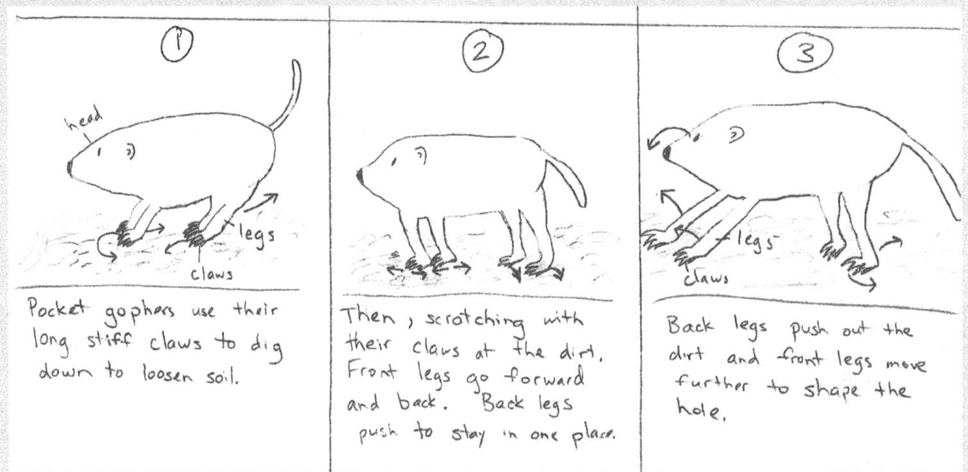

This middle school student's storyboard shows sketches of three moments in the digging action of a pocket gopher.

Ms. Ren circulates and reminds students to consult the word wall for nouns, adjectives, and verbs that describe structures, properties, and functions, and she encourages students to stand up and act out their animal's digging motion to get a better feel for the phenomenon they are trying to model in words and sketches. After each student team has analyzed one digging animal, Ms. Ren facilitates a gallery walk so students can compare and contrast the digging adaptations of different animals. They summarize their findings by charting similarities and differences across the organisms (Figure 14.3).

Figure 14.3. Chart comparing digging structures and functions

Similarities	Differences
• Long claws used for scratching dirt • Front legs shorter than back legs • Front legs for moving dirt; back legs for forward body motion and stability • Can also use nose to push dirt (secondary structure/function for digging)	• Mole front paws push dirt outwards away from body; pangolins dig parallel to their body • Mole paws are like paddles, much wider than arms; pangolin fingers are thin • Pangolin scales protect body against sharp rocks and branches; mole has only hair

This table created by middle school students compares structures and functions of two different digging animals: the mole and pangolin.

Continued

Instructional Application 14.2 *(continued)*

From Animal Inspiration to Robotic Mechanism

Next, students form new teams, and each team debates which animal they see as the most generative inspiration for a biomimetic digging robot. They consider which biological digging structures (e.g., teeth or claws) and which digging functions (e.g., pinching jaws or rotating shoulders) they will most easily be able to translate into mechanical parts and actions with the robotic kits and craft materials provided in the classroom. Consider a team that decides its robot will mimic a pangolin's claws and its hooking and pulling arm movements. To give the team ideas about mechanical hooking and pulling, Ms. Ren suggests that they try manipulating the examples of "cable-driven" cardboard mechanisms that she has already built. In these devices, a motor spins simply to wind a cable or string, which in turn pulls on an object located at a distance from the motor. Exploring prebuilt cable-drives helps the team brainstorm. As they inspect the mechanisms and begin to sketch their own ideas, they say, "We can use a cardboard rectangle to mimic the pangolin's wrist and small pieces of drinking straws cut lengthwise to mimic its claws. For an arm, we could connect the wrist to two more cardboard segments with brads for joints. Then we need something to make the arm hook and pull. We could thread a string from the straws along the cardboard arm back to a motor and have the motor tug the string to move the claws."

"Engineers may decompose the overall action into multiple smaller functions and determine structures that can accomplish each of those functions. Structure and function in engineering helps pursue the goal of designing a system that accomplishes an action or behavior."

After generating multiple possibilities and getting Ms. Ren's feedback on a team consensus plan, each team begins to construct its robot using small motors, carefully shaped cardboard, string, and pieces of craft sticks and drinking straws. Now they use *structure and function* as engineers do: The goal is to make an action happen on the environment, and engineers use knowledge (or construct new knowledge) of structures and their functions in order to design an overall system that carries out that action. Engineers may decompose the overall action into multiple smaller functions and determine structures that can accomplish each of those functions. Structure and function in engineering helps pursue the goal of designing a system that accomplishes an action or behavior.

Figure 14.4 (p. 326) compares a prototype of a digging robot with the mole whose digging structures and functions inspired the robot design. The mole rotates its shoulder joints so its paw and claw structures push plants or soil away from its body and

Continued

Instructional Application 14.2 *(continued)*

out to both sides. Similarly, the digging robot pushes rubble out to the side using craft stick linkages that are mounted to motors programmed to rotate in opposite directions.

Figure 14.4. Real-life mole versus digging robot

Students analyze the structure-function relationships of a biological system and apply their learning to design the structures and functions of a robotic one.

As the students construct their robot prototypes, Ms. Ren circulates and prompts students to explain the reasoning behind their design decisions. She asks questions like, "Why are you attaching that string at that location on the cardboard?" and "Do you think that straw will support the weight of that craft stick chisel?" As she consults with teams, she encourages students to consider how their robot will respond to being jostled around in the test bin or being handled repeatedly through multiple attempts to move rubble. For example, if a motor is knocked slightly out of place, will its digging motion still work? Here, there is an opportunity for the CCC of stability and change to influence students' engineering design thinking. After helping students consider her questions about their robots' robustness, Ms. Ren could have explicitly pointed out that engineers often have to design intentionally for stability because instability and change are the default. The natural world—the environment—constantly acts on human-made products and processes. In engineering design, the goal is often to create a product that can remain at equilibrium even when the natural environment disturbs it.

Continued

Instructional Application 14.2 *(continued)*

Looking for Patterns Across Prototypes

After a few days of constructing, testing, and iterating on their digging robot designs, Ms. Ren's students participate in a design review where they describe how they have mimicked an animal's digging structure and function relationship with robotic hardware and craft materials. They also share information with each other about how much rubble (modeled by shredded paper) their robot was able to move in each test, and how stable their prototype was when jostled around in the earthquake disaster zone (modeled by a large bin). Ms. Ren asks students what similarities they see across the prototypes that moved more rubble than others and whether there is anything in common across the prototypes that withstood the jostling. By encouraging students to look for common features that supported design success for different prototypes, Ms. Ren is highlighting the CCC of patterns, which serves biology and engineering in the same way. Biologists use the concept of patterns as they review data for clusters of information that indicate relationships between variables in living systems, from cells to ecosystems. Similarly, Ms. Ren's middle school engineers are reviewing data from robot testing to determine relationships between structure and function. They are investigating how particular design factors—such as the position of their motors or the length of their craft stick digging arms—are related to the success of their prototypes at meeting criteria. For example, looking across multiple robot prototypes, they might see that when motors were located at the front, digging structures attached to motors were more likely to collapse than when the motors were positioned on the robot's sides. After noticing this pattern, they might wonder about its cause and engage in reasoning about the forces at play in each design variation.

This biomimetic robot design unit provides an opportunity for students to see how they can be more effective engineers by intentionally looking for the CCCs of structure and function, stability and change, and patterns across both biological and technological systems. In multidisciplinary learning experiences such as this one, teachers can support students in valuing the CCCs across disciplines by asking them to reflect on how a CCC helps them generate deeper understandings during a scientific investigation, and then consider how a CCC supports their problem solving during an engineering design challenge.

Summary

Many CCCs and SEPs pervade a well-designed activity. Educators in engineering education, as well as in science education, need to choose specific CCCs to emphasize during a particular activity in order to help students develop facility with them. To make CCCs into key learning objectives, educators could (a) identify the CCCs to be targeted and generate an engineering challenge for which CCCs are particularly salient for that lesson and unit and/or (b) examine a design challenge and decide which CCCs will be particularly helpful in designing effective solutions. Educators should then consider how students might apply CCCs as productive thinking tools. How can CCCs be used in authentic ways to help bridge science and engineering (and math) concepts and strengthen students' reflective engineering design processes?

It is important that CCCs are authentically integrated into the activities and not add-ons or overlays. During activities, students should use CCCs in meaningful ways that tie to DCIs and SEPs. Through such use, students can develop deeper understandings of phenomena, designs, concepts, and practices. By asking students to employ CCCs in meaningful ways to make sense of and solve a problem, lessons can model how engineers and scientists use similar ideas in their work. Using concepts in context for relevant work helps students develop familiarity with them. However, if they are embedded, CCCs might not always be obvious to teachers or students during the activity.

Thus, it is important to explicitly structure opportunities for discourse about CCCs in activities. Instruction needs to make explicit the uses of CCCs and how they are being employed to construct new knowledge, and it must include explicit support for them. As a result of teachers and students participating in such metadiscourse (Reveles, Cordova, and Kelly 2004)—that is, discussion about the ways they are discussing their uses of CCCs—they make public their reflections about CCCs and their use of CCCs in their thinking. Resources, such as those described in the Instructional Applications in this chapter—namely, resources that provide rich questions, worksheets, and reflection questions that model how experiences can be set up, data collection orchestrated, and questions posed—can invite students to wrestle with ideas related to CCCs. For example, students might be asked to indicate how energy is flowing through the system on a diagram of a solar oven, or they might be invited to reflect on the difference between analyzing the structure-function relationships within an animal's anatomy and within a robot's mechanical system. Design notebook templates can first encourage students to conduct and document frequent tests of their designs and then later help them review those results to identify and document particular features that helped a design carry out its core function (Wendell, Andrews, and Paugh 2019).

Across their activities and years of schooling, students also need to realize that there are multiple ways CCCs might be utilized that differ by discipline and subdiscipline. To build this understanding, students also should be given the opportunity to use CCCs in a variety of ways and contexts. The biomimetics example in this chapter demonstrates one context where students can reflect on the similarities and differences between how biologists and engineers use CCCs such as structure and function. CCCs can provide a powerful tool to encourage students to connect their understandings of science principles with evidence of how those principles often shape the world around them through engineering design.

In this chapter, we have explored the relationship between the CCCs and engineering design by illustrating the following key points:

- Engineering is the systematic application of knowledge to design objects, systems, and processes that meet human needs and desires, whereas science aims to describe, explain, and predict the natural world and its physical properties.
- Although the purposes of science and engineering differ, the two disciplines rely on and inform each other.
- CCCs bridge science and engineering by providing tools for thinking more deeply about both natural phenomena and designed artifacts, and educators can make choices about which CCCs to highlight within a particular engineering design experience by considering which science practices and core ideas they most want students to engage with as they work on designs.
- Sometimes CCCs play a different role in science than in engineering. For example, in the biomimetic robotics unit described in this chapter (Instructional Application 14.2 [p. 321]), structure and function in biology helped students more deeply explain existing phenomena (i.e., digging animal adaptations), whereas structure and function in engineering enabled students to design behaviors that didn't yet exist (i.e., digging robot mechanisms). On the other hand, in the solar oven insulation unit (Instructional Approach 14.1 [p. 316]), the CCC of energy and matter played a similar role across science and engineering; it deepened the students' analysis of a range of possible insulating materials and of their designed solar ovens.
- Using real-world engineering problems is a promising way to help students build science and engineering knowledge, familiarity with practices in these fields, and understanding of crosscutting concepts and the depth they add to both inquiry and design.

Acknowledgments

We acknowledge the contributions of Debra Bernstein, Gilly Puttick, Fay Shaw, Michael Cassidy, and Ethan Danahy, collaborators on the Designing Biomimetic Robots project, which is a partnership between Tufts University and TERC, with funding from NSF award #DRL-1742127 (Debra Bernstein, PI). We are also deeply grateful to collaborating teachers and students for all of their efforts to enact parachute design, solar oven design, and biomimetic robot design projects.

References

Cunningham, C. M. 2018. *Engineering in elementary STEM education: Curriculum design, instruction, learning, and assessment.* New York: Teachers College Press.

Cunningham, C. M., and G. J. Kelly. 2017. Epistemic practices of engineering for education. *Science Education* 101 (3): 486–505.

NGSS Lead States. 2013. *Next Generation Science Standards: For states, by states.* Washington, DC: National Academies Press. *www.nextgenscience.org.*

Reveles, J. M., R. Cordova, and G. J. Kelly. 2004. Science literacy and academic identity formulation. *Journal for Research in Science Teaching* 41 (10): 1111–1144.

Wendell, K. B., C. J. Andrews, and P. Paugh. 2019. Supporting knowledge construction in elementary engineering design. *Science Education.* 103 (4): 952–978.

Wendell, K. B., and J. Kolodner. 2014. Learning disciplinary concepts and practices through engineering design. In *Cambridge handbook of engineering education research,* eds. B. Olds and A. Johri, 243–263. New York: Cambridge University Press.

PART IV

Assessment of the CCCs and What Comes Next

Chapter 15

Assessment of Crosscutting Concepts: Creating Opportunities for Sensemaking

Erin Marie Furtak, Aneesha Badrinarayan, William R. Penuel, Samantha Duwe, and Ryann Patrick-Stuart

How do we go about solving a new problem or making sense of a particularly perplexing phenomenon? Sometimes, our science ideas and practices are not enough, and we must engage other ideas and understandings as we explore possible new explanations and mechanisms. In this chapter, we suggest that crosscutting concepts (CCCs) can serve as an avenue for students to make sense of new problems by asking them to engage in *sensemaking*. Doing so helps us open up a new perspective on assessment in the era of reforms aligned with *A Framework for K–12 Science Education* (the *Framework;* NRC 2012); namely, that we can present students with novel situations and then ask them to apply CCCs to help them make sense of that situation. In this chapter, we will address the following questions:

- How do CCCs support students as they make sense of new phenomena and solve novel problems?
- How can we implicitly and explicitly assess student understanding of CCCs?
- How can CCCs inform the way we think about, design, and prepare teachers to enact classroom assessment?

Making Sense of Phenomena and Problems With Crosscutting Concepts

Imagine taking a drive through a valley in the foothills of the Rocky Mountains on a summer afternoon. As you follow the road alongside a cool, gurgling creek, taking in the

Figure 15.1. South- (left) and north-facing (right) slopes in Colorado's Rocky Mountains

surrounding scenery, you might observe that one hillside is covered with shrubs and a few trees, whereas the opposite hillside is covered with more densely packed stands of forest (Figure 15.1). Why might that be? You may already understand how trees grow and how environmental factors may contribute to differential growth; however, the explanation isn't readily apparent. To get to a deeper understanding—or even a better, more specific investigative question—you begin to ask a series of questions that help you make sense of the environment around you in ways that a scientist might.

According to Dr. Helen Quinn—a physicist and science educator who wrote the preface of this book (p. xi) and who chaired the committee that wrote the *Framework* (NRC 2012), which forms the basis of the *Next Generation Science Standards* (*NGSS*; NGSS Lead States 2013)—it's situations like these that best illustrate how crosscutting concepts help scientists make sense of the world around them. That is, when scientists don't readily know an explanation for a phenomenon they observe, they will start asking themselves questions that are grounded in an understanding of concepts like scale, proportion, and quantity; patterns; and cause and effect. For example, when seeing the difference between the north- and south-facing slopes for the first time, a scientist might ask: "What trees and other plants are growing where? What is the relative proportion and quantity of different kinds of trees and plants on the two opposite hillsides? What patterns do I observe where those plants are growing? What is causing those patterns? Are other hillsides in the area like this, or is this a localized phenomenon? How might matter and energy be cycling through this ecosystem?"

Like scientists, students can also pursue questions like these to figure out why different species are growing in different proportions on the north- and south-facing slopes—and the CCCs can help them orient to the phenomenon in productive ways and delve more deeply into possible mechanisms. For example:

- One might observe that the relative *proportion* of trees was greater on one side than on the other.
- One might see a *pattern* in which certain kinds of plants are growing in some places and not in others. In the case of the slopes, a student might observe that ponderosa pine trees were more likely to be growing on the south-facing side but were almost

completely absent on the north-facing side, which had more lodgepole pines and Douglas fir trees.

- Connecting the direction of the slopes of the hillsides to the angle of the Sun would reveal that the south-facing slopes receive more sunlight, and thus water evaporates more quickly, limiting the resources available to plants growing there.

In all of these cases, a grasp of CCCs like patterns, systems and system models, and cause and effect is imperative to framing interesting and scientific lines of investigation and to diving deeper into this puzzling phenomenon to propose mechanisms, hypotheses, arguments, and explanations.

In the hillside example, we can see how different CCCs provide points of entry into a problem situation that students may not have previously encountered, while also creating opportunities for students to demonstrate their knowledge of disciplinary core ideas (DCIs) and engage in science and engineering practices (SEPs). In other words, CCCs represent ways that scientists engage with new phenomena and problems—an essential competency we want students to develop if they are to successfully use their science education in the real world. If the crosscutting concepts are to capture ways that scientists proceed when confronted with new phenomena or problems—and if we want to ensure that students are developing an understanding of the CCCs such that they can confidently confront new situations—we need to reframe assessment as an *opportunity for students to apply crosscutting concepts to make sense of novel phenomena*. In other words, assessing the CCCs requires that we assess students' approach to problems or phenomena when they may have never seen them before or have insufficient understanding of disciplinary core ideas to explain phenomena fully.

Furthermore, we recognize that by applying CCCs in these novel contexts, students can also build on and leverage their experiences from their daily lives, a fundamental element of promoting equitable participation of all students in science learning (e.g., NRC 2014). As Chapter 3 in this book also explains, viewing CCCs as resources that students bring to the classroom creates more access by providing opportunities for students to engage with questions (e.g., looking for patterns, thinking about differences in scale) that they naturally ask in their experiences with the world around them.

This chapter will describe what it means to take a sensemaking perspective on CCCs and how this perspective informs the way we design classroom assessments. We will illustrate how leading with these concepts can be a different and effective approach to assessment design for the science classroom. We'll explain how to create assessment questions that elicit understanding of CCCs, and then we'll illustrate how underlying progressions can support student learning both within and across grade bands. Finally, we will suggest ways to support teacher professional learning to design and use assessment evidence related to CCCs.

What Do We Mean by *Sensemaking*?

A key goal of three-dimensional learning is for students to make sense of phenomena and problems in the world around them and to have opportunities to draw on their everyday experiences and full repertoire of resources as they do so (NASEM 2017). As humans, we are all naturally oriented to making sense of the world around us. In this chapter, we will use the word *sensemaking* to mean the dynamic process by which students build and revise explanations as they try to figure out a scientific phenomenon (Fitzgerald and Palincsar 2019; Odden and Russ 2019). When presented with a new and puzzling situation, students may then work together as they generate possible ideas and explanations related to the phenomenon, which could be the beginnings of an explanation that can resolve the anomaly (Odden and Russ 2019).

CCCs are a critical tool that scientists use to frame and dive deeper into phenomena and problems in which they are interested. In addition, common approaches to assessment posit that students should be presented with new scenarios to which they can apply their developing understandings or that allow them to take new perspectives on questions they are already investigating.

Returning to the idea presented above—that the CCCs are ways scientists might generate approaches to understand a new or unfamiliar phenomenon—we can see how an assessment scenario might create opportunities for students to demonstrate their application of CCCs as they make sense of that phenomenon. Assessments that are intended to assess three-dimensional performances must ask students to contend with compelling phenomena that have a real uncertainty associated with them. Students may use the crosscutting concepts as an entry point into a phenomenon or problem that might otherwise be a roadblock into a proposed investigation, hypothesis, argument, or explanation. In this context, CCCs can be assessed in ways that specifically prompt students to demonstrate their developing ability to use *specific* crosscutting concepts or that allow for flexible use of CCCs as entry points for students to address different phenomena.

For many teachers, the idea of assessing students on phenomena they have not seen before may feel starkly different from the way they usually think about assessment: as a check to see what students have learned so far or as a check for understanding of what has been taught. However, repositioning the assessment of CCCs from a sensemaking perspective gets at the heart of a new way of thinking about assessment—to see the degree to which students are able to apply what they are learning to a novel problem or phenomenon. Ultimately, if we want students to be able to use science in their everyday lives, they must be able to use what they learn in school to explain and understand what they encounter outside of school.

Integrating Uncertainty Into Assessment Design

Assessments designed from a sensemaking perspective may present students with an ambiguous scenario. A scenario presents a phenomenon to students, typically using a combination of textual and visual elements and, in some cases, some form of data such as a graph or table. Ideally, the scenario makes salient what is to be explained—either because it is puzzling, raises questions about taken-for-granted understandings, or has some local or global significance. Providing enough context so students can engage with the problem is key, but this needs to be done in ways that don't overload students with unnecessary details or vocabulary (Fine and Furtak 2020). Then, the assessment can ask students to consider a question that helps them take the next step toward explaining this phenomenon: Would they want to dive into data to look for patterns? At what scale? Do they want to distinguish between causal versus correlational relationships to better argue for mechanism? Multiple CCCs may even be necessary to make sense of some phenomena.

We'll dig more into how to design these kinds of assessments later; for now, let's unpack some different ways of talking about CCCs and assessment: implicit and explicit approaches, formal and informal assessments, and formative and summative uses.

Implicit and Explicit Approaches to Assessing Crosscutting Concepts

There are both implicit and explicit approaches to assessing CCCs. When the primary focus is on disciplinary core ideas and science and engineering practices, the crosscutting concepts are often *implicitly* part of student thinking and performance. From this perspective, a teacher might infer from a student response to a question asking for a mechanistic or causal explanation of a phenomenon that the student has some grasp of cause and effect, because the student's explanation included a cause and an effect and provided sufficient evidence and reasoning to support that claim. However, this approach does not put student use of and development in CCCs front and center. Based on such performances, we can't be sure that students actually understand and can use the CCCs to deepen their understanding or exploration of phenomena. Indeed, one might argue that CCCs have always been implicit, but the perspective of the *Framework* is that students need more explicit support and instruction to be able to better develop the ability to apply CCCs over time such that all learners develop the critical ability to confidently and productively make sense of novel situations.

The alternative approach is to assess crosscutting concepts *explicitly,* or as a distinct element of student understanding. An advantage of this approach is that it more straightforwardly supports the development of a student's ability to apply crosscutting concepts over time. For example, the progressions provided as part of the *Framework* (NRC 2012) show the ways in which student understanding of, say, patterns is expected to develop over grade bands (Table 15.1). In grades 3–5, for instance, students are expected to be able to identify that "similarities and differences in patterns can be used to sort, classify, communicate, and analyze simple rates of change for natural phenomena and designed products." By high school, however, expectations are more sophisticated, and students are expected to observe different patterns at different scales and to identify the ways in which patterns "can provide evidence for causality in explanations of phenomena."

Table 15.1. Progression of the CCC of patterns across grade bands

Grade Band	3-5	6-8	9-12
Students are expected to ...	Identify how patterns of change can be used to make predictions. Use patterns as evidence to support an explanation.	Relate macroscopic patterns to microscopic and atomic-level structures.	Observe different patterns at different scales, and provide evidence for causality in explanations of phenomena.

Source: NSTA 2013.

In addition to promoting the development of student understanding of particular CCCs over time, developing proficiency with CCCs requires presenting students with varied problem contexts in which they have the opportunity to decide which CCC to apply when. This explicit approach can help ensure that all learners are being given the opportunity to use each CCC as a tool in their tool kit for making sense of the world around them and that learners have the ability to use all the CCCs flexibly.

Formal and Informal Assessments

Developing Assessments for the Next Generation Science Standards (NRC 2014) argues that assessment tasks should consist of multiple components, or different types of questions that are linked to a common problem to be solved or phenomenon to be explained. However, this *formal* and preplanned approach is just one way of finding out what students know and are able to do. We acknowledge a full range of activities that teachers and students perform to help teachers gain insights into student thinking. We will call these the *informal*, everyday strategies teachers and students might use to draw out student thinking, such as asking students to write down questions related to a phenomenon they have just encountered or having students plan an investigation to answer a question the class has decided to pursue.

Formal assessments of CCCs might take the form of state tests, districtwide assessments, or classroom assessments that teachers plan to use to get information about students' ability to apply CCCs. These assessments are often used as end-of-instruction opportunities to formally monitor and convey student understanding of CCCs, SEPs, and DCIs together, and may be involved in some kind of systems-level decision-making. In contrast, informal assessments comprise the everyday "noticing and attending" (Coffey et al. 2011; Russ and Luna 2013) teachers might do of student use of CCCs as they engage in regular classroom activities.

Formative and Summative Uses

These different types of assessments—whether formal or informal—can have uses that serve *summative* purposes; that is, they can be used to assign scores or grades to signify amounts of learning. They may also serve *formative* purposes, informing subsequent learning experiences. For example, interim assessments can be designed to provide teachers with reports of student progress, and thus may point to students' specific response patterns on formal assessments that can inform subsequent instruction. This would be a *formative* use of a *formal* assessment.

However, we don't want to diminish the importance of the information teachers can get about student use of CCCs just by walking around and talking with students; that is, listening to their ideas, and then making rough judgments about student progress to inform the next minute, next day, or next week of instruction (Wiliam 2007).

We caution that unless CCCs are explicitly assessed in formal assessments, it's not clear how those assessments might generate information that specifically supports teachers in making inferences about what students are able to do with the CCCs. Similarly, asking clear questions linked to CCCs is more likely to provide teachers with on-the-fly information that's useful for their instruction.

To summarize, we emphasize that CCCs should be assessed explicitly, through a combination of formal and informal approaches, so that we may learn more and more often about students' developing understandings. In addition, assessments of CCCs should also be formative, not just summative, such that they can be used to inform and guide instruction.

Progressions of Crosscutting Concepts Over Time

When designed to explicitly assess CCCs, assessments can also provide the opportunity for students to monitor the progress and development of their ability to use CCCs flexibly to make connections and dive deeply into phenomena. In fact, an advantage of explicitly assessing CCCs is that students' progress on applying them may be tracked over time. The *Framework* lays out the ways in which student use of CCCs should become more explicit over time. Take, for example, the differences between sample grade-level expectations for the CCC of patterns shown in Table 15.1 (p. 338) or the K–12 progression of the CCC of stability and change in Table 15.2.

Table 15.2. Progression of the CCC of stability and change

Primary School (K-2)
• Things may change slowly or rapidly.
• Some things stay the same while other things change.
Elementary School (3-5)
• Change is measured in terms of differences over time and may occur at different rates.
• Some systems appear stable but over long periods of time will eventually change.
Middle School (6-8)
• Stability might be disturbed either by sudden events or gradual changes that accumulate over time.
• Explanations of stability and change in natural or designed systems can be constructed by examining changes over time and processes at different scales, including the atomic scale.
• Small changes in one part of a system might cause large changes in another part.
• Systems in dynamic equilibrium are stable due to a balance of feedback mechanisms.
High School (9-12)
• Much of science deals with constructing explanations of how things change and how they remain stable.
• Systems can be designed for greater or lesser stability.
• Feedback (negative or positive) can stabilize or destabilize a system.
• Change and rates of change can be quantified and modeled over very short or very long periods of time. Some system changes are irreversible.

Source: NRC 2014.

The advantage of these progressions is threefold: First, progressions support teachers and assessment designers in writing assessments that have the appropriate level of sophistication for a particular grade band. Second, they allow teachers to provide helpful feedback to students. Third, they allow teachers to track students' learning over time. We note that supporting student learning through the progression of CCCs is deeply and integrally related to their understandings of DCIs and their grasp of SEPs; that is, as students are able to apply CCCs to explain more sophisticated and complex phenomena, this also supports—and necessitates—deeper understandings of DCIs and abilities to engage in SEPs.

Designing Assessments With the Appropriate Level of Rigor

Working with these progressions can support a teacher's thinking about the rigor of the CCCs and how expectations of the way students work with and apply the CCCs increase as they progress in school. It may be more straightforward to ask simple questions such as, "What is the pattern?" However, as students approach high school, they are expected to work with mathematical representations as they identify patterns, and to observe variations in patterns at different scales.

Providing Feedback

We have found through research that having a clear representation of how student thinking develops in a domain can support teachers in providing students with helpful feedback (e.g., Furtak et al. 2016). For example, a teacher may observe a middle school student, in answering a problem about natural selection, implying that populations remain constant over time. The progression of stability and change suggests that by grades 3–5 and in middle school, students may begin exploring how change can occur at different rates, and even though situations may be stable, they are also capable of rapid change. Thus, appropriate feedback in this situation might be to encourage the student to look at a larger, time-dependent set of data that illustrates how the relative proportions of the number of individuals with particular variations within a population shifts over time or to look at changes in environmental factors that co-occur with rapid changes in those proportions.

Tracking Student Learning Over Time

Finally, a dual advantage of both writing assessments that explicitly assess CCCs, and also using progressions to give students feedback, involves being able to track student progress over time. Our assessment tasks—whether formal or informal—may consist of multiple questions that together help us understand how students are applying CCCs to make sense of a phenomenon. Over the course of a unit, a semester, or an academic year, we can track students' performance and, with support and feedback around the CCCs, students' ability to apply the CCCs could be shown to increase over time.

A project conducted by some of the authors of this chapter is already showing the power of using CCCs to design tasks and support student progress over time. As part of an ongoing Research-Practice Partnership conducted by authors Furtak, Duwe, and Patrick-Stuart, teachers have collaborated to use progressions of CCCs to track how students' ability to apply the CCC of energy and matter becomes more sophisticated across high school physics, chemistry, and biology. By following the CCC into different content areas, teachers have learned how it supports different DCIs in different science domains and how SEPs like Developing and Using Models and Constructing Explanations and Designing Solutions can combine with the CCC and the DCIs to create three-dimensional learning performances. Teachers have developed combinations of assessment items that together support students as they explain novel phenomena, such as using a bucket of ice water to cool cans of soda (chemistry) or modeling how an energy bar can power a runner's muscles (biology).

New Approaches to Assessment Design: Leading With Crosscutting Concepts

As the previous sections have established, leading with CCCs compels us to think carefully about how questions on assessments are worded. Instead of asking students questions about their conceptual knowledge or prompting students to engage in a scientific practice, assessments that focus on CCCs bring patterns, scale, systems, and other concepts to the forefront.

Take, for example, the hillside phenomenon used at the beginning of this chapter. That phenomenon was developed for a bundle of standards that collectively assess interdependent relationships in ecosystems (LS2.A) and weather and climate (ESS2.D), along with the SEPs of Analyzing and Interpreting Data, Developing and Using Models, and Constructing Explanations and Designing Solutions. However, if we want students to lead with CCCs, our questions will be substantively different from those formed to primarily assess DCIs or SEPs. Table 15.3 illustrates how, following from this bundle of performance expectations (PEs), we can identify different questions that focus students on looking for patterns and identifying components of the system or how those components interact.

Writing these kinds of questions, however, is not routine for those of us working in science education who have learned and taught in settings that prioritize checking to see if students have developed adequate conceptual understanding. This approach is also counter to ways we might write questions if, for example, students have had ample opportunities to learn about and work with interactions between biotic and abiotic factors in ecosystems and have linked those interactions to regional climates and the amount of energy reaching north- and south-facing slopes. Some of us might think it unfair to test students on unfamiliar content.

However, from the perspective of broadening equity and access for students from diverse backgrounds in learning science, everyday, accessible, and novel phenomena are key. For example, in a study conducted by Furtak, Duwe, and Patrick-Stuart, high school

Table 15.3. Example questions for hillside phenomenon with links to different PEs and the three-dimensions

Question	Performance Expectation	Crosscutting Concept	Disciplinary Core Idea	Science and Engineering Practice
What patterns do you observe in the plants that are growing on the north- and south-facing slopes? How do you explain the patterns?	MS-LS2-2: Construct an explanation that predicts patterns of interactions among organisms across multiple ecosystems.	Patterns	LS2.A: Ecosystems: Interactions, Energy, and Dynamics	Constructing Explanations and Designing Solutions
What are the ecosystems you observe on the hillsides? What are the components of the hillside ecosystems? How do the biotic and abiotic factors within the ecosystems interact?	MS-ESS2-6: Develop and use a model to describe how unequal heating and rotation of the Earth cause patterns of atmospheric and oceanic circulation that determine regional climates.	Systems and System Models	ESS2.D: Weather and Climate	Developing and Using Models
Draw a diagram that shows how changes to an abiotic factor of one of the hillside ecosystems affect other biotic factors.	MS-LS2-1: Analyze and interpret data to provide evidence for the effects of resource availability on organisms and populations of organisms in an ecosystem.	Cause and Effect	LS2.A: Ecosystems: Interactions, Energy, and Dynamics	Analyzing and Interpreting Data

biology students living outside Denver were provided with the opportunity to think about how altitude might affect rates of respiration. They were encouraged to think through their own and their families' experiences with altitude. However, if we only assess students' ability to display knowledge acquired in school in forms and media that are familiar to them from dominant cultural and linguistic communities, then we run the risk of further privileging students from those dominant communities who have had greater access to—and in many cases success in—those traditional settings. Conversely, if we reposition student knowledge through the lens of making sense of everyday experiences, we provide more opportunities for a broader range of students to show us what they know.

As a support for writing questions of this type, the third author of this chapter, William Penuel, worked with colleagues across several linked initiatives to develop a set of "Crosscutting Concepts Assessment Prompts" designed to help teachers ask questions that foreground the different crosscutting concepts in their assessment design (see Table 15.4). The prompts are a set of partially and fully developed questions that are associated with each of the seven crosscutting concepts in the *Framework* (NRC 2012) and Appendix G from the *NGSS* (NGSS Lead States 2013). The Crosscutting Concepts Assessment Prompts consists of sets of prompts for each CCC. Each set works with the assumption that students have been provided with some kind of context: a set of data, a phenomenon or problem scenario, or a model. The contexts were derived from the *Framework*, which provided some—limited—guidance as to the kinds of phenomena and problem contexts for which the crosscutting concepts might be used.

Table 15.4. Sample crosscutting concepts assessment prompts

Crosscutting Concept	Features of Phenomenon/ Problem/Scenario	Sample Questions
Patterns	Ask after presenting students with observational data as part of a phenomenon ...	• What patterns do you observe in the data presented in the [table, chart, graph, model output]? • What does the pattern of data you see allow you to conclude about _____? • Does the pattern in the data support the conclusion that is related to ______? Why or why not? • What mathematical representations of the data could help you identify patterns in the data? • What observations could you make next to help explain the pattern in the data? • What kind of mathematical function best fits the pattern of data you see? • For bivariate data: How strong is the correlation between *x* and *y*? (Calculate the correlation coefficient.)
Cause and Effect	When a system or situation presented in the scenario involves complex or relational causality (e.g., as in ecosystems and co-evolution) ...	• Draw a diagram that shows how changes to one component of the system affect components that are not directly connected to that component. • If [change to one component of a complex system] occurred, what do you predict would happen to [component that has an indirect rather than direct connection to the first component]? • What feedback loops are causing this system to be in [balance/equilibrium]? • How can a small change to ______ have a big effect on _______?
Scale, Proportion, and Quantity	When eliciting students' ability to change scales to investigate phenomena that are too large or small to see or too long or short to observe directly ...	• Why could [people in the scenario] see [object] when they observed it [under a microscope/with a telescope] but not when they looked just with their eyes? • How could we test whether ______ is changing, even though it looks like it is not? • Which of the patterns presented in the scenario do you think could be observed at a [faster/slower, smaller/larger] scale? Why?

Continued

Table 15.4. *(continued)*

Crosscutting Concept	Features of Phenomenon/ Problem/Scenario	Sample Questions
Systems and System Models	When the model is of a designed system ...	• Create a set of instructions for building [system] that another student can follow. • If you could control *X* in the system, would it stop *Y*? Why or why not? • How could you test whether this system satisfies the design constraints described in the scenario?
Energy and Matter	When eliciting understanding of the cycling of matter, ask students ...	• Where is matter that enters [system] coming from? • What happens to matter as it moves within [system]? • Where does matter that leaves [system] go? • Draw a picture showing the stocks and flows of matter in [system]. • What evidence is there that matter is conserved in this cycle?
Structure and Function	After presenting students with a novel system to investigate that they have not explored before ...	• What function do you think [structure] serves in this system? How could we find out? • This system performs [functions]. How do you think the structures support or enable those functions? • This organism engages in [behavior] to [function]. How might [structure] help explain how it is able to perform [behavior]?
Stability and Change	When the scenario presents a system or phenomenon with feedback loops ...	• How does [process or mechanism A] affect [process or mechanism B]?

Note: See *stemteachingtools.org* for the complete set of prompts (Penuel and Van Horne 2016).

As an example, for the CCC of patterns, students might be presented with data from an experimental study focused on isolating causal variables as part of the phenomenon—for instance, the effect of varying factors that could affect the strength of an electromagnetic field. As part of the assessment, students design and carry out an investigation, creating a data table that represents the strength of the field (unit) and the number of coils in a solenoid. Assessment questions for these data might build on the CCC question prompts for analyzing collected data, which include, "What does the pattern of data you see allow you to conclude from the experiment?" or "Does the pattern in the data support the conclusion that ________ is caused by _________? Why or why not?" When adapted into a specific assessment item, the questions might say, "Does the pattern in the data support the conclusion that *increased strength of the field* is caused by *increasing the number of coils*? Why or why not?"

The CCC prompts are grounded in the idea that it is useful to draw on multiple metaphors for using the crosscutting concepts to support student sensemaking (Rivet et al. 2016). Some prompts ask students to "look for" or "notice" particular aspects of a phenomenon presented to them, drawing on the metaphor of crosscutting concepts as a lens. (For example, one prompt for patterns asks, "What does the pattern of data you see allow you to conclude from the experiment?" Likewise, one prompt for stability and change asks, "How could we test whether it is changing, even though it looks like it is not?") Some prompts present crosscutting concepts as opportunities to make connections to phenomena students have studied before. Still other prompts ask students to use crosscutting concepts as tools. (For instance, one prompt for scale, proportion, and quantity asks, "Is the relationship between these variables linear, exponential, or something different altogether? How does the pattern in the data support your conclusion?") Finally, some of the prompts focus on crosscutting concepts as rules of the game—that is, they prompt students to consider mechanisms that can be investigated across a range of phenomena. (For instance, one prompt for cause and effect asks, "What [properties, entities, or rules] that aren't described explain what you see happening [in the scenario]?)

Let's look at a couple of examples of prompts designed using this basic format. These prompts were both created from scratch and formed by modifying existing items. The first example is asked at the end of a lesson in which students have come to a conclusion about the causes of a rapid increase in the population of buffalo on the Serengeti in the 1960s, after the elimination of a disease that suppressed their numbers. The next lesson will present students with an opportunity to use a simulation that includes not only buffalo but also wildebeest to explore how, if at all, their population dynamics are interrelated. Students will need to specify models to test, and so the item prepares them for a discussion at the beginning of class the next day. In addition to prepping students for the upcoming class, the item illustrates the way crosscutting concepts can serve as a lens to prepare students for future learning about a novel investigative phenomenon by engaging them with it before they actually encounter it in class (Schwartz and Bransford 1998). The disciplinary core ideas relevant to explaining the phenomenon are fully developed in subsequent lessons, and the assessment question below is designed to prepare them for this learning:

> What evidence could you examine to help decide whether the change in the wildebeest population from 1975 to 2000 was part of a large change in the Serengeti system?
>
> A. We could look at more detailed information about the numbers of wildebeest.
> B. We could see whether there was a disturbance to the ecosystem in 1975.
> C. We could see if the Rinderpest came back.
> D. We could examine patterns of change in the size of populations of other animals and plants living on the Serengeti.

In this instance, students might respond to the multiple-choice question and then, either in writing or in conversation with their peers and teacher, talk about what they would expect to see in the data that could help them decide if the change in the wildebeest population was part of a large change in the ecosystem.

Using the CCC prompts has also supported teachers in modifying existing assessments based on existing items. For example, teachers and instructional coordinators from the Aurora Public School (APS) district in Colorado and researchers from the University of Colorado gathered on a recent summer afternoon to design common assessments for the teachers in grades 6–10 to use with their students in the upcoming academic year. The ninth-grade assessment writing team, consisting of a district instructional coordinator, a high school physics teacher, and a university-based researcher, worked with HS-PS2-1: Analyze data to support the claim that Newton's second law of motion describes the mathematical relationship among the net force on a macroscopic object, its mass, and its acceleration. This performance expectation explicitly cites the CCC of cause and effect; however, teachers also found utility thinking about the CCC of patterns when writing assessment items.

The foundation for the teachers' task design came from a released item from the State of Kentucky Science Assessment System (Kentucky Department of Education n.d.). Working with the CCC prompts, the item writing team adapted the existing item into a multicomponent task related to the question "How do airbags protect passengers during a crash?" The evidence statements in the *NGSS* for HS-PS2-1 help us see the ways in which the CCC is applied to this performance expectation (Figure 15.2).

Figure 15.2. Evidence statements for HS-PS2-1

Observable features of the student performance by the end of the course:			
1	Organizing data		
	a	Students organize data that represent the net force on a macroscopic object, its mass (which is held constant), and its acceleration (e.g., via tables, graphs, charts, vector drawings).	
2	Identifying relationships		
	a	Students use tools, technologies, and/or models to analyze the data and identify relationships within the datasets, including:	
		i.	A more massive object experiencing the same net force as a less massive object has a smaller acceleration, and a larger net force on a given object produces a correspondingly larger acceleration; and
		ii.	The result of gravitation is a constant acceleration on macroscopic objects as evidenced by the fact that the ratio of net force to mass remains constant.
3	Interpreting data		
	a	Students use the analyzed data as evidence to describe* that the relationship between the observed quantities is accurately modeled across the range of data by the formula $a = F_{net}/m$ (e.g., double force yields double acceleration, etc.).	
	b	Students use the data as empirical evidence to distinguish between causal and correlational relationships linking force, mass, and acceleration.	
	c	Students express the relationship $F_{net}=ma$ in terms of causality, namely that a net force on an object causes the object to accelerate.	

The writing team found that, although the crosscutting concept of cause and effect yielded important questions related to the fundamental relationship of $F = ma$, it was also useful to write items that related to students simply making sense of data from multiple airbag trials prior to interpreting it. The Crosscutting Concepts Assessment Prompts for patterns includes sets of item stems that focus on causal relationships, and the team used the set of data while examining graphs of the data.

The team found that leading with questions about CCCs allowed them to ask questions that explicitly assessed the CCCs (e.g., "What patterns do you observe in the data?"). It also provided important points of entry into reasoning with the DCI of Newton's second law (e.g., "Which of the three airbag designs best protect drivers from rapid acceleration?"), as well as the SEPs of Analyzing and Interpreting Data and Engaging in Argument From Evidence (e.g., "Using the data provided, construct a written argument for which airbag design provides the most protection to a driver. Support your claim with evidence from the data table and/or graph.").

Ultimately, we encourage teachers to use these prompts early on in tasks as a kind of "on ramp" to help students notice key features of the scenario presented that could be productive starting points for sensemaking. As the blanks in many prompts indicate, most prompts cannot be simply inserted into an extended task without considering how it must be tailored to the phenomenon or problem at hand.

Professional Learning to Support Assessment of CCCs

We envision that all the preceding supports (i.e., the CCC question prompts and progressions) may be used as resources for teacher collaboration and professional learning within and across grade bands. Here, we provide two examples—first, using progressions of CCCs to support teacher professional learning across grade bands; and second, using supports for writing questions related to CCCs to support teacher assessment design.

Example One: The CCCs and Professional Learning About Assessment

Authors Furtak, Patrick-Stuart, and Duwe have collaborated since 2014 in a long-term partnership between the Aurora Public Schools and the School of Education at the University of Colorado, Boulder. APS is a physics-first school district, with high school students taking physics in ninth grade, chemistry in tenth grade, and biology in eleventh grade. When this repositioning of the science courses took place, teachers were challenged to rethink the way the different disciplines were taught with this new sequence. As Patrick-Stuart put it, "What do we get by teaching biology *after* teaching physics and chemistry?"

Early in the partnership, while working with small groups of science teachers to identify priority standards, Duwe and Patrick-Stuart looked deeply at matrices provided at the *NGSS*@NSTA hub (*https://ngss.nsta.org*) to unpack the ways that crosscutting concepts progressed over different grade bands (NRC 2012). Through joint reading of studies that unpacked the way that energy and matter cycling cut across the disciplines (e.g., Park and Liu 2016), the partnership focused its efforts on coordinated professional learning and assessment codesign around the crosscutting concept of energy and matter, thus examining how crosscutting concepts can promote both horizontal and vertical coherence within an educational system (Shepard, Penuel, and Pellegrino 2018). To support ongoing transition to the *NGSS*, the partnership also expanded this goal to integrate modeling energy and matter cycling within systems as cutting across high school grade bands.

Early professional learning experiences for teachers created space for those with different disciplinary backgrounds to engage in a modeling and explanation activity (Eisenkraft 2016). Posed with the challenge to model energy transfers and transformations when a piece of puffed corn snack was burned, teachers first individually constructed models and then shared their models with science teachers from other disciplines before negotiating and constructing a group model (Figure 15.3). The teachers were surprised at the differences between the models, and the components they included, across content areas. The model in Figure 15.3 features a list of the different ways the science teachers were thinking about and discussing forms of energy in this system, and it also includes a number of different representations that capture these discussions—diagrams, arrows showing energy transfers and transformations, and even a graph. Those teaching the physical sciences focused more on thermal energy and combustion, and those from the biological sciences highlighted energy transfers and transformations.

Figure 15.3. Sample model created by cross-disciplinary teams of science teachers modeling energy

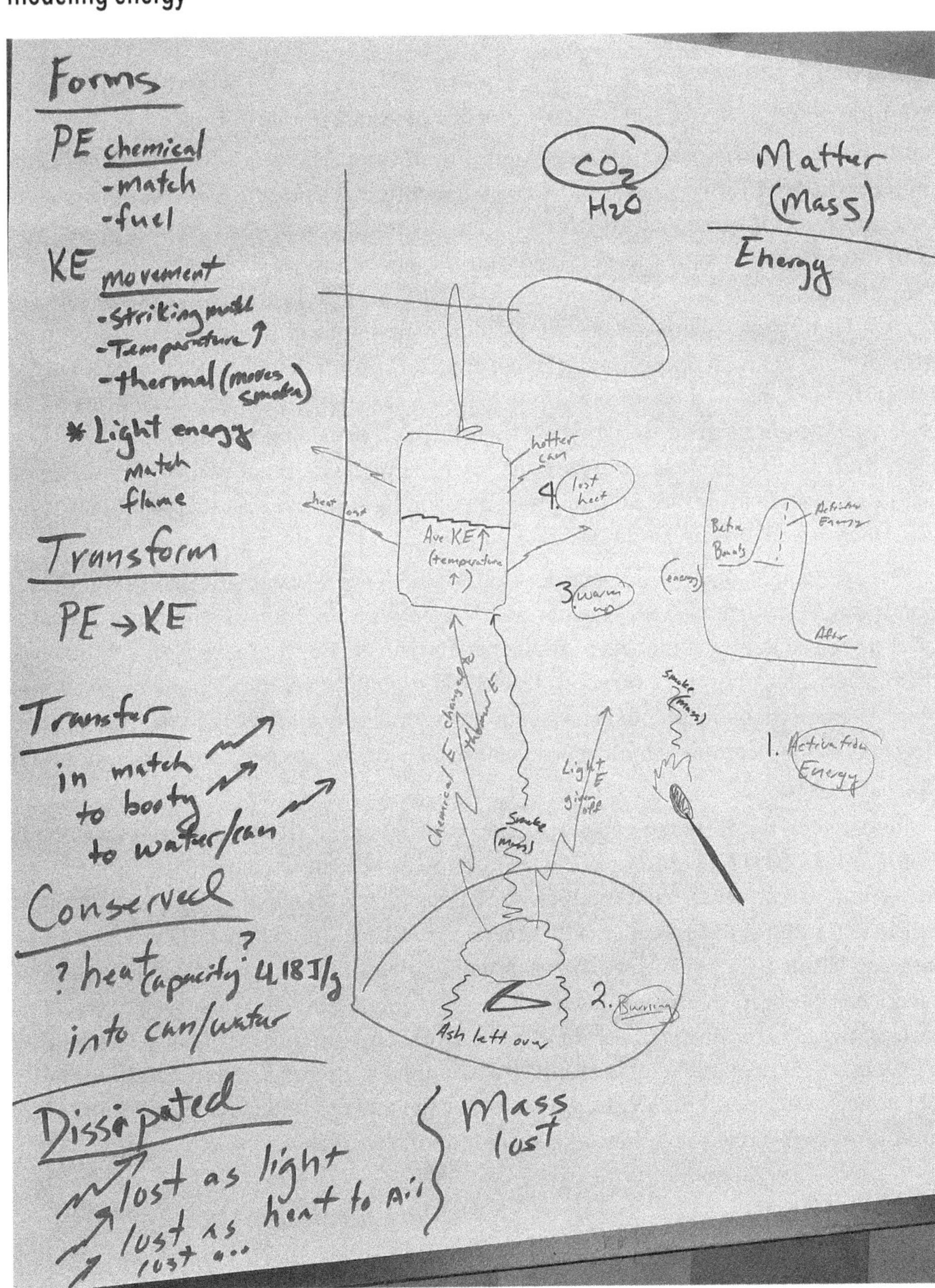

Conversations such as these can create space for teachers to think about how they are providing explicit opportunities for students to develop and use CCCs. In addition, it can help teachers see the ways that explicit emphasis on CCCs can increase the rigor of expectations for how students apply those CCCs on assessments.

Example Two: Using CCCs to Support Teacher Assessment Design

Author Penuel has worked with colleagues to develop and test a workshop series to prepare teachers to integrate use of the Crosscutting Concepts Assessment Prompts tool into three-dimensional assessments. The workshop series is part of the inquiry-Hub research-practice partnership between the University of Colorado, Boulder, and Denver Public Schools, with involvement of BSCS Science Learning (*www.colorado.edu/program/inquiryhub/professional-learning*). The principal aim of the workshop series is to provide teachers with opportunities to use the CCC prompts tool and a parallel tool for integrating science and engineering practices into assessments (*http://stemteachingtools.org/brief/30*) to help them design their own three-dimensional multicomponent tasks. Over the course of the series, teachers analyze existing tasks, practice using the scaffolds with existing scenarios that present phenomena to be explained, and develop their own assessments and scoring guides for assessments.

Over the course of the series, teachers follow and learn a systematic process for developing three-dimensional assessments that begins with analysis of standards and proceeds to selection of phenomena or problems that can anchor the tasks. Rather than have them select their own phenomenon, teachers are given a set of phenomena to choose from, because choosing phenomena can be a lengthy process. They then use a phenomenon to develop a complete multicomponent task using the suggested questions for both SEPs and CCCs.

Research on the workshop shows several benefits of the workshop series and points to the value of the CCC prompts in supporting teachers in developing better three-dimensional assessments. First, studies show that the workshop series helps teachers notice key features of assessments that make them three-dimensional (Penuel, Wingert, and Van Horne 2018). This is significant because integrating CCCs and SEPs into assessments has not been "standard practice." In addition, independent analyses of teachers' tasks show growth after participating in the workshop on teachers' ability to attend to CCCs in assessments (Penuel et al. 2019). Evaluation data indicate that nearly a quarter of teachers who attend the workshop series say the tools for integrating CCCs and SEPs are the most valuable thing they take away from the workshops.

Practical Steps: Recommendations for Assessing the Crosscutting Concepts

1. **Plan for eliciting and using evidence from explicit assessments of CCCs.** Explicit assessments are needed to understand how student understanding is developing. Making a plan for how you will use evidence to inform instruction will help you make good formative use of data. Consider:
   - What are the different ways I can elicit the CCCs using both formal and informal prompts within my classroom?
   - What are some informal ways I can elicit student understanding of CCCs?
   - What are some ways I can address student difficulties that arise in understanding CCCs that might be revealed in my assessments?
2. **Unpack the CCCs using the progressions.** The progressions house a wealth of information about what students are expected to understand about the CCCs at different grade bands, as well as how they progress over time. Consider:
   - For prompts I design, how can I use the progressions to ensure I am eliciting grade-level appropriate understandings of the CCCs?
   - How is the grade-level expectation for this CCC different from the grade bands above and below?
   - How do I ensure the prompt is grade-appropriate for the other dimensions?
3. **Decide what you want students to show you they know and can do with a CCC.** As the CCCs are unpacked, it becomes clear that there are many different aspects to even a single CCC. Decide which part you want to focus on now. Consider:
   - What CCCs have students been using in the classroom up to this point?
   - Is it more important that students practice and demonstrate using a specific CCC or that they show they can flexibly use one of many CCCs to engage in sensemaking in ways they choose?
   - What would it look like to elicit understanding of a focal CCC for a phenomenon students haven't yet encountered?
   - What does this imply for how to elicit this thinking—what kinds of information and cueing should students be provided with?
4. **Create intentional uncertainty in the assessment scenario.** Based on which aspects of student thinking are the primary focus, direct students' attention to some uncertainty in the phenomenon or problem that is most closely related to the CCC you seek to assess. Consider:
   - What is puzzling, intriguing, culturally relevant, or globally significant about this phenomenon?

- How can the story, images, and data presented cue students to start using the targeted CCC to present that uncertainty and get students to think more deeply?

5. **Provide opportunities for professional learning of assessing the CCCs.** Consider:

   - What ideas do teachers bring about CCCs and assessment that can be elicited at the outset of a professional learning series?
   - How can learning opportunities be sequenced to build teachers' confidence and understanding over time?
   - What tools are needed to help teachers analyze, adapt, and design assessments?

Summary

In this chapter, we have sought to illustrate how CCCs are essential tools for students to make sense of novel phenomena and how that perspective can inform the design of assessments. We find that the key opportunity presented by the CCCs is to allow students points of entry to explain new and puzzling situations, which, in turn, can provide teachers with vital information about what students know and can do.

We collectively emphasize that crosscutting concepts in the *NGSS* must be highlighted in assessment design, in professional learning experiences with teachers, and, integrally, in teaching and learning experiences with students. If we expect students to learn to identify and leverage crosscutting concepts, we must use the language of the CCCs in our interactions with them and encourage students to see these connective threads—as well as their progress in applying them over time—as they engage with phenomena in the process of science learning.

One thing is clear: We want to prepare students to be effective sensemakers and problem-solvers. Doing so requires that we give all learners opportunities to develop the full suite of tools they may need to make sense of novel phenomena through crosscutting concepts and to have procedures in place to track how they develop their ability to apply crosscutting concepts over time. We also need to prepare students to engage with intentionally simplified and scaffolded "school science," as well as the increasingly uncertain and ambiguous scenarios they will encounter in the real world. Finally, we want to be able to (1) signal and provide feedback about progress toward the features of learning and performance that will be necessary for student success and (2) help make sure that all learners are supported in learning the CCCs.

Because it is becoming increasingly clear that a firm grasp of the crosscutting concepts is imperative for confident learners who are prepared to dive deeply into new phenomena and problems they encounter, foregrounding the crosscutting concepts is a necessary component of three-dimensional assessments. Failure to do so limits how well educators can support learners in becoming adept and flexible sensemakers and problem-solvers. However, a conscious inclusion of the CCCs in instruction and assessment promotes equity for all learners. That is, explicitly highlighting and providing feedback will enable all learners to develop better understandings of CCCs.

Acknowledgments

This material is based in part on work supported by the National Science Foundation (Grant Nos. DRL-1561751 and DRL-1748757).

References

Coffey, J. E., D. Hammer, D. M. Levin, T. Grant. 2011. The missing disciplinary substance of formative assessment. *Journal of Research in Science Teaching* 48 (10): 1109–1136.

Eisenkraft, A. 2016. Teaching about energy as a crosscutting concept. In *Teaching About Energy Across the Sciences*, ed. J. Nordine, 39–60. Arlington, VA: NSTA Press.

Fine, C., and E. M. Furtak. 2020. Toward framework-based design criteria for classroom assessment of emergent bilingual learners. *Science Education* 104 (3): 393–420.

Fitzgerald, M. S., and A. S. Palincsar. 2019. Teaching practices that support student sensemaking across grades and disciplines: A conceptual review. *Review of Research in Education* 43 (1): 227–248.

Furtak, E. M., K. Kiemer, R. K. Circi, R. Swanson, V. de León, D. Morrison, and S. C. Heredia. 2016. Teachers' formative assessment abilities and their relationship to student learning: findings from a four-year intervention study. *Instructional Science* 44 (3): 267–291.

Kentucky Department of Education. n.d. Science assessment system through course task: Airbag effectiveness. *https://education.ky.gov/curriculum/conpro/science/Documents/Airbag_Effectiveness_TCT.pdf.*

National Academies of Sciences, Engineering, and Medicine (NASEM). 2017. *Seeing students learn science: Integrating assessment and instruction in the classroom.* Washington, DC: National Academies Press.

National Research Council (NRC). 2012. *A framework for K–12 science education: Practices, crosscutting concepts, and core ideas.* Washington, DC: National Academies Press.

National Research Council (NRC). 2014. *Developing assessments for the* Next Generation Science Standards. Washington, DC: National Academies Press.

National Science Teaching Association (NSTA). 2013. NGSS@NSTA Hub. *https://ngss.nsta.org.*

NGSS Lead States. 2013. *Next Generation Science Standards: For states, by states.* Washington, DC: National Academies Press. *www.nextgenscience.org.*

Odden, T. O. B., and R. S. Russ. 2019. Defining sensemaking: Bringing clarity to a fragmented theoretical construct. *Science Education* 103 (1): 187–205.

Park, M., and X. Liu. 2016. Assessing understanding of the energy concept in different science disciplines. *Science Education* 100 (3): 483–516.

Penuel, W. R., and K. Van Horne. 2016. Prompts for integrating crosscutting concepts into assessment and instruction. *STEM Teaching Tools. http://stemteachingtools.org/assets/landscapes/STEM-Teaching-Tool-41-Cross-Cutting-Concepts-Prompts.pdf.*

Penuel, W. R., A. S. Lo, J. K. Jacobs, A. Gardner, M. A. M. Stuhlsatz, and C. Wilson. 2019. Tools for supporting teachers to build quality 3D assessment tasks. Paper presented at the Annual International Meeting of NARST, Baltimore, MD.

Penuel, W. R., K. Wingert, and K. Van Horne. 2018. Preparing teachers to notice key dimensions of next generation science assessment tasks. In *13th international conference of the learning sciences, Volume 2*, eds. J. Kay and R. Luckin, 1215–1217. London: International Society of the Learning Sciences.

Rivet, A. E., G. Weiser, X. Lyu, Y. Li, and D. Rojas-Perilla. 2016. What are crosscutting concepts in science? Four metaphorical perspectives. In *Transforming Learning, Empowering Learners: The International Conference of the Learning Sciences (ICLS) 2016, Volume 2*, eds. C. K. Looi, J. L. Polman, U. Cress, and P. Reimann, 970–973. Singapore: International Society of the Learning Sciences.

Russ, R. S., and M. J. Luna. 2013. Inferring teacher epistemological framing from local patterns in teacher noticing. *Journal of Research in Science Teaching* 50 (3): 284–314.

Schwartz, D. L., and J. D. Bransford. 1998. A time for telling. *Cognition and Instruction* 16 (4): 475–522.

Shepard, L. A., W. R. Penuel, and J. W. Pellegrino. 2018. Using learning and motivation theories to coherently link formative assessment, grading practices, and large-scale assessment. *Educational Measurement: Issues and Practice* 37 (1): 21–34.

Wiliam, D. 2007. Keeping learning on track: Classroom assessment and the regulation of learning. In *Second handbook of mathematics teaching and learning*, ed. J. F. K. Lester, 1053–1098. Greenwich, CT: Information Age Publishing.

Chapter 16

The Role of Crosscutting Concepts in Teacher Sensemaking and Empowerment

Emily C. Miller and Tricia Shelton

We begin this chapter by acknowledging that teachers inhabit a precarious place in discussions around STEM education. As teachers, we are well positioned to play a major role in addressing inequities in STEM education; however, our agency to do so is often limited. Standards, assessments, and often curricula are handed to us without our input and usually without permission to modify the documents according to local contexts or the students in front of us. Our trajectory of professional learning often consists of replicating experts' ideas of classroom implementation.

We propose that teacher engagement in collaboration, reflection, and ultimately teacher learning—all elements of sensemaking—benefit the *building of theory* around three-dimensional learning and crosscutting concepts (CCCs). Teacher intellectual work is fruitful for the conceptualization of student science learning for the field. This chapter argues that CCCs and the tenuous understanding that surrounds them present the perfect opportunity for fostering structural change. The field's demand to better understand CCCs in the classroom can be leveraged for teachers to make critical choices, actualize those choices, and interrogate larger social structures that have impeded teacher choice.

We ask this question: As teachers are uniquely and intimately placed *within the dynamic and impromptu action* of student science learning, could the academic field expand to encompass and capitalize on teachers' perspectives and experiences? We believe this expansion would enable a deeper understanding and improve the application of CCCs of the *Next Generation Science Standards* (*NGSS*; NGSS Lead States 2013).

Organization of This Chapter

First, we define key terms to aid the exploration of our question. We define *sensemaking* as collaboration and reflection, which, together, give rise to teacher learning. Next, we introduce the dimension CCCs and argue that this dimension of the *NGSS* creates opportunity for teacher input and opportunity for teachers to benefit from providing that input. We present two vignettes to illustrate some possible scenarios for teachers to build pedagogical and theoretical understanding about CCCs. The first vignette highlights teacher sensemaking about their own learning of CCCs in order to ask questions about student learning, and the second focuses on teacher sensemaking around artifacts and lesson plans. In the two vignettes, all teacher participants engage in sensemaking through collaboration, reflection, and learning. We argue that the sensemaking activity is empowering for teachers—as well as facilitators, administrators, and researchers. Last, we consider the implications for CCCs, teacher collaboration, reflection, and learning in moving toward equity in STEM education and across sociostructural barriers.

The *Next Generation Science Standards* and Upheaval

The *NGSS* shifts ideas about science teaching and learning and provides the perfect state of upheaval that enables sensemaking and creative solutions (Maitlis 2005). The *NGSS* builds on documents that move toward conceptualizing the practice side of science understanding. They move away from emphasizing facts and procedures and toward engagement in practices and lenses for viewing the natural world, which scientists employ. The *NGSS* is designed to create opportunities for students to actively engage in knowledge construction—to be "doers" of science, rather than "receivers." In short, the *NGSS* provides an opportunity for educators at all levels to develop learning environments in which students have epistemic agency—to be positioned as able to shape the knowledge and practices of their classroom community—in their knowledge construction.

However, the *NGSS* have not yet been implemented to provide an opportunity for *teachers* to be positioned as shapers of pedagogical knowledge and practices in their science classrooms. There is a "chain of delivery" for the approach and agency implicated by the *NGSS*—from academic researchers to standards developers to curriculum developers to teachers to students. To fully achieve the aims of the *NGSS*, all the links within this chain must be afforded some level of agency, and we can recognize that teachers are a vital link within this chain of *NGSS* delivery.

In this chapter, we present a view of teacher sensemaking about teacher practice during *NGSS* enactment, where there is less emphasis on the "right answer" and more emphasis on making use of teachers' experiences, diverse perspectives, and intellectual resources. We define *scientific sensemaking* as the dialogic activity of searching for meaning and coherence using science and engineering practices (SEPs) for explaining phenomena and solving problems (Ford 2012; Gupta and Elby 2011). For teachers, sensemaking in professional learning is often directed and channeled. Of course, teachers also engage in collaboration, reflection about practice, and learning without the presence of a professional learning facilitator. Teacher collaboration, reflection, and learning are three elements of teacher sensemaking—the dialogic activity of searching for meaning and coherence to solve pedagogical problems—which build on our knowledge of how people change practices in sociocultural contexts.

Teacher Collaboration

Schoolwide change initiatives often constrain the topics of teacher collaboration from the top down, while teachers continue to collaborate on what they find important and engaging in brief moments of free time. Teacher collaboration is considered an essential component in effective schoolwide improvement of professional learning and student performance (e.g., Buxton, Carlone, and Carlone 2005). Collaboration involves organized, intensive dialogue that is increasingly precise around teacher practices, students' needs, and related goals. According to Warren et al. (2001), collaborative activity among teachers and students requires recognizing all participants as intellectuals who are co-learning difficult topics. Similarly, collaboration among teachers, and between teachers and researchers, requires the same components: Teachers and researchers share respect and responsibilities toward co-learning in difficult problem spaces. Is it possible to imagine a space for teachers that contains authentic collaboration based on intellectually demanding, critical, and sometimes uncomfortable dialogue that evaluates ongoing practice?

Teacher Reflection

Teacher reflection includes in-the-moment pedagogical decisions during teaching, as well as metacognitive, retrospective analysis after teaching (Larrivee 2008). Instead of focusing solely on more immediate results, teacher reflection is a lengthy and more reiterative process by which teachers work and rework curricular approaches. Effective communities create environments that bring about and sustain the personal and professional discomfort that accompanies and reinforces reflective practice and cultural change (Alaimo 2008). As such, reflective practice seeks to examine uncertainties surrounding critical questions about equity and diversity. In terms of reflective collaborative practice, authentic reflection requires all participants to explore vulnerabilities—this includes teachers, and if present, researchers and administrators (Carlone, Haun-Frank, and Kimmel 2010). We see the definition of *reflection* as inclusive of sensemaking because it involves the solving of pedagogical problems.

Teacher Learning

Simply put, teacher learning is change in practice based on reflection, possibly over sustained periods of time and collaborative inquiry (van Es 2012). Teacher learning involves dynamic planning, inquiry, and action processes. When people are professionally and personally vulnerable as intellectual equals, we believe learning occurs as (1) deep collaborative and reflective practice, leading to transformed teacher practice in classrooms and (2) transformative activity that empowers both teachers and students.

Overview of Two Vignettes: Introduction to the *Next Generation Science Standards* and Crosscutting Concepts

This chapter uses two vignettes based on real events. They highlight teacher collaboration and reflection using CCCs as the lever for soliciting teacher intellectual resources and transforming traditional practices of the educational community.

A primary goal of the *NGSS* is to shift science teaching and learning through clarifying that knowing and doing science involves three dimensions. The *NGSS* describes engaging in scientific thinking as a process that consists of applying disciplinary core ideas (DCIs), science and engineering practices, and crosscutting concepts. The three dimensions work together to fully describe the process of scientific thinking. With the *NGSS*, students develop an understanding of how to explain and describe phenomena and solve design problems. This application of three dimensions is a departure from previous standards, which included more emphasis on scientific end-points such as facts, procedures, and theories. The *NGSS* focuses on bringing authentic scientific processes into the classroom, helping students understand how knowledge is attained and critically evaluated, thereby more closely resembling scientific inquiry in the science, technology, engineering, and mathematics (STEM) subjects and careers. Arguably, the three dimensions build on reform documents from earlier standards, including Science for All Americans (AAAS 1990) and the *National Science Education Standards* (NRC 1996). The novel aspect of the *NGSS* is the theory of three-dimensional learning and a greater emphasis on the CCCs and on bringing the processes of science into classrooms.

Scientists and engineers work to understand and explain phenomena and solve problems by leveraging CCCs as a lens through which to ask questions about the data and to consolidate and coordinate group perspectives. Although the utility of CCCs by scientists has been documented and assessed (NRC 2012), their potential to enhance learning in the classroom has not been rigorously evaluated and tested in the same way as the other two dimensions of the *NGSS* (Osborne, Rafanelli, and Kind 2018). CCCs are described as, at best, under-operationalized (Osborne, Rafanelli, and Kind 2018). Assuming CCCs do not provide the unifying framework across science disciplines, as it has been argued (e.g., Osborne, Rafanelli, and Kind 2018) or simply demand further operationalization, we explore the potential for CCCs to enrich conversations, sensemaking, and theory building by teachers.

As stated earlier, in regard to sensemaking, teachers must be engaged as co-intellectuals in meaningful endeavors where uncertainty and discomfort are valued in the problem space. We use the following vignettes to show that, within the endeavor of meeting the *NGSS*, the CCCs can be leveraged to promote a sensemaking environment for teachers. We offer the following two very different vignettes to illustrate the rich engagement that is possible with teachers' sensemaking around changes in teacher practice and building pedagogical theory.

In both vignettes, we stray from the traditional design of teacher sensemaking in professional learning and professional learning community (PLC) settings, instead mirroring the sensemaking discussions we aim for with students in our classrooms. The sensemaking focuses on soliciting and leveraging intellectual resources and experiences for an authentic purpose. Thus, both vignettes illustrate designed environments for collaborative sensemaking about three-dimensional learning with diverse groups of teachers. In addition, in both vignettes, we positioned teachers as the experts. For proof of concept, we propose that

1. teachers can use practice to build theory,
2. the activity of building theory is empowering to teachers, and
3. the resulting theory is useful for the field.

Introduction to Vignette 16.1

The first vignette involves elementary and middle school teachers from one state involved in the collaborative activity of figuring out CCCs and their usefulness as tools in explaining and predicting phenomena. One of the authors, as a researcher/practitioner, partnered with a science consultant from a state department of instruction. This vignette is loosely based on work we did in the state to consider science questions such as modeling progressions, language in science, and CCCs. The following vignette includes quotes that are representative of the discussion. Both authors agree that partnering with practitioners provides excellent opportunities to further evaluate the use of CCCs for sensemaking in the classroom while also empowering teachers and potentially increasing *NGSS* uptake. However, we saw the paucity of research on CCCs as an opportunity for practitioner empowerment. By leveraging teacher practice within a partnership, we brought teachers into the larger conversation as valued and expert contributors.

VIGNETTE 16.1
Teacher Collaborative Sensemaking About the Crosscutting Concepts' Context

Approach

Our aim was to explore the potential for CCCs to enrich conversations and sensemaking in classroom settings by working with groups of teachers representing districts in urban, suburban, and rural settings. We expected the conversation to be practice-oriented; just as sensemaking brings questions to the forefront and embraces areas of confusion, we expected to embrace our own uncertainties. Teachers spend a great deal of time with the *NGSS* trying to understand others' interpretations of the documents and apply those interpretations to contexts. We intended to interrogate the ideas that had been proposed so far by standards committees, who tend to be somewhat more removed from day-to-day classroom practice, and to seek diverse and creative thinking that was grounded in practice and exploration of science teaching, reflection on teacher practice, and student learning.

To accomplish this end, my partner and I (Miller)—who are, as stated, two science teachers from the middle and elementary level—designed an Earth science phenomenon experience for the teachers to engage in. We asked the teachers, "What do the CCCs mean in the effort to explain and predict the phenomenon? How do the CCCs transform science learning? What questions can we take up and explore in our classrooms?"

We recruited eight teachers from four different districts across Wisconsin in urban, rural, and small-town settings. These community settings further represented some of the state's geographic and socioeconomic variability: a south-central city, a small resort town in the far north, a small city on the eastern lakeshore, and west-central farmland towns. Finally, we sought representation of three elementary grade bands: K–1, 2–3, and 4–5. The *NGSS* includes only two grade bands at the elementary level, but we felt that the differences between grades K and 2 are broad enough to support enriched conversation.

Phenomenon: Flooding Event

My teaching partner and I planned an initial engagement in a phenomenon to inform our discussion of CCCs and how to support students with CCCs. We selected a phenomenon based on the potential for motivating flexible thinking—the purposeful use of the dimensions as having affordance for approaching the unfamiliar phenomenon

Continued

Vignette 16.1 *(continued)*

(Miller et al. 2018). The phenomenon is a multilayered and rich event. By "multilayered," we mean there are many DCIs and CCCs that can be used for sensemaking; by "rich," we mean there are various possible accurate understandings of that event that range from straightforward to sophisticated (Miller et al. 2018).

The flooding event is an Earth science phenomenon, but physical science and life science can be applied to the attraction of the particles to the water, the behavior of water, and the saturation of a natural ecosystem. I had previously used the experience with teachers. In the past, this phenomenon enabled conversation about science that incorporated life experiences, as well as cultural and linguistic intellectual resources from school, family, and community.

We began the experience with a picture of a flooded field and a girl playing in a field in a raincoat. We asked the teachers to describe what they noticed and what they wondered about the picture. We followed up with prompts, including, "Tell me more about what you are thinking," and "Why are you asking that question?" All of the teachers offered various evidences of science understanding through the questions they asked, all the while underscoring the diversity in our group:

- "Are the hills around the field close by?"
- "Is the girl surprised by the flooding?"
- "Are there puddles in some areas and not in others?"
- "Is there a river nearby or water seeping underground?"
- "How many different kinds of plants are in the field?"
- "I wonder if the field is fallow, or is it usually used for grazing?"

It was clear that, even in this initial engagement with the phenomenon, teachers brought in various experiences to make sense of the photo and ask questions. We recorded every question that the teachers asked.

We asked the driving question we had developed for the phenomenon-driven activity: "How are the Earth materials and water interacting so that it floods?" We were interested in having the group develop predictive models; thus, we created an opportunity to consider how CCCs were used in figuring out the phenomenon and to explore opportunities for student learning.

We passed around plates with sand, potting soil, and soil collected from a yard. The teachers examined the materials and described what they noticed. We asked the teachers to make predictions about "water-holding" capacity for each sample and offer explanations for why some materials hold water and others allow water to pass through quickly.

Continued

Vignette 16.1 *(continued)*

Predictive Models for Which Material Will Hold the Most Water

After discussing their initial thinking, we asked the teachers to develop predictive models to show their thinking about which material would have the greatest capacity to "hold" water. We asked them to show the mechanisms that created that capacity.

Teachers worked on their models and came up with various predictions. Sharia and her partner Pam suggested that the sand would block the water from seeping through. Pam explained her reasoning. She cited experiences on the waterfront of Lake Michigan and knowledge of the streams that lead to the lake. "You know how you can watch a wave pass over the sand? It rolls over the sand, and it moves some of the sand in the water, but slides right over the beach." She showed with her hands the waves retreating from the beach. "The sand is almost like a protective coat of paint! It doesn't let the water pass through." I asked if their point of view made sense to the others. Kristin volunteered, "Well, we do understand that point of view because it does look like the wave rolls over and over and over, and the sand on the beach barely moves. But we were thinking this was more of a function of the structure of the particles of sand. We were wondering if the shape matters. The particles are smaller, and as we saw in the evidence, they are more similar to one another than the particles in the soil and in the potting soil …" At this point, another teacher jumped in, "Can we get consensus on what is meant by particles? I mean, thinking about the classroom, we would definitely want a definition."

The team came to agreement on the word *particles* and continued working through their reasoning. "If there are many more similar particles on one hand, compared to jagged and mismatched pieces on the other, we were thinking of a jigsaw puzzle. Nothing would get past the particles that fit together so perfectly; at the same time, the jagged pieces would have more spaces between the particles, not only to let water pass through but also to hold the water trapped between them," remarked the fourth-grade teacher, Amber.

After a long discussion about the predictions of these teacher partnerships, which were similar even as the reasoning differed, the partnership from the small town brought up a different prediction altogether. They thought the function of the particles was most important. They reasoned that the potting soil was supposed to hold water—that the particles were meant to absorb water and hold on to moisture. Thus, it was the potting soil that would hold the most water. The sand did not hold plants, and plants had a hard time growing in sand, so this meant there was evidence that the water passed through the sand, probably the fastest.

The final group to share focused on the number of particles in one space. They imagined that the material with the largest particles also had the fewest particles,

Continued

Vignette 16.1 *(continued)*

and this material would be the easiest for water to pass through. (Another teacher interjected with, "But the organic material in the soil will expand, won't it?") For comparison, they brought up a crowd of people standing in a space such as a street mall. If you want to get through the crowd of people, it would be more difficult with more people gathered there; if there are fewer people, it is easier to pass through.

Collaborative Discussion About CCCs

All partnerships applied different CCCs, which allowed different ideas to come out, and the discussion was very rich. We presented our sensemaking activity as an explicit opportunity to reflect on the practice of CCCs and to inform the resulting teacher practice, as well as the three-dimensional assessments for the statewide discussion.

Carina suggested, "I am imagining ... that when scientists get together, let's say to figure out if an area is going to flood, they may have the same or different focuses. And I understand that there is value in having the discussion about what questions are important. But there's a good purpose for NOT constraining the conversation at first. Maybe with a wide-open approach to the CCCs, the scientists get more ideas on the table, just like we did!"

Pam added excitedly, "Also, and really importantly, the different CCCs that came out represented the diversity in the room." Sharia agreed, saying, "Right, it was the different views we come with that caused us to approach the problem differently. You guys didn't have as much of a system's view of water and land, because you don't see the interaction between the lake and the land every day, like we do. So, if I had insisted, as the teacher, that the class adopt a certain CCC from the beginning, I would have excluded some of you!"

Many in the room nodded. Mary said that she saw the usefulness of the open-ended, less structured approach to CCCs as giving rise to more ideas—and ideas that they might not have anticipated.

Kristin offered a connection to literacy: "It's sort of like phonics versus whole language. We can insist that the students adopt a certain phoneme, even if they don't use that phoneme because they speak Spanish, or we can emphasize the purpose of reading and writing, and return to the phonemes after the purpose has been realized. In this case, the purpose of explaining an event is the most important aspect, and then the pieces, or the phonemes, are analyzed with the teacher's help. This is really interesting because we can scaffold both phonemes and particular CCCs, but then we maybe end up with a less rich text, like 'Dan the man has a tan van,'" she laughed.

Continued

Vignette 16.1 *(continued)*

The teachers in the group decided to return to their classrooms with two lessons that approached the use of CCCs differently. One would be an open-ended lesson similar to the one we employed together, and the other a scaffolded lesson, where they specifically directed thinking based on one CCC. They would return to the group with data on how many students were engaged, and how the CCCs supported the understanding of the natural phenomenon.

Interest in Assessment

When some of the teachers reconvened, they engaged in pedagogical argumentation about the use of the CCCs. Two participants thought the more structured approach to a CCC both required and built language. They used their classroom examples to show that when they explicitly taught the language of the CCC, they would see it in the students' writing. Both teachers who promoted this line of reasoning had examples from students' science writing to back up their claim.

Amber, the kindergarten teacher, said she found that a more open-ended approach to the CCCs enabled creative thinking. She described a lesson where the students were trying to make an ice cube melt. Because she did not give them very many instructions, and they had an open-ended problem, the students all tried different techniques to solve the problem. They knocked ice cubes against each other, put them in water, wrapped an ice cube in a paper towel, broke the ice cube apart into pieces, and rubbed it in their hands. When she asked her kindergarteners to explain their thinking, the partners used many of the CCCs in their explanations: cause and effect, energy and matter, stability and change, and systems and system models. She was thrilled with the lesson. She said that all of her "kinders" gained from testing and reporting out on the different ideas. She also thought there was a benefit to sharing the explanations because the explanations supported the larger group in understanding a variety of CCCs.

My co-presenter, Kyle, and Mary built on Amber's story to discuss assessments of the CCCs. They seemed almost wary about having the audacity to suggest what they would like to see in *NGSS* assessments, laughing and looking around conspiratorially. "I think it shouldn't be vocabulary in the assessment. I would like to see an assessment where there is a scenario like the ice cube problem and the student who is taking the test would have to identify the CCC. That is a skill, right? Figuring out what CCC is being applied?" Mary spoke up, "I would almost prefer to see an open-ended question, like with the ice cube, where you would see the students try different ideas and then attach those approaches to the CCCs." Pam responded, "Yes, it would be good if it isn't a test where the student is steered to say one right answer—that would take the authenticity and engagement out of science!"

Conclusions From Vignette 16.1

This vignette illustrates that (1) teachers can use practice to build theory of three-dimensional teaching practice, (2) the activity of building theory is empowering to teachers, and (3) the resulting theory is useful for the field. Teachers identified two separate pedagogical theories in teaching CCCs and discussed the pros and cons of each application, either structured work around one CCC or more open-ended inquiry that is then connected to a wider range of CCCs. They determined some of the costs and benefits of each approach and proposed ideas about related approaches on assessment, as well. The group of diverse teachers collaborated to come up with various ways to approach CCCs. They developed emerging approaches after comparing their own uses of the CCCs for the flood event, and they also tested ideas for using the CCCs in the classroom. The teachers were not required to try out these questions; rather, they were inspired to reflect on their teaching, and they were interested in coming back together to hear what others found.

Through being part of the initiative to develop their own questions and build their own theoretical conjecture (Sandoval 2004), teachers were empowered to action. They used the CCCs to discuss equity in their own classrooms and consider the diversity of voices that were being portrayed through an open-ended activity. It is important to note that many of the ideas that the collaborative group wrestled with were precisely the questions discussed among administrators, science curriculum designers, and large-scale reformers at the state and national agency levels. However, other remarks, such as preference for open-ended questions about CCCs on assessments, diverged from reform documents such as the EQuIP rubric (*www.nextgenscience.org/resources/equip-rubric-science*).

If similar collaborative efforts were to occur across multiple scales—from local to national—researchers, teachers, and students would benefit from intellectual and social rewards while also helping to improve science education. Researchers would provide expertise to develop curriculum; teachers would provide expertise, observations, and feedback around what does and does not work in real-world classroom contexts; and students would benefit from curriculum rooted in both sound science and proven practice.

Introduction to Vignette 16.2

The learning community in this second vignette comprised eight educators who were working together to deepen their understanding of *NGSS* design and pedagogy to support its implementation. An essential element of the community was an environment of trust for dialogue and reflection around their own classrooms using third-point references, like *A Framework for K–12 Science Education* (the *Framework*; NRC 2012). This community could be replicated in many settings, including school-based PLCs and online communities.

This vignette portrays the empowerment of teachers through discussion of teaching practice using CCCs as the instigator and using reflection as the primary lever in shifting instructional approaches in implementing the *NGSS*. Educators at the high school level engage in a sensemaking discussion about CCCs through inquiry around lesson plans and student work. This discussion includes many science educators in a professional learning project about the *NGSS*. This vignette emphasizes reflective practice but is collaborative in nature, and the space is questioning preconceptions about pedagogy. The colleagues analyze one teacher's lesson plan about life science structures and processes and the related student work, looking for evidence of CCCs. They ask consequential questions about appropriate scaffolding and the explicitness versus implicitness of CCCs as tools for making sense of phenomena. Ultimately, they develop an understanding of CCCs as tools that enable the exploration of phenomena. This second vignette contrasts with the first because it is focused only on teacher practice and demonstrates how sensemaking elicits thinking and then overturns preconceived ideas about CCCs. This vignette is also an amalgamation of numerous actual events and conversations that included the authors.

Eight science educators participated on a collaborative call to provide feedback to a high school teacher, one of this chapter's authors, who designed a unit aligned to the *NGSS*. After teaching the unit, the teacher sought reflective collaborative discussion for multiple reasons. The teacher sought to improve student sensemaking in her classroom and shift her teaching practice. One of the goals of the group was to facilitate shared learning about the *NGSS*, especially the CCCs. The following conversation focused on opportunities for students to develop and use the CCCs to make sense of phenomena and develop solutions to problems. The work specifically targeted performance expectations, and, as in the earlier vignette, left open the possibility of developing additional CCCs that were not part of the targeted performance expectations. We selected this vignette because it illustrates an evolutionary step in thinking about the CCCs—from simply an organizing principle in the *Framework* to an effective connecting tool and *a tool for organizing and deepening one's thinking*, not just across disciplines but also across a unit. Moreover, the group expanded their thinking to include the use of CCCs in making sense of phenomena and developing science understanding.

VIGNETTE 16.2
Teacher Reflection on Evidence of Crosscutting Concepts in Practice

Approach

In this discussion across multiple school districts, collaborative conversations between the teachers and researchers in the PLC focused on *implicit* versus *explicit* use of the CCCs by students and *implicit* versus *explicit* evidence of opportunities for students to use CCCs within lesson and unit plans. Nevertheless, the teachers and researchers struggled to determine what *explicitness* around the CCCs looked like and how they could provide scaffolding for students. They asked, "What does scaffolding of CCCs look like in instruction?" The following phone call among the eight science educators details a reflective effort with two objectives: (1) finding evidence of the explicit use of the CCCs from the *instructional plan,* provided in advance of the call, and (2) finding evidence of the explicit use of the CCCs from *student work.*

High School Football Water Intoxication Case Resulting in the Phenomenon of Multiple System Failure

The high school teacher who requested the call had designed and taught a unit featuring a case in which a high school football player died from drinking too much water. The teacher chose the phenomenon of water intoxication because it was an accessible, interesting, and relatable topic for high school students during the hot months of football season in the South. The phenomenon made students wonder how this event could happen. The answer to this question would involve multiple layers of complexity. For example, students would need to use multiple elements of various disciplinary core ideas to explain what happened to cause the football player to die and evaluate the mechanistic accounts based on reasoning and evidence. The three-week unit targeted two performance expectations—HS-LS1-2 and HS-LS1-3—and it called for two CCCs: systems and system models and stability and change. The teacher and writer of the unit felt that the CCC of structure and function would also help the students make sense of the phenomenon. This unit would serve as the first unit of the year and lay the foundation of sensemaking in the classroom.

Anna began the call: "In order to explain the phenomenon, students are seeking answers to their own questions by blending these three dimensions: the disciplinary core idea LS1, the science and engineering practices of Developing and Using Models and Planning and Carrying Out Investigations, and the crosscutting concept of systems and system models."

Continued

Vignette 16.2 *(continued)*

Brigit disagreed: "I focused my evidence on a different crosscutting concept. My evidence references the lessons toward the beginning of the unit, where students were expected to frame their thinking using the crosscutting concept of structure and function in order to develop an understanding of the hierarchical nature of systems, or LS1. However, I found that there was no evidence of differentiated instruction around the crosscutting concepts in the instructional plan."

A rich and reflective discussion followed Brigit's statement. The participants asked, "What does it mean for students to use and develop crosscutting concepts to explain phenomena?" and "What does it mean to provide instructional supports or differentiation for student use of the crosscutting concepts?" To resolve these issues, Brigit suggested discussing student work to add to the conversation about whether there was evidence of student use of CCCs to figure out the phenomenon and whether or not instructional supports or differentiation were present. At this point, the call facilitator asked the teacher and writer of the unit to share student work. In sharing and discussing the student work, interesting clarifications around CCCs developed.

John said, "I realize something important about my own approach to this question. When I looked for evidence of opportunity for students to develop and use the CCCs in the instructional plan, I was looking for buzzwords about the targeted CCC of systems and system models. I was looking for this same evidence when I noticed the instructional plan called for students to use a CCC that is not called out in the targeted performance expectations: structure and function. So, I notice *how* I was looking for support of the CCCs. I was using my district definition of what counts as differentiation. In my district, we support students instructionally in certain ways, like providing graphic organizers or planning tiered learning where students are offered varying levels of complexity or depth of knowledge."

John paused and continued slowly and thoughtfully forming his words, "Now that we have discussed evidence of student learning, I have changed my thinking in a big way! I can see that the evidence of crosscutting concept use by students and the differentiation that happened was on a *metacognitive* level. For example, the student work shows that students were using the crosscutting concept of structure and function to organize and deepen their thinking about the science ideas being built. The student work is where we look for this. It shows that all students were able to use the organizing framework of structure and function to think about the phenomenon and learn the target science ideas."

Anna jumped in. She was also excited by John's finding and saw an opportunity for calling out diversity. "The in-the-moment questions chosen by the teacher to support student sensemaking were what differed among student work. This was the

Continued

Vignette 16.2 *(continued)*

differentiation! We also can confidently claim that the student work showed individual student thinking because all of the models were a bit different. The models all showed that students used the crosscutting concepts, but students represented this thinking differently. For example, one student used scale to show how one system malfunctioned, contributing to the football player's death, whereas another student used the crosscutting concept of structure and function to explain the importance of all systems working together for the organism to function. The teacher asked different questions to support students in organizing their thinking, using the crosscutting concepts in a way that made sense to them while also moving everyone toward the explanation of the phenomenon."

John agreed, saying, "The teacher questions and support for the process, like sentence stems and prompts, clearly gave students a structure to organize their thinking throughout the unit, but it was *their* thinking." He continued, "I have a whole new way to think about the crosscutting concepts now—a metacognitive lens on both the student and teacher level."

All of the participants discussed thinking about CCCs on this metacognitive level. The group consensus was that it was critical to consider student work when looking for CCCs in a lesson. Their original doubt disappeared, and they agreed there was evidence that the students developed and used the targeted CCC of systems and system models, as well as the additional CCC of structure and function, chosen because of its sensemaking power in the context of this phenomenon. The teachers agreed that differentiated instructional supports delivered through in-the-moment teacher questioning helped the students attend to their thinking through the lens of a CCC and make sense of the phenomenon.

This collaborative call and the discussion with colleagues led to additional learning, reflection, and changes in the teachers' science instruction as they began a new semester with new students. The teachers wanted to test out this new thinking in the new semester. The goal was to determine if instructional supports for students in the development and use of multiple CCCs would empower them to use the CCCs *by choice* to make sense of phenomena or to design solutions to problems.

The call concluded with some further discussion about trying out instructional supports for CCCs. The teacher and writer of the unit shared some of the instructional supports she found effective. She said that early in the focal lesson, she provided opportunities for students to hear her *think aloud* about her use of the CCC of structure and function and how it helped her understand a system and the interactions of the system's components. For example, she described to the students the types of questions she asked herself (questions framing thinking using the CCCs)

Continued

Vignette 16.2 *(continued)*

and how these questions helped her think about the phenomenon and how it happens. Her students then had an opportunity to use a similar approach to develop and use these CCCs as a tool to make sense of a phenomenon.

She described how this think-aloud practice, in combination with her in-the-moment questions to students, had supported her students in sensemaking using the crosscutting concepts, as evidenced by the student work. Her intentional questions during this student sensemaking time supported students with varying levels of sophistication in their understanding of and proficiency around the CCCs, and, thus, the questions moved their learning forward. She said the students organized evidence using the CCCs in their explanations. They connected the evidence to build an evolving science understanding and to explain how the phenomenon happened.

Developing Theory: Crosscutting Concepts as Thinking Tools

The teachers returned to teaching with changed understandings about CCCs after the professional learning phone call. They were armed with their new plan to test with their students. Their goal was to support students in the development and use of multiple CCCs and to empower them to use the CCCs to make sense of phenomena or to design solutions to problems. The teachers decided to focus on the following key area:

> *Crosscutting concepts connect disciplines, help students organize and think about evidence and phenomena, and get students to independent thinking.*

When the group reconvened, similar to the structure of the earlier call, the teachers agreed to determine the degree of success in employing CCCs by evaluating lesson plans and student work and also by paying close attention to student discussions. Right away, Allen provided an example of his high school students understanding CCCs as tools for thinking. He said, "One of my students described her understanding of crosscutting concepts as supporting conversation among her peers. I heard her say, 'If we get stuck in our argument and the talking stops, we can just choose a new crosscutting concept question or stem and throw it out to the group to think through our evidence, and we will be talking again—and look smart!' This was evidence to me that students had internalized and valued the use of crosscutting concepts as a way to frame thinking about phenomena and the evidence that helps them figure out the science explaining how or why the phenomena occurs."

The collaborative teacher group reflected on student articulations like the one above, evidence of student work, and observations of student learning in the next semester. They determined that many students gained independence through using

Continued

Vignette 16.2 *(continued)*

CCCs as thinking tools, but not all students. Therefore, the next investigative question for the educators became "What would be needed for this to happen for all students?" This equity focus led to an investigation of the *types* of questions the teachers were asking. Ultimately, they concluded that questions, both teacher-posed and student-initiated, can serve multiple purposes in a science classroom. When used in teacher-student and peer-to-peer interactions, questions can do the following:

1. They can be vehicles to encourage clarification of science ideas and deepen science understanding.
2. They can be used to show students that their input is valued and to support a community that builds shared practices and engages in authentic collaboration within the classroom.

Conclusions From Vignette 16.2

This second vignette underscored how reflective practice and collaborative conversation can change teacher practice, as well as disrupt structures that traditionally limit interpretive and pedagogical intellectual work to experts, to the exclusion of teachers. The teachers demonstrated the agency to try out their own theories and generalize for others. They willingly adapted their practice in order to return to the dialogue, collecting evidence and evaluating their practice as the process goal.

The teachers in the vignette used their practice to build theory, adding value to the ongoing discourse among researchers and curriculum developers. The teachers leveraged the common nexus of policy, research and practice, and working together toward common goals in science education. This example demonstrates how the educational policy and academic fields benefit from soliciting, collecting, and incorporating this dynamic knowledge emerging from the *impromptu action* of student science learning. It also emphasizes that teachers are equipped to lead educational change.

How the Two Vignettes Compare Across Collaboration, Reflection, and Learning

The two vignettes illustrate instances of sensemaking, which we define as the process of collaboration, reflection, and teacher learning that leads to changes in practice. Both vignettes underscore the benefit of working directly with teachers in collaborative settings around intellectually demanding topics, where teachers' ideas and experiences can be harnessed for authentic purposes.

The first vignette was *collaboration*-forward. This project led to a demand for teacher reflections, as well as in-the-moment consideration of the relationship between teacher practice and learning. Teachers returned with observations around changes in their practice; the discussion was invaluable in improving application of CCCs in the classroom. The vignette provided representations of teachers with different personalities, cultural backgrounds, and teaching experiences. These differences enhanced the collaborative activity, helping demonstrate both the value of different perspectives on CCCs in making sense of the phenomenon and the implications for equity (Lee and Luykx 2006). They also described an enhanced understanding of the value of student diversity, diversity of ideas, and the benefit of persistently soliciting different perspectives on CCCs from teachers and students alike. The two days of the project helped teachers view their own diverse knowledge as being valuable to the collective effort of knowledge building.

The second vignette was *reflection*-forward, with a retrospective analysis of how the relationship between teacher practice and student learning drove the conversation. The group inspected a lesson plan for evidence of the use of CCCs. Then they assessed student work to evaluate the students' uptake of the dimension of CCCs. This group formed a consensus that CCCs might be better thought of as ways to frame thinking about phenomena in order to connect evidence and evolving science understanding. In addition, the group of educators used analysis of student work, lesson plans, and student articulations to determine that differentiation through in-the-moment teacher questioning and peer-to-peer questions based on the CCCs is essential. Both outcomes informed the academic and practice communities. They ultimately solved an ongoing instructional problem of differentiation through their inquiry: teacher support of equity through questioning.

Teachers as Theory Builders

Further thoughtful discussion needs to be conducted on teacher sensemaking that privileges the collaboration between teachers for the purposes of theory building. Coburn (2003) describes discrete stages in teacher sensemaking as they interact with the *NGSS*, constructing understanding, gatekeeping, and negotiating practical and technical details. It is important to note that teachers can also build theory to inform both the reform documents and the implementation of reforms. Sensemaking can be undertaken individually, and the dialogic process involves the teacher and the writer of the document; however, we emphasize community-based sensemaking among teachers.

In order for teachers to be theory builders, time and space for reflection are essential. Professional learning communities are collaborative teams designed to foster reflective teacher-centered conversations. They place teacher learning at the nexus of practice and reflection and are focused on student learning. PLCs can be more or less structured by school-based protocol and schoolwide initiatives. There is documented success for PLCs in meeting such goals (Vescio, Ross, and Adams 2008). To make space for teachers to both contribute to and inform theory building, it is important to ensure that PLC structures are not excessively constrained with predetermined objectives. A key approach in PLCs would be to engage teachers as co-intellectuals in more open-ended, challenging, and rich endeavors. Refining practice around the CCCs offers exactly this opportunity.

We see the potential for teachers to be positioned as collaborators in theory building, with the express goal to reflect on reform documents to change practice. Underconceptualized CCCs provide the perfect springboard for engaging teachers at local levels as theory builders and agents in pedagogical reflective practice. Collaboration and reflection support learning as change in participatory practice, bolstering all levels of the chain of reform and supporting deeper learning at community and institutional levels. Unfortunately, much of the potential that teachers have to offer can remain untapped because top-down structures constrain and, to some extent, undervalue the input of classroom practitioners. The examples presented in this chapter show how the research community could reverse this by treating teachers perhaps less as subjects of study and more as long-term collaborators and coauthors in developing theory. The affordances of the *Framework* and the *NGSS* give teachers more voice and power than ever before to lead change and contribute to the building of theory. This is a trend that is promising and should be continued during rollout. Bringing teachers, researchers, and professional learning facilitators together, as equals and as sensemakers, will generate both local and broader systemic transformation.

References

American Association for the Advancement of Science (AAAS). 1990. *Science for all Americans.* New York: Oxford University Press.

Alaimo, S. P. 2008. Nonprofits and evaluation: Managing expectations from the leader's perspective. *New Directions for Evaluation* 2008 (119): 73–92.

Buxton, C. A., H. B. Carlone, and D. Carlone. 2005. Boundary spanners as bridges of student and school discourses in an urban science and mathematics high school. *School Science and Mathematics* 105 (6): 302–312.

Carlone, H. B., J. Haun-Frank, and S. C. Kimmel. 2010. Tempered radicals: Elementary teachers' narratives of teaching science within and against prevailing meanings of schooling. *Cultural Studies of Science Education* 5 (4): 941–965.

Coburn, C. E. 2003. Rethinking scale: Moving beyond numbers to deep and lasting change. *Educational Researcher* 32 (6): 3–12.

Ford, M. J. 2012. A dialogic account of sense-making in scientific argumentation and reasoning. *Cognition and Instruction* 30 (3): 207–245.

Gupta, A., and A. Elby. 2011. Beyond epistemological deficits: Dynamic explanations of engineering students' difficulties with mathematical sense-making. *International Journal of Science Education* 33 (18): 2463–2488.

Larrivee, B. 2008. Development of a tool to assess teachers' level of reflective practice. *Reflective Practice* 9 (3): 341–360.

Lee, O. and A. Luykx. 2006. *Science education and student diversity: Synthesis and research agenda.* New York: Cambridge University Press.

Maitlis, S. 2005. The social processes of organizational sensemaking. *Academy of Management Journal* 48 (1): 21–49.

Miller, E., E. Manz, R. Russ, D. Stroupe, and L. Berland. 2018. Addressing the epistemic elephant in the room: Epistemic agency and the *Next Generation Science Standards*. *Journal of Research in Science Teaching* 55 (7): 1053–1075.

National Research Council (NRC). 1996. *National science education standards.* Washington, DC: National Academies Press.

National Research Council (NRC). 2012. *A framework for K–12 science education*. Washington, DC: National Academies Press.

NGSS Lead States. 2013. *Next Generation Science Standards: For states, by states.* Washington, DC: National Academies Press. *www.nextgenscience.org.*

Osborne, J., S. Rafanelli, and P. Kind. 2018. Toward a more coherent model for science education than the crosscutting concepts of the *Next Generation Science Standards*: The affordances of styles of reasoning. *Journal of Research in Science Teaching* 55 (7): 962–981.

Sandoval, W. A. 2004. Developing learning theory by refining conjectures embodied in educational designs. *Educational Psychologist* 39 (4): 213–223.

van Es, E. 2012. Using video to collaborate around problems of practice. *Teacher Education Quarterly* 39 (2): 103–116.

Vescio, V., D. Ross, and A. Adams. 2008. A review of research on the impact of professional learning communities on teaching practice and student learning. *Teaching and Teacher Education* 24 (1): 80–91.

Warren, B., C. Ballenger, M. Ogonowski, A. S. Rosebery, and J. Hudicourt-Barnes. 2001. Rethinking diversity in learning science: The logic of everyday sense-making. *Journal of Research in Science Teaching* 38 (5): 529–552.

Chapter 17

A Call to Action for Realizing the Power of Crosscutting Concepts

Jeffrey Nordine, Okhee Lee, and Ted Willard

Based on our interactions with practitioners, we hear questions about crosscutting concepts (CCCs) from the field. Some of the commonly asked questions include the following:

- What does use of CCCs look like in the classroom?
- How do I integrate CCCs with the other dimensions during instruction?
- How does students' understanding of CCCs change over the course of their K–12 education?
- How do we assess students' understanding of CCCs?

We conclude this book by sharing our thoughts on the current state of the research and policy on crosscutting concepts and inspiring practitioners for a "call to action" to contribute research and policy based on classroom implementation. As research on CCCs is emerging as envisioned in *A Framework for K–12 Science Education* (the *Framework*; NRC 2012) and the *Next Generation Science Standards* (*NGSS*; NGSS Lead States 2013), the K–12 community of practitioners has unique opportunities to provide leadership in defining what high-quality CCC-informed instruction looks like and to contribute to the emerging research area and policy initiative.

In this chapter, we first situate CCCs in the larger context of the rapid advancements of scientific discoveries, engineering ingenuities, and technological innovations. Then we describe the role of CCCs to serve as conceptual tools across science and engineering disciplines, especially as lenses to ask questions about phenomena or problems. Next, we stress that all students bring their intuitive understanding of CCCs from their homes and communities and that teachers need to guide all students in using CCCs intentionally over the course of instruction. Finally, we propose a call to action for practitioners to play key roles in this emerging area.

Like many things in the world, science and technology are changing fast. The wealth of knowledge about the natural world has exploded in recent decades, and many of the technological devices in our lives today are simply too complicated for a single person to fully understand. In addition to generating vastly deeper knowledge, scientists and engineers are increasingly working across traditional disciplinary boundaries (Ledford 2015). Science and engineering fields like geochemistry, biotechnology, and network neuroscience are providing insights and innovations that would have been hard to imagine just a decade or two ago. It is not just that science fields are becoming more interdisciplinary—the problems facing society also require thinking across the traditional science and engineering boundaries. For example, evaluating the causes and remedies of climate change involves thinking across the life, Earth, physical, and space sciences, but nobody can be an expert in all of these fields. The most pressing challenges facing society today require an ability to think and reason about problems that involve more scientific and technical knowledge than any one person can reasonably possess. So what is to be done?

Enter CCCs. As conceptual tools that are used across all sciences, they are particularly useful as lenses to ask questions about phenomena and problems with which we are unfamiliar. No matter how complicated battery technology becomes as batteries begin to incorporate increasingly exotic materials and chemical interactions, they will always obey the conservation laws (energy and matter: flows, cycles, and conservation). Even though many new cars—particularly electric and hybrid cars—are simply too complicated for a home mechanic anymore, causal thinking (cause and effect: mechanism and explanation) may help troubleshoot minor automotive problems so more expensive repairs might be avoided. When used in conjunction with one another, CCCs can often provide deeper insight into phenomena by providing complementary perspectives that both broaden and sharpen our view on the rapidly changing world that students will inherit.

Not too long ago, the interdisciplinary science fields like nanotechnology and quantum biology didn't exist. Just 20 years ago, it would have been hard to imagine that people would soon make a living as vloggers, data miners, or smartphone app developers, but today these jobs are ubiquitous. In the coming decades, students will encounter situations and challenges they could not have prepared for during their time in school, yet they will need to face these challenges using the conceptual tools they have learned to use. As useful as CCCs are within science and engineering, their utility extends beyond these fields. A student who is well equipped to use the lenses of CCCs to gain valuable perspectives on problems we cannot yet anticipate will have a valuable set of tools to make sense of the novel challenges of the future.

The *Framework* and *NGSS* expect that all students develop the ability to use CCCs effectively. Traditionally, CCCs were not made explicit to all students and were reserved "as background knowledge for students in 'gifted,' 'honors,' or 'advanced' programs" (NGSS Lead States 2013, p. 30). In contrast, the *Framework* and *NGSS* expect that CCCs are made accessible to all students. As all students come to the science classroom with intuitive ideas about CCCs, teachers need to leverage these intuitive ideas and guide students in using CCCs to make sense of phenomena or problems.

As potentially powerful as the CCCs are for thinking about novel problems, the research base regarding how students build ideas about CCCs over time and what types of instructional strategies is still far from robust. The *Framework* recognizes this, noting, "The research base on learning and teaching CCCs is limited. For this reason, the progressions we describe should be treated as hypotheses that require further empirical investigation" (NRC 2012, p. 84).

Although research into the teaching and learning of CCCs is ongoing, the vision of three-dimensional learning is here now. Researchers and policymakers have outlined a vision for what science and engineering teaching and learning should look like in order to prepare 21st-century learners, so the K–12 community of science teachers has the opportunity to provide leadership in defining what high-quality CCC-informed instruction looks like in practice. In fact, since the publication of the *Framework* and *NGSS*, many classrooms across the country have begun to provide examples of how use of the CCCs can enhance student learning.

This book is a starting point, and we hope it is also a call to action. The research and policy communities desperately need insights and experiences that can only be gained through classroom practice, and we expect that readers will expand on the ideas contained within this book. Teachers could and should develop concrete examples of truly three-dimensional science instruction (which includes meaningful opportunities for students to learn and use CCCs). This learning is critical to developing a better understanding of how CCCs can help strengthen science learning, broaden access to science for all students, and prepare the next generation of scientists and citizens who are well prepared to tackle the novel problems and challenges in the world they will inherit.

At this point, we would like to talk to the readers directly. As you adopt and adapt these ideas into your own classroom, we encourage you to consider sharing the results of your work with the broader science education community. There are powerful venues to share your learning. For example, you can write articles for one of the National Science Teaching Association (NSTA) journals or write blog posts. Social media provides a number of opportunities. There are forums on a variety of topics in the NSTA Learning

Center where you can share ideas or ask questions. Many state science teacher associations have Facebook groups, and there is also a national Facebook group called NGSS Educators. On Twitter, there are a number of chats, such as #NGSSchat, where educators virtually gather together at a particular time to discuss a particular topic. You should not overlook more traditional face-to-face gatherings, such as state science teacher conferences and NSTA's national conferences, area conferences, and STEM Forum. Attending (and presenting!) at these events provides an opportunity to meet colleagues and share the story of your classroom's use of CCCs. The *Framework* makes clear that student learning should be a communal activity. Teacher learning should be communal, as well.

A central goal of this book is to clarify the nature and potential of CCCs to strengthen science learning, and we hope it has provided a window into the design and implementation of effective three-dimensional science instruction that leverages the power of CCCs. We summarize core messages of the book as follows:

- CCCs serve as conceptual tools, when used along with science and engineering practices and disciplinary core ideas, that provide lenses to ask questions and make sense of phenomena or problems.
- CCCs should be made accessible to all students as both an entry into science learning and an outcome of science learning.
- As CCCs are not well established in the field, there is an opportunity for practitioners to be leaders by illustrating what high-quality CCC-informed science instruction looks like and contributing to emerging research and policy initiatives.

We hope this book serves as an instructional guide to capitalize on CCCs for science teaching and learning with all students.

References

Ledford, H. 2015. How to solve the world's biggest problems. *Nature* 525 (7569): 308–311.

National Research Council (NRC). 2012. *A framework for K–12 science education: Practices, crosscutting concepts, and core ideas.* Washington, DC: National Academies Press.

NGSS Lead States. 2013. *NGSS Appendix D: All standards, all students: Making the Next Generation Science Standards accessible to all students.* Washington, DC: National Academies Press. *www.nextgenscience.org.*

Image Credits

Chapter 1

Figure 1.1: William L. Farr, Wikimedia Commons, CC BY-SA 4.0, *https://commons.wikimedia.org/wiki/File:Ranchland_with_Texas_bluebonnets_(Lupinus_texensis)_in_western_Kerr_County,_Texas,_USA_(17_April_2015).jpg.*

Figure 1.2: Authors

Chapter 2

All images come from the authors.

Chapter 4

Figure 4.1 (a): R.E.H., Wikimedia Commons, Public Domain, *https://commons.wikimedia.org/wiki/File:EB1911_Zebra_-_Equus_zebra_white_background.jpg.*

Figure 4.1 (b): G. H. Ford, Wikimedia Commons, Public Domain, *https://commons.wikimedia.org/wiki/File:PoeciloconcerFasciatusFord.jpg.*

Figure 4.1 (c): Pearson Scott Foresman, Wikimedia Commons, Public Domain, *https://commons.wikimedia.org/wiki/File:Sandpiper_(PSF).png.*

Figure 4.2: Covitt, B. A., J. M. Dauer, and C. W. Anderson. 2017. The role of practices in scientific inquiry. In *Helping students make sense of the world using next generation science and engineering practices*, ed. C. V. Schwarz, C. Passmore, and B. J. Reiser, 59–83. Arlington, VA: NSTA Press.

Figure 4.3: U.S. Geological Survey, Wikimedia Commons, Public Domain, *https://commons.wikimedia.org/wiki/File:Snider-Pellegrini_Wegener_fossil_map.gif.*

Figure 4.4 (a): U.S. Geological Survey, Wikimedia Commons, Public Domain, *https://commons.wikimedia.org/wiki/File:Map_plate_tectonics_world.gif.*

Figure 4.4 (b): U.S. Geological Survey, Wikimedia Commons, Public Domain, *https://commons.wikimedia.org/wiki/File:Oceanic.Stripe.Magnetic.Anomalies.Scheme.svg.*

Figure 4.5: Authors

Figure 4.6 (a): 2012rc, Wikimedia Commons, Public Domain, *https://commons.wikimedia.org/wiki/File:Periodic_table_large.png.*

Figure 4.6 (b): V. G. Ceballos, Wikimedia Commons, CC BY-SA 4.0, *https://commons.wikimedia.org/wiki/File:Tetrapoda_Cladogram.jpg.*

Figure 4.7 (a): U.S. Geological Survey, Wikimedia Commons, Public Domain, *https://commons.wikimedia.org/wiki/File:Bryce_strat.jpg.*

Figure 4.7 (b): U.S. Department of Agriculture, Public Domain, *www.nrcs.*

usda.gov/wps/portal/nrcs/detail/soils/use/worldsoils/?cid=nrcs142p2_054002.

Figure 4.8 (a): Hamed Rajabpour and Nariman Ghorbani, Wikimedia Commons, CC BY-SA 4.0 *https://commons.wikimedia.org/wiki/File:Lunar_Phases.jpg.*

Figure 4.8 (b): Nicholas Caffarilla, Wikimedia Commons, CC BY-SA 3.0 *https://commons.wikimedia.org/wiki/File:Anise_Swallowtail_Life_Cycle.svg.*

Figure 4.8 (c): NOAA, Public Domain, *https://water.weather.gov/ahps2/hydrograph.php?gage=acyn4&wfo=phi.*

Figure 4.9 (a): Authors

Figure 4.9 (b): U.S. Global Change Research Program, Public Domain, *www.globalchange.gov/browse/multimedia/global-temperature-and-carbon-dioxide.*

Figure 4.10: Original drawings by Craig Douglas for Authors

Figure 4.11: Pima County Department of Environmental Quality. 2012. South-Side Private Well Monitoring Program, Tucson International Airport Area Superfund Site. 2012 Annual Report. Tucson, AZ.

Figure 4.12: Authors

Figure 4.13: Authors

Figure 4.14: National Parks Service, Public Domain, *www.nps.gov/isro/learn/nature/wolves.htm.*

Figure 4.15: Authors

Figure 4.16: Authors

Chapter 5

All images come from the authors.

Chapter 7

All images come from the authors.

Chapter 8

Figure 8.1: *Carbon TIME* project, CC BY-NC-SA 4.0, *https://carbontime.bscs.org.*

Classroom Snapshot 8.1 Image: Authors

Classroom Snapshot 8.2 Images: Authors

Figure 8.2: *Carbon TIME* project, CC BY-NC-SA 4.0, *https://carbontime.bscs.org.*

Chapter 9

Bridge: Authors

Capybara Skeleton: Museum of Veterinary Anatomy FMVZ USP, Wikimedia Commons, CC BY-SA 4.0, *https://commons.wikimedia.org/wiki/File:Capybara_skeleton.jpg.*

Figure 9.1: UpstateNYer, Wikimedia Commons, CC BY-SA 3.0, *https://commons.wikimedia.org/wiki/File:USSupremeCourtWestFacade.JPG.*

Figure 9.2: FEMA News Photo, Wikimedia Commons, Public Domain, *https://commons.wikimedia.org/wiki/File:FEMA_-_262_-_Photograph_by_FEMA_News_Photo_taken_on_10-01-1989_in_California.jpg.*

Figure 9.3: Authors

Figure 9.4 (left and right): Jomae Sica

Figure 9.5 (left): Cyril Mehodius Jansky, 1914, Public Domain, *https://commons.wikimedia.org/wiki/File:Industrial_lifting_magnet.jpg.*

Figure 9.5 (right): Authors

Chapter 10

Figure 10.1: Authors

Chapter 11

Figure 11.1: Authors

Figure 11.2: Authors

Figure 11.3: Authors

Figure 11.4: Jim Clark, Lattice Enthalpy (Lattice Energy), *chemguide.co.uk/physical/energetics/lattice.html.*

Chapter 12

Figure 12.1: American Association for the Advancement of Science (AAAS)/ Project 2061. 2017. *Toward high school biology: Understanding growth in living things, teacher edition,* p. xi. Arlington, VA: NSTA Press.

Figure 12.2: American Association for the Advancement of Science (AAAS)/ Project 2061. 2020. *Matter and energy for growth and activity, teacher edition,* p. viii. Arlington, VA: NSTA Press.

Figure 12.3: Anonymous student work

Figure 12.4: Jo Ellen Roseman

Figure 12.5: American Association for the Advancement of Science (AAAS)/Project 2061. 2017. *Toward high school biology: Understanding growth in living things teacher edition*, pp. 292–293. Arlington, VA: NSTA Press.

Figure 12.6: Jo Ellen Roseman

Figure 12.7: American Association for the Advancement of Science (AAAS)/ Project 2061. 2020. *Matter and energy for growth and activity, teacher edition,* p. 304. Arlington, VA: NSTA Press.

Figure 12.8: Anonymous student work

Figure 12.9: American Association for the Advancement of Science (AAAS)/ Project 2061. 2020. *Matter and energy for growth and activity, teacher edition,* p. 185. Arlington, VA: NSTA Press.

Figure 12.10: Anonymous student work

Figure 12.11: Authors

Chapter 14

Figure 14.1: Authors

Figure 14.2: Authors

Figure 14.3: Authors

Figure 14.4 (left): GalleryOfHope, Wikimedia Commons, CC BY 4.0, *https://commons.wikimedia.org/wiki/File:Top_Of_Mole_Head.jpg.*

Figure 14.4 (right): Authors

Chapter 15

Figure 15.1: Authors

Figure 15.2: NGSS Lead States. 2013. *Next Generation Science Standards: For states, by states.* Washington, DC: National Academies Press. *www.nextgenscience.org.*

Figure 15.3: Authors

Index

Page numbers printed in **boldface type** refer to figures or tables.

Index

Index

Index

Index